Lecture Notes in Electrical Engineering

Volume 461

About this Series

"Lecture Notes in Electrical Engineering (LNEE)" is a book series which reports the latest research and developments in Electrical Engineering, namely:

- Communication, Networks, and Information Theory
- Computer Engineering
- Signal, Image, Speech and Information Processing
- Circuits and Systems
- Bioengineering

LNEE publishes authored monographs and contributed volumes which present cutting edge research information as well as new perspectives on classical fields, while maintaining Springer's high standards of academic excellence. Also considered for publication are lecture materials, proceedings, and other related materials of exceptionally high quality and interest. The subject matter should be original and timely, reporting the latest research and developments in all areas of electrical engineering.

The audience for the books in LNEE consists of advanced level students, researchers, and industry professionals working at the forefront of their fields. Much like Springer's other Lecture Notes series, LNEE will be distributed through Springer's print and electronic publishing channels.

More information about this series at http://www.springer.com/series/7818

Wookey Lee · Wonik Choi
Sungwon Jung · Min Song
Editors

Proceedings of the 7th International Conference on Emerging Databases

Technologies, Applications, and Theory

Springer

Editors

Wookey Lee
Department of Industrial Engineering
Inha University
Incheon
Korea (Republic of)

Wonik Choi
Department of Information
and Communication Engineering
Inha University
Incheon
Korea (Republic of)

Sungwon Jung
Department of Computer Science
and Engineering
Sogang University
Seoul
Korea (Republic of)

Min Song
Department of Library
and Information Science
Yonsei University
Seoul
Korea (Republic of)

ISSN 1876-1100 ISSN 1876-1119 (electronic)
Lecture Notes in Electrical Engineering
ISBN 978-981-10-6519-4 ISBN 978-981-10-6520-0 (eBook)
https://doi.org/10.1007/978-981-10-6520-0

Library of Congress Control Number: 2017953433

Printed on acid-free paper

This Springer imprint is published by Springer Nature
The registered company is Springer Nature Singapore Pte Ltd.
The registered company address is: 152 Beach Road, #21-01/04 Gateway East, Singapore 189721, Singapore

Preface

Please accept our warmest welcome to the seventh International Conference on Emerging Databases: Technologies, Applications, and Theory (EDB 2017) which was held in Busan, Korea, on August 7–9, 2017. The KIISE (Korean Institute of Information Scientists and Engineers) Database Society of Korea hosts EDB 2017 as an annual forum for exploring technologies, novel applications, and researches in the fields of emerging databases. We have thrived to make EDB 2017 the premier venue for researchers and practitioners to exchange current research issues, challenges, new technologies, and solutions.

The technical program of EDB 2017 has embraced a variety of themes that fit into seven oral sessions and one poster session. We have selected 26 regular papers and 9 posters with high quality. The following sessions represent the diversity of themes of EDB 2017: "NoSQL Database," "System and Performance," "Social Media and Big Data," "Graph Database and Graph Mining," and "Data Mining and Knowledge Discovery." In addition to the oral and poster sessions, the technical program has provided one keynote speech by Dr. Mukesh Mohania (IBM Academy of Technology, Australia), two invited talks by Prof. Alfredo Cuzzocrea (University of Trieste, Italy) and Prof. Carson Leung (University of Manitoba, Canada), and one tutorial by Prof. Jae-Gil Lee (KAIST, Republic of Korea).

We would like to give our sincere thanks to all our colleagues who served on the Program Committee members and external reviewers. The success of EDB 2017 would not have been possible without their dedication. We would like to thank Bong-Hee Hong (Pusan Nat'l Univ., Korea), Young-Kuk Kim (Chungnam Nat'l Univ., Korea), Young-Duk Lee (Korea Data Agency, Korea), Hiroyuki Kitagawa (Tsukuba University, Japan), and Sean Wang (Fudan University, China) (Honorary Co-Chairs); Jinho Kim (Kangwon Nat'l Univ., Korea) and Wookey Lee (Inha Univ., Korea) (General Co-Chairs); and Youngho Park (Sookmyung Women's Univ., Korea), Wonik Choi (Inha Univ., Korea), and James Geller (NJIT, USA) (Organization Committee Co-Chairs) for their advices and supports. We are also grateful to all the members of EDB 2017 for their enthusiastic cooperation in organizing the conference.

Last but not least, we would like to give special thanks to all of the authors for their valuable contributions, which made the conference a great success.

Sungwon Jung
Min Song
Program Committee Co-chairs

Contents

Optimizing MongoDB Using Multi-streamed SSD

Trong-Dat Nguyen(✉) and Sang-Won Lee

College of Information and Communication Engineering, Sungkyunkwan University, Suwon 16419, Korea
{datnguyen,swlee}@skku.edu

Abstract. Data fragmentation in flash SSDs is a common problem that leads to performance degradation, especially when the underlying storage devices become aged by heavily updating workloads. This paper addresses that problem in MongoDB, a popular document storage in the current market, by introducing a novel stream mapping scheme that based on unique characteristics of MongoDB. The proposal method has low overhead and independent with data models and workloads. We use YCSB and Linkbench with various cache sizes and workloads to evaluate our proposal approaches. Empirical results shown that in YCSB and Linkbench, our methods improved the throughput by more than 44% and 43.73% respectively; reduced 99th-percentile latency by up to 29% and 24.67% in YCSB and Linkbench respectively. In addition, by tuning the leaf page size in B+Tree of MongoDB, we can significantly improve the throughput by 3.37x and 2.14x in YCSB and Linkbench respectively.

Keywords: Data fragmentation · Multi-streamed SSD · Document store · Optimization · MongoDB · WiredTiger · Flash SSD · NoSQL · YCSB · Linkbench

1 Introduction

Flash solid state drives (SSDs) have several advantages over hard drives e.g. fast IO speed, low power consumption, and shock resistance. One unique characteristic of NAND flash SSDs is "erase-before-update" i.e. one data block should be erased before writing on new data pages. *Garbage collection (GC)* in flash SSD is responsible for maintaining free blocks. Reclaiming a non-empty data block is expensive because: (1) erase operation itself takes orders of magnitude slower than read and write operations [1], and (2) if the block has some valid pages, GC first copy back those pages to another empty block before erasing the block.

Typically, locality of data access has a substantial impact on the performance of flash memory and its lifetime due to wear-leveling. IO workload from client queries has skewness i.e. small proportion of data that has frequently accessed [10,11,15]. In flash-based storage systems, hot data identification is the process of

W. Lee et al. (eds.), *Proceedings of the 7th International Conference on Emerging Databases*, Lecture Notes in Electrical Engineering 461,
https://doi.org/10.1007/978-981-10-6520-0_1

distinguishing *logical block addresses (LBAs)* that have frequently accessed data (*hot data*) with the others less frequently accessed data (*cold data*). Informally, *data fragmentation* in flash SSD happens when writing data pages with different lifetimes to a block in an interleaved way. In that case, one physical block includes hot data and cold data which in turn increases the overhead of reclaiming blocks significantly. Prior researchers solved the problem by identifying hot/cold data either based on history address information [12] or based on update frequency [13,14]. However, those approaches had a degree of overhead for keeping track of metadata in DRAM as well as CPU cost for identifying hot/cold blocks. Min et al. [16] designed a Flash-oriented file system that groups hot and cold segments according to write frequency. In another approach, TRIM command is introduced to aid upper layers in user space and kernel space notifying flash FTL which data pages are invalid and no longer needed, thus reducing the GC overhead by avoiding unnecessary copy back of those pages when reclaiming new data blocks [18].

Recently, NoSQL solutions have become popular and been alternatives to traditional relational database management systems (RDBMSs). Among many NoSQL solutions, MongoDB[1] is one of the representative document stores with WiredTiger[2] as the default storage engine that shares many common characteristics with traditional RDBMS such as transaction processing, multi-version concurrency control (MVCC), and secondary index supporting. Moreover, there is a conceptual mapping between MongDB's data model and traditional table-based data model in RDBMS [7]. Therefore, MongoDB is interested not only by developers from industrial but also from researchers in academia. Most of the researchers compared between RDBMSs and NoSQLs [3,4], addressing data modeling transformation [8,9] or load-balanced sharding [5,6].

Performance degradation due to data fragmentation also exists in NoSQL solutions with SSDs as the underlying storage devices. For example, NoSQL DBMSs such as Cassandra and RocksDB take the log-structured merge (LSM) tree [21] approach have different update lifetime for files in each level of LSM tree. Kang et al. [10] proposed a *Multi-streamed SSD (MSSD)* technique to solve data fragmentation in Cassandra. The key idea is assigning different *streams* to different file types then groups data pages with similar update lifetimes into same physical data blocks. Extended from the previous research, Yang et al. Adopting file-based mapping scheme from Cassandra to RocksDB is inadequate because in RocksDB, there are concurrency compaction threads that compact files into several files. Therefore writes on files with different lifetime are located in the same stream. To address that problem, Yang et al. [11] extended the previous mapping scheme with a novel stream mapping with locking scheme for RocksDB.

To the best of our knowledge, no study has investigated on data fragmentation problem in MongoDB using multi-streamed SSD technique. Nguyen et al. [17] exploited TRIM command to reduce overhead in MongoDB. However, TRIM command does not entirely solve data fragmentation [11]. WiredTiger uses

[1] https://www.mongodb.com/mongodb-architecture.
[2] http://source.wiredtiger.com/2.7.0/index.html.

B+Tree implementation for its collection files as well as index files. However, the page sizes of internal pages and leaf pages in collection files are not equal i.e. 4 KB and 32 KB respectively. Meanwhile, the smaller page size is known to work better for flash SSD because it can help reducing *write amplification* ratio [2], so we can further improve throughput in WiredTiger by tuning smaller leaf page size.

In this paper, we propose a novel boundary-based stream mapping to exploit the unique characteristic of WiredTiger. We further extend the boundary-based stream mapping by introducing an on-line high efficient stream mapping based on data locality. We summarize our contributions as below:

- We investigated WiredTiger's block management in detail and pointed out two causes for data fragmentation: (1) writing on files have different lifetimes, and (2) there is internal fragmentation in collection files and index files. We adopt a simple stream mapping scheme based on those observations that map each file types with different streams. Further, we proposal a novel stream mapping scheme for WiredTiger based on the boundaries on collection files or index files. This approach improves the throughput in YCSB [19] and Linkbench [20] up to 44% and 43.73% respectively, improved the 99th-percentile latency in YCSB and Linkbench up to 29% and 24.67% respectively.
- We suggested a simple optimization of changing the leaf page size in B+tree from its default value 32 KB to 4 KB. In combination with the multi-streamed optimization, this simple tuning technique improved the throughput by threefold and 2.16x for YCSB and Linkbench respectively.

The rest of this paper is organized as follow. Section 2 explains the background of multi-streamed SSD and MongoDB in detail. Proposal methods are described in Sect. 3. We explain the leaf page size optimization in Sect. 4. Section 5 discusses evaluation results and analysis. Lastly, the conclusion is given in Sect. 6.

2 Background

2.1 Multi-streamed SSD

Kang et al. [10] originally proposed the idea of mapping streams to different files so that data pages with similar update lifetime are grouped in the same physical block. Figure 1 illustrates how different between regular SSD and multi-streamed SSD (MSSD) work. Suppose that the device had eight logical block addresses (LBAs) and divided into two groups: hot data (LBA2, LBA4, LBA6, LBA8), and cold data are remains. There are two write sequences for both regular SSD and MSSD. The first sequence is written continuously from LBA1 to LBA8, and then the second write sequence only includes hot LBAs i.e. LBA6, LBA2, LBA4, and LBA8.

In regular SSD, after the first write sequence, LBAs are mapped to block 0 and block 1 according to write order regardless of hot or cold data.

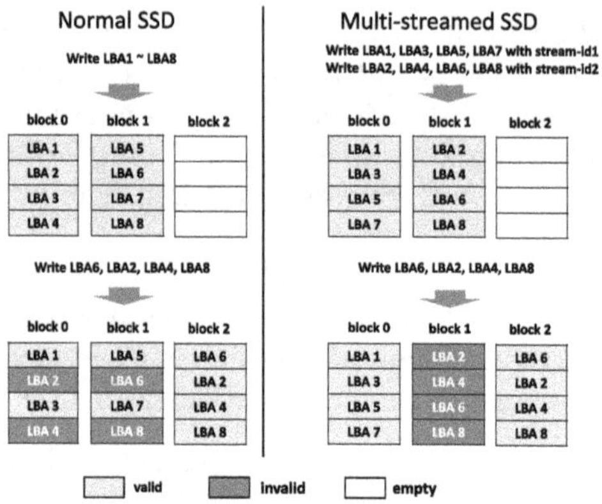

Fig. 1. Comparison between normal SSD and multi-streamed SSD

When the second write sequence occurs, new coming writes are done in empty block 2, corresponding old LBAs become invalid in block 0 and block 1. If the GC process reclaims block 1, there is an overhead for copying back LBA5 and LBA7 to another free block before erasing block 1.

The write sequences are similar in MSSD; however, in the first write sequence, LBAs assigned to a corresponding stream according to their hotness values. Consequently, all hot data grouped into block 1. After the second write sequence finished, all LBAs in block 1 become invalid and erasing block 1 in such case is quite fast, due to the copying back overhead is eliminated.

2.2 MongoDB and WiredTiger

MongoDB and RDBMS. Document store shares many similar characteristics to traditional RDBMS such as transaction processing, secondary indexing, concurrency controlling. MongoDB has emerged as standard document stores in NoSQL solutions. There is a conceptually mapping between the data model in RDBMS and the one in MongoDB. While *database* concept is same for both models; *tables*, *rows*, and *columns* in RDMBS can be seen as *collections*, *documents*, and *document fields* in MongoDB, respectively. Typically, MongoDB encodes documents as BSON[3] format and uses WiredTiger as the default storage engine since the version 3.0. WiredTiger uses B+Tree implementation for collection files as well as index files. In collection file, maximum page sizes are 4 KB and 32 KB for internal pages and leaf pages respectively. From now on, we use WiredTiger and MongoDB interchangeably unless there is some specific distinguishes.

Block Management. Understanding internal block management of WiredTiger is the key to optimizing the system using MSSD approach.

[3] http://bsonspec.org/spec.html.

WiredTiger uses *extents* to represent location information of data blocks in memory i.e. offsets and sizes.

Each checkpoint keeps track of three linked lists of extents for managing allocated space, discard space, and free space, respectively. WiredTiger keeps only one special checkpoint called *live checkpoint* in DRAM that includes block management information for the current working system. When a checkpoint is called, before writing the current live checkpoint to disk, WireTiger fetches the previous checkpoint from the storage device to DRAM; then merges its extent lists with the live checkpoint. Consequently, reused allocated space from the previous checkpoint after the merging phase finished. During the checkpoint time, WiredTiger discards unnecessary log files and the reset the log write offset to zero.

An important observation is that, once a particular region of the storage device is allocated in a checkpoint, it is reused again in the next checkpoint. That forms an internal fragmentation in the storage device that leads to the high overhead of GC process if the underlying storage is SSD. Next section discusses this problem in detail.

3 Boundary-Based Stream Mapping

3.1 Asymmetric Amount of Data Written to File Types

The amount of data written to files is a reliable criterion to identify the bottleneck of the storage engine and the root cause of data fragmentation that leads to high overhead in GC process.

Table 1. The proportions of data written to file types with various of workloads

Benchmark	Operation ratio C:R:U:D	Colls	Pri. Indexes	2nd indexes	Journal
Y-Update-Heavy	0:50:50:0	93.6	n/a	n/a	6.4
Y-Update-Only	0:0:100:0	89.6	n/a	n/a	10.4
LB-Original	12:69:15:4	58.6	3.1	37.22	1.08
LB-Mixed	12:0:84:4	66.1	0.5	31.13	2.27
LB-Update-Only	0:0:100:0	67.6	0.02	30.2	2.18

Table 1 shows the proportions of data written to collection files, index files and journal files under various workloads with different operations i.e. create, read, update, delete (CRUD). For simple data model, YCSB workload A (Y-Update-Heavy), and YCSB only update workload (Y-Update-Only) are carried out. To experiment more complex data model, we use original Linkbench workload (LB-Original), mixed operations workload (LB-Mixed), and only update Linkbench workload (LB-Update-Only). Write ratios to other files e.g. metadata, system data, are too small i.e. less than 0.1%, that can be excluded from the table.

As observed from the table, write distributions to file types are different depending on the CRUD ratios of the workload. In YCSB benchmark, since the data model is simple key-value with only one collection file and one primary index file, almost writes are on collection file, and there is no update on primary index file. In Linkbench, however, collection files and secondary index files are hot data which have frequency accessed, primary index files and journal files are cold data that receive the low proportion of writes i.e. less than 5% in total. This observation implicates that difference written ratios in file types lead to hot data and cold data locate in the same physical data block in SSD that result in high overhead in GC process as explained in the previous section.

To solve this problem, we use a simple file-based optimization that assigns different streams for different file types. To minimize the overhead of the system, we assign a file to a corresponding stream only when open that file. Table 3 in Sect. 5 describes the detail of stream mapping in file-based method.

3.2 Multi-streamed SSD Boundary Approach

We further analyze the write patterns of WiredTiger to improve the optimization. We define write region (*region* in short) is the area between two logical file offsets that data is written on in a duration of time. Figure 2 illustrates the written patterns of different file types in the system under Linkbench benchmark with LB-Update-Only workload in two hours using *blktrace*. The x-axis is the elapsed time in seconds, the y-axis is the file offset. *DirectIO* mode is used to eliminate the effect of Operating System cache. Collection file and secondary index file have heavily random write pattern on two regions i.e. *top* and *bottom* that separated by a boundary in dashed line as illustrated in Fig. 2(a), and (c). In the other hand, the primary index file and journal file follow sequence write patterns as illustrated in Fig. 2(b), and (d) respectively.

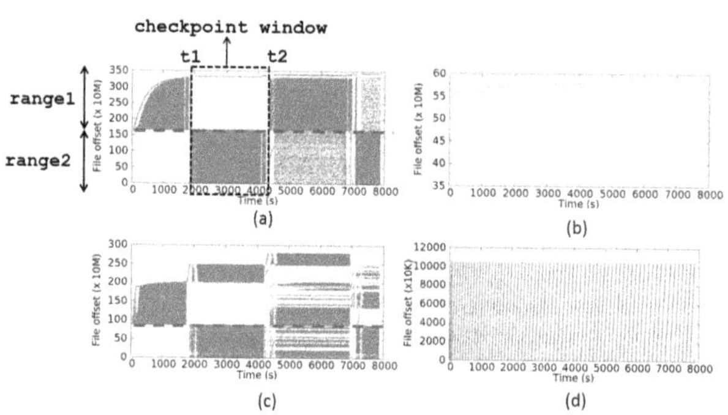

Fig. 2. Write patterns of various file types in WiredTiger with Linkbench benchmark, (a) Collection file, (b) primary index file, (c) secondary index, and (d) journal file

Algorithm 1. Boundary-based stream mapping

```
1: Require: boundary of each collection file and index file has computed
2: Input: file, and offset to write on
3: Output: sid - stream assign for this write
4: boundary ← getboundary(file)
5: if file is collection then
6:     if offset < boundary then
7:         sid ← COLL_SID1
8:     else
9:         sid ← COLL_SID2
10: else if file is index then
11:     if offset < boundary then
12:         sid ← IDX_SID1
13:     else
14:         sid ← IDX_SID2
15: else                                  ▷ Other files i.e. metadata
16:     sid ← OTHER_SID
```

One important observation is that, at a given point of time, the amount of data written to two regions i.e. *top* and *bottom* is asymmetric and switches after each checkpoint. In this paper, we call that phenomenon is *asymmetric regions writing*. Due to the asymmetric regions writing phenomenon, for a given file, there is an *internal fragmentation* that dramatically affects to the overhead of GC in SSDs. Obviously, file-based optimization is inadequate to solve the problem. In this approach, writes on one file are mapped with one stream, thus inside that stream, internal fragmentation still occurs. Therefore, we proposal a novel stream assignment named *boundary-based* stream mapping. The key idea is using the file boundary that separates the logical address of a given file to the top region and the bottom region. As described in Algorithm 1, firstly, the boundary of each collection and index file is computed as the last file offset after the load phase finished. Then in query phase, before writing a block data on a given file, the boundary is retrieved again as in line 4, based on the boundary values and the file types, the stream mapping is carried out as in line 7, 9, 12, 14, and 16. After stream id is mapped, the write command to the underlying file is given as *posix_fadvise(fid, offset, sid, advice)*, where *fid* is file identify, *offset* is the offset to write on, *sid* is stream id mapped and *advice* is passed as a predefined constant.

4 B+Tree Leaf Page Size Tuning

WiredTiger uses B+Tree to implement collection files and index files. Accesses to internal pages are usually more frequent than leaf pages. Due to page replacement policy in the buffer pool, internal pages are kept in DRAM longer than leaf pages. So there is an asymmetric amount of data written to components of B+Tree as presented in Table 2. We keep track of the number of writes on each component of B+Tree by modifying the original source code of WiredTiger. Exclude from typical components i.e. root page, internal page, and leaf page, *extent page* is a special type that only keeps metadata for extent lists in a checkpoint. For only update workload, while writes only occur on collection file in YCSB, both collection files and index files in Linkbench are updated. Because root pages and internal pages are accessed more frequent than leaf pages, WiredTiger keeps them in DRAM as long as possible and mostly writes them to disk at the checkpoint time belong with extent pages. In the other hand, leaf pages are flushed out not only at the checkpoint time but also at the normal thread through evicting dirty pages from buffer pool in the reconciliation process. Therefore, more than 99% of the total writes occur on leaf pages.

Table 2. Percentage of writes on page types in collection files and index files in YCSB and Linkbench

Benchmark	Collection (%)				Index (%)			
	Root page	Int. page	Leaf page	Ext. page	Root page	Int. page	Leaf page	Ext. page
YCSB	$5e^{-5}$	0.27	99.72	$9.3e^{-5}$	0	0	0	0
Linkbench	$7e^{-5}$	0.61	39.86	$14e^{-5}$	$13e^{-5}$	0.28	59.23	$26e^{-5}$

Note that the default sizes for internal pages and leaf pages are 4 KB and 32 KB respectively. Large leaf page size leads to high write amplification such that some bytes update from workload lead to whole 32 KB data page written out to disk. It becomes worse with heavy random update workload such that almost 99 percent of writes occur on leaf pages. We suggest a simple but effective tuning that decreases the leaf page size from its default 32 KB to 4 KB. This changing improves the throughput significantly as discussed in the next section.

5 Evaluation and Analysis

5.1 Experimental Settings

We conducted the experiments with YCSB 0.5.0[4] and LinbenchX 0.1[5] (an extended version of Linkbench that supports MongoDB) as the top client layer and used various of workloads as illustrated in Table 1. We use 23 million 1-KB documents in YCSB and *maxid1* equal to 80 million in Linkbench respectively. In the server-side, we adopt a stand-alone MongoDB 3.2.1[6] server with WiredTiger as storage engine. Cache sizes vary from 5 GB to 30 GB, other settings in WiredTiger are kept as default. To enable multi-streamed technique, we use a modified Linux kernel 3.13.11 along with customized Samsung 840 Pro as in [11].

[4] https://github.com/brianfrankcooper/YCSB/releases/tag/0.5.0.
[5] https://github.com/Percona-Lab/linkbenchX.
[6] https://github.com/mongodb/mongo/archive/r3.2.1.tar.gz.

To exclude the network latency, we setup the client layer and the server layer on the same commodity server with 48 cores Intel Xeon 2.2 GHz processor, 32 GB DRAM. We execute all benchmarks during 2 hours with 40 client threads.

5.2 Multi-streamed SSD Optimization Evaluation

To evaluate the effect of our proposal multi-streamed SSD based methods we conducted experiments with different stream mapping schemes as shown in Table 3. In the original WiredTiger, there is no stream mapping; thus all file types use stream 0 that reserve for files in Linux Kernel as default. In file-based stream mapping, we used total four streams that map each stream to a file type. In boundary-based stream mapping, metadata files and journal files are mapped in the same way with the file-based approach. The different is that there is two streams map with collection files, one for all top regions and another for the bottom regions. We map streams for index files in the same manner without considering primary index files or secondary index files.

Figure 3 illustrates the throughput results for various benchmarks and workloads. Note that in Linkbench benchmark with maxid1 equals to 80 million, the total index size is quite large i.e. 33 GB that requires the buffer pool size large enough to keep almost index files in DRAM. In addition, in LB-Original workload that exist read operations, pages tend to fetched in and flush out buffer

Table 3. Stream mapping schemes

Method	Kernel	Metadata	Journal	Collections	Indexes
Original	0	0	0	0	0
File-based	0	1	2	3	4
Boundary-based	0	1	2	3,4	5,6

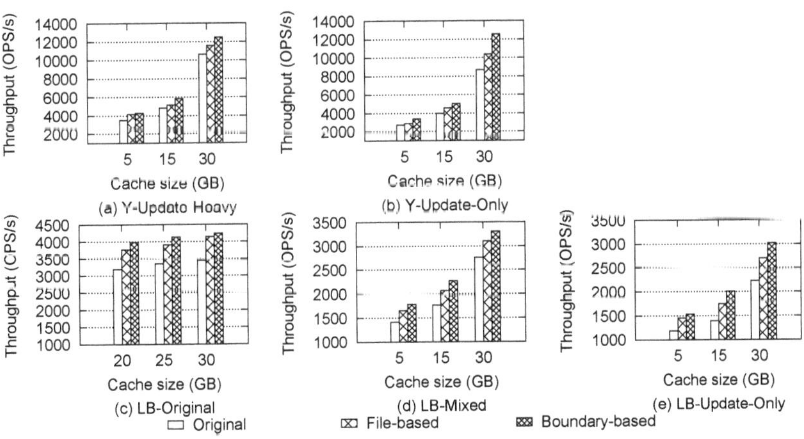

Fig. 3. Throughput of optimized methods compared with the original

pool more frequent hence we use large cache sizes i.e. 20 GB, 25 GB, and 30 GB in LB-Original workload. In general, multi-streamed based methods have greater throughput than the original. The more frequent writing the workload has, the more benefit multi-streamed based approaches gain.

In YCSB benchmark, boundary-based shows the throughput improve up to 23% at 5 GB cache size and 44% at the cache size is 30 GB in Y-Update-Heavy and Y-Update-Only workload respectively. For Linkbench benchmark, the boundary-based method has throughput improve up to 23.26%, 28.12%, and 43.73% for LB-Original, LB-Mixed, and LB-Update-Only respectively. In the YCSB benchmark, the percentage of throughput improvement of the boundary-based method has remarkable gaps compared with file-based that up to approximate 14% and 24.4% for Y-Update-Heavy and Y-Update-Only respectively. However, those differences become smaller in Linkbench that just 6.84%, 11% and 18.8% for LB-Original, LB-Mixed, and LB-Update-Only respectively. For NoSQL applications in distributed environment, it is also important to consider the 99th-percentile latency of the system to ensure clients have acceptable response times. Figure 4 shows the 99th-percentile latency improvements of multi-streamed based methods compared with the original. Overall, similar with throughput improvement, 99th-percentile latency correlates with the overhead of the GC hence the better one method solve data fragmentation, the lower 99th-percentile latency it reduces.

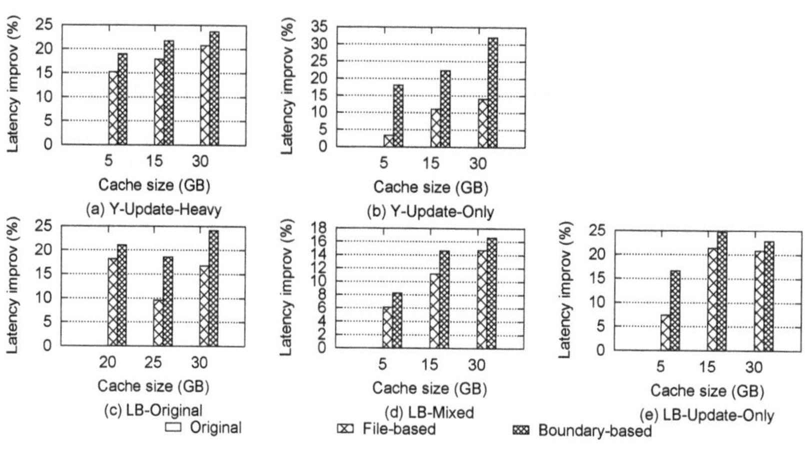

Fig. 4. Latency of optimized methods compared with the original

Boundary-based is better than file-based method. In YCSB, compared with the original WiredTiger, the boundary-based method reduces the 99th-percentile latency 29.3%, 29% for Y-Update-Heavy, Y-Update-Only, respectively. In Linkbench, it reduces up to 24.13%, 16.56%, and 24.67% for LB-Original, LB-Mixed, and LB-Update-Only respectively. Once again, boundary-based benefits in reducing the latency in simple data model i.e. YCSB decrease a little in complex data model i.e. Linkbench.

5.3 Leaf Page Size Optimization Evaluation

To evaluate the impact of leaf page size we conducted the experiment on original WiredTiger as well as our proposal method with YCSB benchmark and Linkbench using various workloads and cache sizes for 32 KB leaf page size (default) and 4 KB leaf page size. For the space limitation in the paper, we only show the throughput results of the heaviest write workloads i.e. Y-Update-Only and LB-Update-Only as in Fig. 5. Overall, compared with the original WiredTiger 32-KB as the based line, Boundary-based-4 KB leaf page shows dramatically improvement of throughput that up to 3.37x and 2.14x for Y-Update-Only and LB-Update-Only respectively. In YCSB, with the same method, changing leaf page size from 32 KB to 4 KB increase the throughput sharply triple or double. In Linkbench, however, reducing leaf page size from 32 KB to 4 KB has the maximum throughput improvement are 1.96x, 1.4x, and 1.5x for the original method, the file-based method, and the boundary-based method respectively. Note that in Linkbench, small leaf page size optimization lost its effect with small cache size i.e. 5 GB. The reason is with the same maxid1 value, reducing the leaf page size from 32 KB to 4 KB increases the number of leaf pages and the number of internal pages in the B+Tree that lead to collection files and index files become larger and require more space from the buffer pool to keep the hot index files in DRAM.

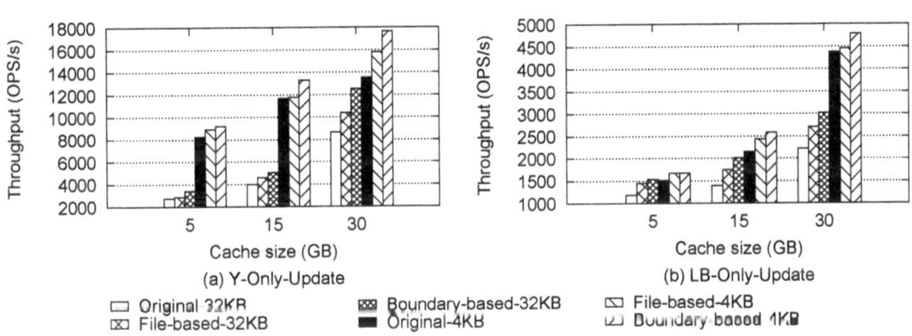

Fig. 5. Throughput of optimized methods compared with the original

6 Conclusion

In this paper, we discussed data fragmentation in MongoDB in detail. The file-based method is the simplest one that solves the data fragmentation due to the different lifetime of writes on file types but remains internal fragmentation caused by asymmetric regions writing. For simple data model in YCSB, the boundary-based approach is adequate to solve the internal fragmentation that shows good performance improvement but lost its benefits with the complex data model in Linkbench. In addition, reducing the maximum leaf page size in collection files or index files from 32 KB to 4 KB can gain significant improvement in throughput in both YCSB and Linkbench. In general, our proposal approaches can adopt to any storage engine that has similar characteristics with WiredTiger i.e. asymmetric files writing and asymmetric region writing. Moreover, we expect to further optimize the WiredTiger storage engine by solving the problem of boundary-based with complex data model i.e. Linkbench in the next research.

Acknowledgments. This research was supported by the MSIP (Ministry of Science, ICT and Future Planning), Korea, under the "SW Starlab" (IITP-2015-0-00314) supervised by the IITP (Institute for Information & communications Technology Promotion).

References

1. Lee, S.W., Moon, B., Park, C., Kim, J.M., Kim, S.W.: A case for flash memory SSD in enterprise database applications. In: Proceedings of the 2008 ACM SIGMOD International Conference on Management of Data, pp. 1075–1086 (2008). doi:10.1145/1376616.1376723
2. Lee, S.W., Moon, B., Park, C.: Advances in flash memory SSD technology for enterprise database applications. In: Proceedings of the 2009 ACM SIGMOD International Conference on Management of data, pp. 863–870 (2009)
3. Aboutorabi, S.H., Rezapour, M., Moradi, M., Ghadiri, N.: Performance evaluation of SQL and MongoDB databases for big e-commerce data. In: International Symposium on Computer Science and Software Engineering (CSSE), pp. 1–7. IEEE, August 2015. doi:10.1109/CSICSSE.2015.7369245
4. Boicea, A., Radulescu, F., Agapin, L.I.: MongoDB vs oracle-database comparison. In: EIDWT, pp. 330–335, September 2012
5. Liu, Y., Wang, Y., Jin, Y.: Research on the improvement of MongoDB auto-sharding in cloud environment. In: 7th International Conference on Computer Science and Education (ICCSE), pp. 851–854. IEEE (2012). doi:10.1109/iccse.2012.6295203
6. Wang, X., Chen, H., Wang, Z.: Research on improvement of dynamic load balancing in MongoDB. In: 2013 IEEE 11th International Conference on Dependable, Autonomic and Secure Computing (DASC), pp. 124–130. IEEE, December 2013. doi:10.1109/DASC.2013.49
7. Zhao, G., Huang, W., Liang, S., Tang, Y.: Modeling MongoDB with relational model. In: 2013 Fourth International Conference on Emerging Intelligent Data and Web Technologies (EIDWT), pp. 115–121. IEEE (2013). doi:10.1109/EIDWT.2013.25
8. Lee, C.H., Zheng, Y.L.: SQL-to-NoSQL schema denormalization and migration: a study on content management systems. In: 2015 IEEE International Conference on Systems, Man, and Cybernetics (SMC), pp. 2022–2026. IEEE, October 2015. doi:10.1109/SMC.2015.353
9. Zhao, G., Lin, Q., Li, L., Li, Z.: Schema conversion model of SQL database to NOSQL. In: 2014 Ninth International Conference on P2P, Parallel, Grid, Cloud and Internet Computing (3PGCIC), pp. 355–362. IEEE, November 2014. doi:10.1109/3PGCIC.2014.137

10. Kang, J.U., Hyun, J., Maeng, H., Cho, S.: The multi-streamed solid-state drive. In: 6th USENIX Workshop on Hot Topics in Storage and File Systems (HotStorage 14) (2014)
11. Yang, F., Dou, K., Chen, S., Hou, M., Kang, J.U., Cho, S.: Optimizing NoSQL DB on flash: a case study of RocksDB. In: Ubiquitous Intelligence and Computing and 2015 IEEE 12th International Conference on Autonomic and Trusted Computing and 2015 IEEE 15th International Conference on Scalable Computing and Communications and Its Associated Workshops (UIC-ATC-ScalCom), pp. 1062–1069 (2015). doi:10.1109/uic-atc-scalcom-cbdcom-iop.2015.197
12. Hsieh, J.W., Kuo, T.W., Chang, L.P.: Efficient identification of hot data for flash memory storage systems. ACM Trans. Storage (TOS) **2**(1), 22–40 (2006). doi:10.1145/1138041.1138043
13. Jung, T., Lee, Y., Woo, J., Shin, I.: Double hot/cold clustering for solid state drives. In: Advances in Computer Science and Its Applications, pp. 141–146. Springer, Heidelberg (2014). doi:10.1007/978-3-642-41674-3_21
14. Kim, J., Kang, D.H., Ha, B., Cho, H., Eom, Y.I.: MAST: multi-level associated sector translation for NAND flash memory-based storage system. In: Computer Science and its Applications, pp. 817–822. Springer, Heidelberg (2015). doi:10.1007/978-3-662-45402-2_116
15. Lee, S.W., Moon, B.: Design of flash-based DBMS: an in-page logging approach. In: Proceedings of the 2007 ACM SIGMOD International Conference on Management of Data, pp. 55–66. ACM, June 2007. doi:10.1145/1247480.1247488
16. Min, C., Kim, K., Cho, H., Lee, S.W., Eom, Y.I.: SFS: random write considered harmful in solid state drives. In: FAST, p. 12, February 2012
17. Nguyen, T.D., Lee, S.W.: I/O characteristics of MongoDB and trim-based optimization in flash SSDs. In: Proceedings of the Sixth International Conference on Emerging Databases: Technologies, Applications, and Theory, pp. 139–144. ACM, October 2016. doi:10.1145/3007818.3007844
18. Kim, S.H., Kim, J.S., Maeng, S.: Using solid-state drives (SSDs) for virtual block devices. In: Proceedings Workshop on Runtime Environments, Systems, Layering and Virtualized Environments, March 2012
19. Cooper, B.F., Silberstein, A., Tam, E., Ramakrishnan, R., Sears, R.: Benchmarking cloud serving systems with YCSB. In: Proceedings of the 1st ACM Symposium on Cloud Computing, pp. 143–154 (2010). doi:10.1145/1807128.1807152
20. Armstrong, T.G., Ponnekanti, V., Borthakur, D., Callaghan, M.: LinkBench: a database benchmark based on the Facebook social graph. In: Proceedings of the 2013 ACM SIGMOD International Conference on Management of Data, pp. 1185–1196. ACM, June 2013. doi:10.1145/2463676.2465296
21. O'Neil, P., Cheng, E., Gawlick, D., O'Neil, E.: The log-structured merge-tree (LSM-tree). Acta Informatica **33**(4), 351–385 (1996). doi:10.1007/s002360050048

Quadrant-Based MBR-Tree Indexing Technique for Range Query Over HBase

Bumjoon Jo and Sungwon Jung(✉)

Department of Computer Science and Engineering,
Sogang University, 35 Baekbeom-ro, Mapo-gu, Seoul 04107, Korea
{bumjoonjo, jungsung}@sogang.ac.kr

Abstract. HBase is one of the most popular NoSQL database systems. Because it operates on a distributed file system and supports a flexible schema, it is suitable for dealing with large volumes of semi-structured data. However, HBase only provides an index built on one dimensional rowkeys of data, which is unsuitable for the effective processing of multidimensional spatial data. In this paper, we propose a hierarchical index structure called a Q-MBR (quadrant based minimum bounding rectangle) tree for effective spatial query processing in HBase. We construct a Q-MBR tree by grouping spatial objects hierarchically through Q-MBRs. We also propose a range query processing algorithm based on the Q-MBR tree. Our proposed range query processing algorithm reduces the number of false positives significantly. An experimental analysis shows that our method performs considerably better than the existing methods.

Keywords: HBase · NoSQL · Spatial data indexing · Q-MBR tree · Range query

1 Introduction

In recent years, a number of studies have attempted to use Hadoop distributed file system and MapReduce framework to deal with spatial queries on big spatial data [1–3]. However, these methods suffer from a large amount of data I/O during query processing because of a lack of spatial awareness of the underlying system. In order to overcome this problem, there have been attempts to distribute spatial data objects over cloud infrastructure by considering their spatial proximity [4–7]. SpatialHadoop [4] and Hadoop-GIS [5] observe the spatial proximity of data, and store adjacent data into same storage block of Hadoop. They provide global index to retrieve relevant blocks for query processing and also provide local index to explore data in each blocks. Dragon [6] and PR-Chord [7] use similar indexing techniques on P2P environment. The limitation of these methods is that they are vulnerable to update. When the updating is issued, distribution of data is changed and entire index structure should be modified.

As an alternative to methods based on the Hadoop system, there have been several studies have enhanced the spatial awareness of NoSQL DBMS, especially HBase [8–10]. HBase provides an effective framework for fast random access and updating of data on a distributed file system. Because HBase only provides an index built on one dimensional rowkeys of data, most studies attempt to provide a secondary index of

W. Lee et al. (eds.), *Proceedings of the 7th International Conference on Emerging Databases*, Lecture Notes in Electrical Engineering 461,
https://doi.org/10.1007/978-981-10-6520-0_2

spatial data in the HBase table format. However, because these are not designed to fully utilize the properties of HBase, inefficient I/O occurs during spatial query processing.

In this paper, we propose indexing and range query processing techniques to efficiently process large spatial data in HBase. Our proposed indexing method adaptively divides the space into quadrants like a quad tree, by reflecting the data distribution, and creates an MBR in each quadrant. These MBRs are used to construct a secondary index to access spatial objects. The index is stored as an HBase table, and accessed in a hierarchical manner.

This paper is organized as follows. Section 2 describes our data partitioning method, named Q-MBR, and the index structure that employs it. Section 3 describes the algorithms for insertion and range query using the Q-MBR tree. In Sect. 4, we experimentally evaluate the performance of our index and algorithms. Finally, Sect. 5 concludes the paper.

2 Spatial Data Indexing Using Quadrant-Based MBR

2.1 Data Partitioning with Quadrant-Based MBR

We split the space using a quadrant based minimum bounding rectangle, named a Q-MBR. To construct the Q-MBR, we divide the space into quadrants, and create an MBR for the spatial objects in each quadrant. If the number of spatial objects in an MBR exceeds a split threshold, then the quadrant is recursively divided into smaller-sized sub-quadrants and MBRs are created for each sub-quadrant. Note that this partitioning method can create an MBR containing only a single spatial object. Figure 1 shows an example of the Q-MBR. The table shown in the figure is a list of Q-MBRs generated by the points on the left side of the figure. In this example, we assume that the capacity of the Q-MBR is four.

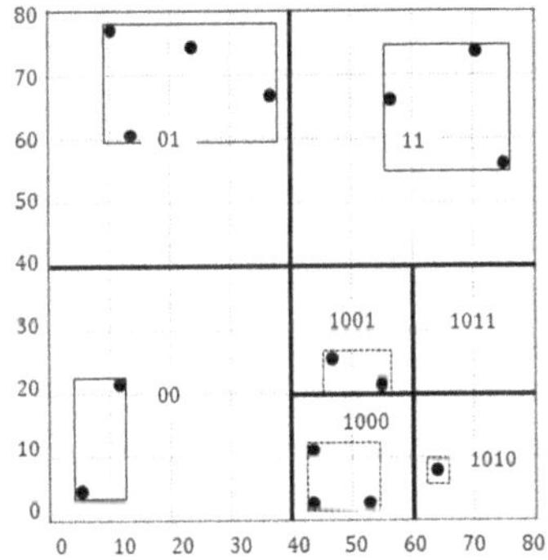

Quadrant	MBR
00	(5,5), (22,12)
01	(10,60), (36,78)
10	(43,2), (64,26)
1000	(13,3), (53,11)
1001	(46,22), (55,26)
1010	(64,8), (64,8)
11	(56,57), (75,74)

Q-MBR

Fig. 1. An example of quadrant-based MBR

As shown in Fig. 1, Q-MBR contains both information about the quadrant and the MBR. The reason for maintaining information on both is to store the Q-MBR in the HBase table and use it as the building block of our hierarchical index structure.

The quadrant information is used as the rowkey, in order to reduce the cost of updating. If the precise MBR information is used as a rowkey, then we frequently have to create a new rowkey, because MBR information is update-sensitive. In the worst case, we create a new rowkey and redistribute every spatial object when each update occurs. Hence, the MBR information is stored in a column, which is relatively inexpensive to update. This MBR information is used for distance calculations in spatial query processing.

2.2 Hierarchical Index Structure

The spatial objects in Q-MBR are accessed in a hierarchical manner through an index tree. This index tree, named a Q-MBR tree, is implemented in an HBase table format. The structure of a Q-MBR tree is similar to that of a quad-tree. The properties of a Q-MBR tree are as follows. First, while related techniques sort spatial objects in z-order and group objects according to the auto-sharding of the table, Q-MBR can group spatial objects into smaller units as the user requires. The next property is that a Q-MBR tree does not require an additional index structure, such as a BGRP tree or the R+ tree of KR+ tree, in order to build and maintain itself. The structure of a Q-MBR tree node is described in Table 1.

Table 1. Structure of a Q-MBR tree node

Type	Component	Description
Internal node	Quadrant	The binary values of quadrant information
	MBRs of children	The coordinates of the lower left and the upper right corners of the MBR
	Number of objects	Number of objects in sub-tree
Leaf node	Quadrant	The binary values of the quadrant information
	Data objects	The list of spatial objects in this node
	Number of objects	Number of objects in this node

An internal node consists of the quadrant information of the node, the MBRs of the child nodes and the number of objects included in their sub-tree. The quadrant information of the node can be represented by binary values. When we split a node, the newly created sub-quadrants can be enumerated according to the z-order. For example, if partitioning occurs at the root node, then the sub-quadrants are named using two-bit values, such as 00, 01, 10 and 11. If the sub-quadrant is recursively partitioned, then the name of the sub-quadrant is created by concatenating the name of their parent with the newly created two-bit name.

A leaf node consists of a quadrant, a list of spatial objects, and the number of objects in this leaf node. The quadrant information and number of objects are similar to their counterparts for an internal node. The difference lies in the list of spatial objects. HBase provides a function of data filtering in order to only transmit data of interest to a client. We define the filtering function as computing the distance between a query point and a spatial object. Therefore, the list of spatial objects contains their coordinates and ids. Figure 2 shows an example of a hierarchical Q-MBR index structure for spatial objects shown in Fig. 1. In this example, we assume that capacity of a leaf node is four.

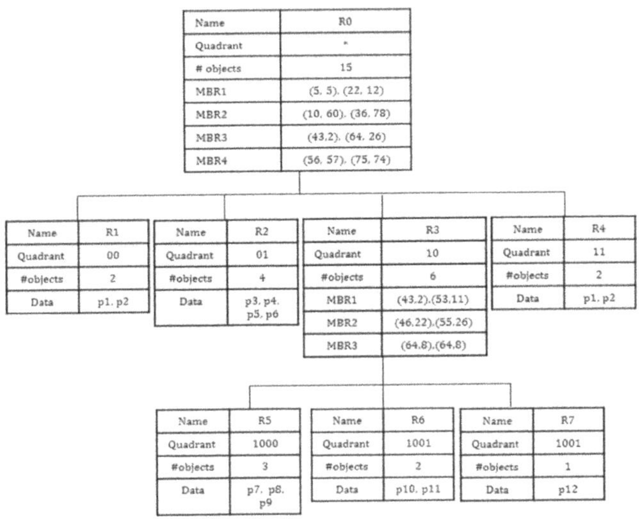

Fig. 2. An example of a Q-MBR tree

2.3 Representation of a Q-MBR Tree in HBase

In order to store a Q-MBR tree in an HBase table, it is necessary to design a schema that supports effective I/O considering the characteristics of HBase. In particular, because leaf nodes storing a group of spatial object have a large number of entries, designing table schema for efficiently loading leaf nodes from the table is important to improve the overall performance of index traversing. Due to the flexibility in schemas of HBase, a table of HBase can take one of two forms: tall-narrow and flat-wide. A tall-narrow table has a large number of rows with few columns, and a flat-wide table consists of a small number of rows with many columns.

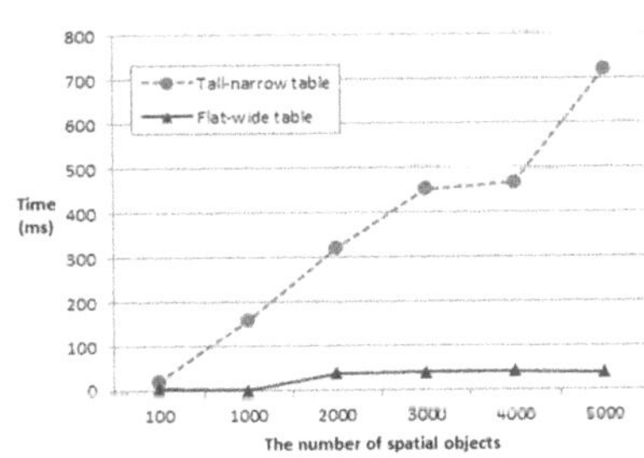

Fig. 3. Response time for loading spatial objects from an HBase table

The format of wide table is more appropriate for loading a large-sized leaf node from the HBase table. Because the spatial objects stored in a single row have the same rowkey, the time required for data fetching from a wide table is shorter. Figure 3 shows the response times for loading spatial objects from a tall-table and a wide-table. The x-axis of the graph indicates the number of spatial objects loaded from the HBase table at each time. In the case of the tall-narrow table, a desired number of rows are read at a time through a *scan* operation. In the flat-wide table, the number of spatial objects read at a time is stored in a single row, and these are obtained through a *get* operation. As shown in the figure, loading objects from the wide table delivers a better performance. Based on this observation, we store the spatial objects in each leaf node in a single row, and add a new column entry whenever a spatial object is inserted.

Row key (Quadrant)	Meta family		MBR family				Data Family			
	Is Leaf (Boolean)	# objects	00	01	10	11	d1	d2	d3	d4
Root	0	15	(5,5), (22,12)	(10,60), (36,78)	(43,2), (64,26)	(56,57), (75,74)				
00	1	2					p_1	p_2		
01	1	4					p_3	p_4	p_5	p_6
10	0	6	(43,2), (53,11)	(46,22), (55,26)	(64,8), (64,8)					
1000	1	3					p_7	p_8	p_9	
1001	1	2					p_{10}	p_{11}		
1010	1	1					p_{12}			
11	1	3					p_{13}	p_{14}	p_{15}	

Fig. 4. Table for a Q-MBR tree node

Figure 4 presents the table of the Q-MBR tree for the example in Fig. 2. An internal node, such as the root or 10, has a column family of MBRs that indicates the MBR information of its children. On the other hand, leaf nodes contain a column family of data objects, which maintains a list of spatial objects. For the purpose of illustration, the column qualifier of each spatial object is enumerated from d1 to d4. However, in order to use column filtering, each spatial object should have a unique column qualifier consisting of their coordinate values. The splitting threshold of a leaf node is determined according to the batch size of the RPC in the HBase system. The batch size is the unit size of a transmission in the HBase system. This can be defined according to the requirements of the user.

3 Algorithms of Spatial Data Insertion and Range Query Processing

3.1 Insertion Algorithm for a Q-MBR Tree

Algorithm 1 presents the insertion algorithm for a Q-MBR tree. The algorithm inserts a spatial object into the leaf node whose quadrant covers the location of the spatial object. To find the appropriate leaf node, the algorithm retrieves the Q-MBR tree using the quadrant information for each node. This searching process is described in lines 2 to 7. The MBR information of children and the number of spatial objects are updated when the internal nodes are traversed. After updating the information of the current traversed node, the algorithm calculates the rowkey of the next node, and loads this from the table for the next iteration. If the appropriate leaf node is found, then the spatial object is added to the *dataFamily* of the leaf node. When the number of spatial objects in a leaf node exceeds the split threshold, the function SplitNode() is called to split the leaf node. The function SplitNode(*node*) returns an internal node that is the result of partitioning. The splitting process creates new children by dividing the quadrant into four sub-quadrants, and redistributes the spatial objects into newly created leaf nodes.

Algorithm 1. Insert data point for a Q-MBR tree

Input : Spatial object ***p***, split threshold ***S***
Output : The updated Q-MBR tree after insertion

```
node N←root node of the Q-MBR tree
while(N is not a leaf node)
   For(each child C of N ) do
      if(quadrant of C covers p)
         update MBR of C in N
         increase object counter of N
         N ←C
if(object counter of N < S ) then
      add an spatial object p to DataFamily of N
      increase object counter of N
else if(size of N >= S ) then
   N =SplitNode(N)
   for(each child C of N) do
      if(quadrant of C covers p)
         update MBR of C in N
         add an spatial object p to DataFamily of C
         increase object counter of C
End
```

Figure 5 present an example of insertion. Suppose that the spatial object p1, marked with a star in the figure, is inserted in the Q-MBR tree from Fig. 5(a). The algorithm starts with an examination of the root node R0. Because the quadrant of R2 covers the location of p1, the algorithm updates the MBR of R2 in the root node, and loads R2 from the index table. The next step is to insert the object p1 into R2. However, the number of objects in R2 exceeds the split threshold after this insertion. Therefore, R2 is split into sub-quadrants, and the spatial objects in R2 are redistributed to the children. As a result, the new leaf nodes R8, R9 and R10 are inserted into the index table, as shown in Fig. 5(b).

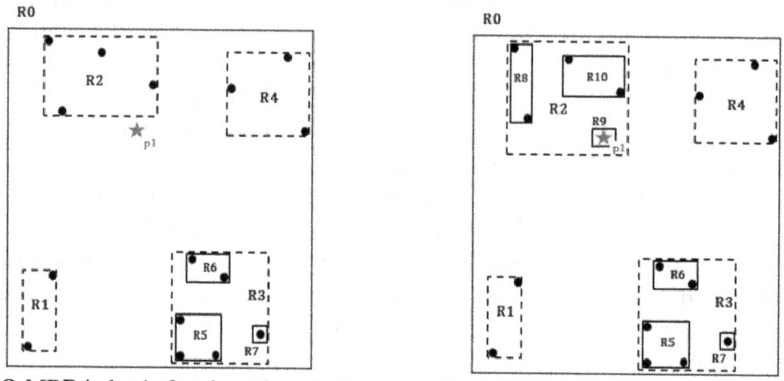

(a) Q-MBR index before insertion of p1 (b) Q-MBR index after insertion of p1

Fig. 5. An example of insertion algorithm

3.2 Range Query Algorithm for a Q-MBR Tree

The range query receives a query point ***q*** and query radius ***r*** as input, and returns a set of data points whose distance from the query point is less than ***r***. Our algorithm for processing a range query is presented in Algorithm 2. The proposed algorithm explores the Q-MBR tree in BFS (breadth-first-search) order, and reads as many rows from the index table as possible at each time, in order to reduce the number of data requests to the region server. Two sets, named ***Nt*** and ***Rk*** in the algorithm, are maintained for this processing. The first, ***Nt***, stores the nodes that are required to be traversed in the current iteration. ***Rk*** is a set of rowkeys to be loaded from the index table for the next iteration of the algorithm. The algorithm is terminated if there are no more nodes in either of the sets.

The algorithm starts by inserting rowkey of the root node into ***Rk***, and loading it from the index table. If the current traversed node is an internal node, then the algorithm calculates the minimum distance between the MBRs of its children and the query point ***q***. The rowkeys of the child nodes with distance less than the query radius ***r*** are inserted into ***Rk***. After all of the nodes in ***Nt*** have been traversed, the algorithm loads nodes from ***Rk*** from the index table, and stores the result into ***Nt*** for the next iteration. When the current traversed node is a leaf node, the algorithm calculates the distance between the spatial objects and the query point in order to answer the query. If the distance between a spatial object ***p*** and the query point ***q*** is less than or equal to the query radius ***r***, then the algorithm inserts ***p*** into the result set ***R***.

Algorithm 2. Range query algorithm

Input : Query point ***q*** and range ***r***
Output : A set of spatial objects within a given query range

```
Nt ←∅

Rk ←∅

Result set R ←∅
insert rowkey of root node into Rk
while Rk is not empty do
   Nt=Nt ∪{Prefetch nodes in Rk}
   while Nt is not empty do
      for (each node n ∈Nt )do
         if(n is a leaf node) then
            for (each object p ∈ n) do
               if(distance(p,q) <=r) then
                  R=R∪{p}
         else if(n is not a leaf node) then
            for (each child c of n) do
               if(overlap(MBR of c, q, r)) then
                  insert rowkey of c into Rk
Return R
```

Figure 6 shows an example of a range query. The algorithm starts by inserting therowkey of R0 into Rk and loading it into *Nt*. There are two children, R2 and R3, which overlap with the query range. Therefore, the rowkeys of R2 and R3 are inserted into ***Rk*** at the first iteration, and loaded together from the index table. Similarly, the rowkeys of R9 and R6 are inserted and loaded at the second iteration. Because R9 and R6 are leaf nodes, the next loop inspects the data objects of R9 and R6 in order to answer the query. As a result, the result set R contains the two points, p1 and p2, and the algorithm is terminated, because there are no more nodes to traverse.

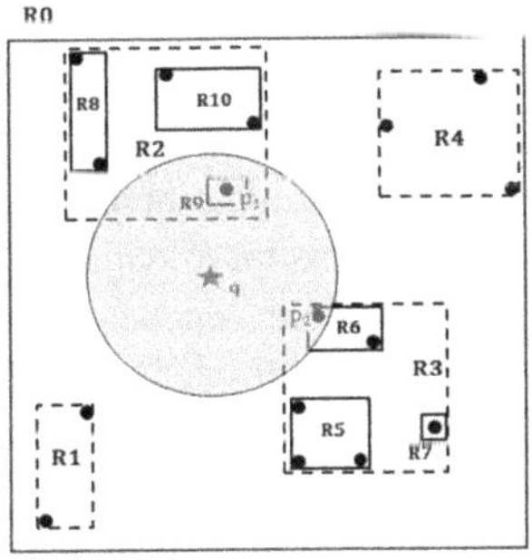

Fig. 6. An example of a range query

4 Performance Analysis

4.1 Experimental Setup and Datasets

We use synthetically generated databases, to control the size and distribution of the data. The first synthetic database contains two-dimensional uniformly distributed data, and the second contains two-dimensional data that follows a normal distribution. We implemented our method on a pseudo-distributed HBase cluster of four nodes, using HBase 0.98.0 and Hadoop 2.4.0 as the underlying system. Our experiments were performed on a physical machine that consists of a 2.5 GHz quad-core, 32 GB memory, and a 1 TB HDD, and runs 64bit Linux.

For all of the experiments, we compare our method (labeled as Q-MBR tree in the graphs) with MD-HBase [8] (labeled as MD-HBase in the graphs), and KR+ tree [9]. The average response time of 100 random queries is used for comparison. We set the parameters of MD-HBase and KR+ tree according to the analysis of [9]. MD-HBase must determine the capacity of the grid cell to group spatial objects. We set the threshold to 2500. The parameters of KR+ tree consist of the lower and upper bounds of the rectangle, and the order of the grid. We set the boundary of rectangle to (100, 50), and the order to eight. Q-MBR tree also uses the capacity of the Q-MBR as the parameter. In varying the capacity, there is a trade-off between the complexity of the index and the selectivity. We set the capacity of Q-MBR to 1024 after measuring the performance of a range query for one million datasets.

4.2 Performance Evaluation for Range Query

Effect of Query Radius

For these experiments, the database size was fixed at 10 million. Figure 7 plots the response times of range queries with a query radius increasing from 0.5% to 5% of the space. As shown in Fig. 7, Q-MBR tree outperforms MD-HBase and KR+ tree. In particular, when the size of the retrieved data increases, Q-MBR tree achieves a better performance than the other two methods because the time for loading data objects from the table is shorter. Although the table structure of KR+ tree is similar to that of Q-MBR tree, the performance of KR+ tree is inferior to that of Q-MBR tree. The reason for this is that the range query algorithm of KR+ tree is based on a key table produced by grid partitioning.

Effect of Database Size

For this set of experiments, the database size was increased from one million to 10 million points. The query radius was set as constant. Figure 8 shows that as the database size increases, the response time also increases for all methods. However, as can be seen from the figure, the rate of this increase in the response time of the Q-MBR tree is lower than for the other two methods. Because the three parameters required by KR+ tree are sensitive to the data distribution, this method shows the worst performance in this experiment.

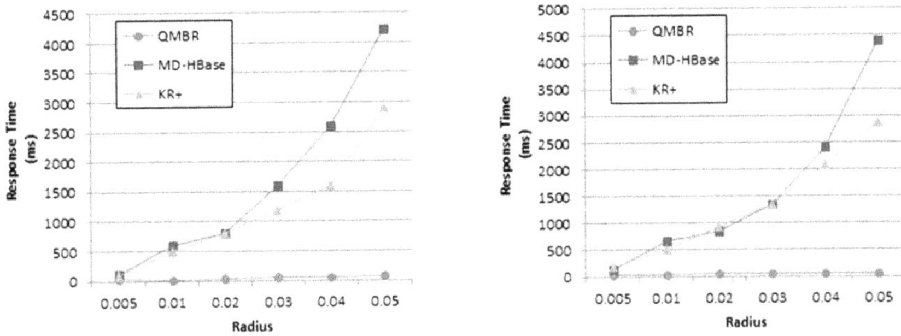

(a) Dataset with uniform distribution (b) Dataset with normal distribution

Fig. 7. Effect of query radius on the response time

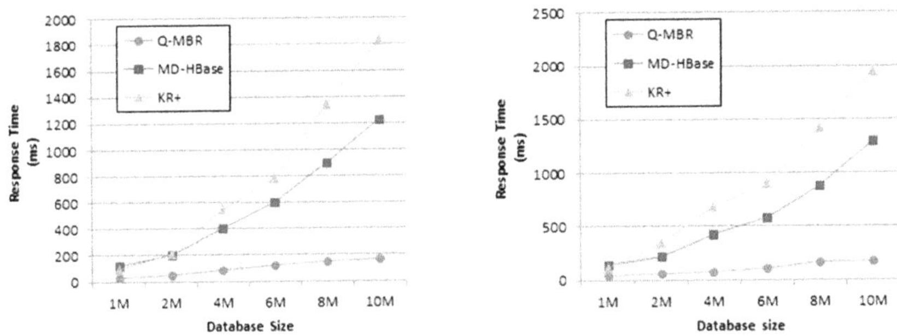

(a) Dataset with uniform distribution (b) Dataset with normal distribution

Fig. 8. Effect of database size on the response time

5 Conclusion

In this paper, we have presented Q-MBR tree, an efficient index scheme for handling large scale spatial data on an HBase system. The proposed scheme recursively divides the space into quadrants, and creates MBRs in each quadrant in order to construct a hierarchical index. Q-MBR provides better filtering power for processing spatial queries than existing schemes. A Q-MBR tree is stored in a flat-wide table, in order to enhance the performance of index traversal. Algorithms for range queries using Q-MBR tree have also been presented in this paper. Our proposed algorithms significantly reduce the query execution times, by prefetching the necessary index nodes into memory while traversing the Q-MBR tree. Experimental results demonstrate that our proposed algorithms outperform those of the existing two methods, MD-HBase and KR + tree. We are currently developing an effective kNN query algorithm suitable for Q-MBR tree.

References

1. Cary, A., Sun, Z., Hristidis, V., Rishe, N.: Experiences on processing spatial data with MapReduce. In: SSDBM, Lecture Notes in Computer Science Scientific and Statistical Database Management, pp. 302–319 (2009)
2. Wang, K., Han, J., Tu, B., Dai, J., Zhou, W., Song, X.: Accelerating spatial data processing with MapReduce. In: IEEE 16th International Conference on Parallel and Distributed Systems, pp. 229–236 (2010)
3. Zhang, S., Han, J., Liu, Z., Wang, K., Feng, S.: Spatial queries evaluation with MapReduce. In: Eighth International Conference on Grid and Cooperative Computing, pp. 287–292 (2009)
4. Eldawy, A., Mokbel, M.F.: SpatialHadoop: a MapReduce framework for spatial data. In: 2015 IEEE 31st International Conference on Data Engineering, pp. 1352–1363 (2015)
5. Aji, A., Wang, F., Vo, H., Lee, R., Liu, Q., Zhang, X., Saltz, J.: Hadoop GIS. a high performance spatial data warehousing system over MapReduce. Proc. VLDB Endowment **6** (11), 1009–1020 (2013)
6. Carlini, E., Lulli, A., Ricci, L.: Dragon: multidimensional range queries on distributed aggregation trees. Future Gener. Comput. Syst. **55**, 101–115 (2016)
7. Li, J.F., Chen, S.P., Duan, L.M., Niu, L.: A PR-quadtree based multi-dimensional indexing for complex query in a cloud system. In: Cluster Computing, pp. 1–12 (2017)
8. Nishimura, S., Das, S., Agrawal, D., Abbadi, A.E.: MD-HBase: design and implementation of an elastic data infrastructure for cloud-scale location services. Distrib. Parallel Databases **31**(2), 289–319 (2012)
9. Van, L.H., Takasu, A.: An efficient distributed index for geospatial databases. In: Lecture Notes in Computer Science Database and Expert Systems Applications, pp. 28–42 (2015)
10. Wei, L., Hsu, Y., Peng, W., Lee, W.: Indexing spatial data in cloud data managements. Pervasive Mob. Comput. **15**, 48–61 (2014)

Migration from RDBMS to Column-Oriented NoSQL: Lessons Learned and Open Problems

Ho-Jun Kim, Eun-Jeong Ko, Young-Ho Jeon, and Ki-Hoon Lee(✉)

School of Computer and Information Engineering,
Kwangwoon University, 20 Kwangwoon-ro, Nowon-gu,
Seoul 01897, Republic of Korea
kihoonlee@kw.ac.kr

Abstract. Migration from RDBMS to NoSQL has become an important topic in a big data era. This paper provides a comprehensive study on important issues in the migration from RDBMS to NoSQL. We discuss the challenges faced in translating SQL queries; the effect of denormalization, secondary indexes, and join algorithms; and open problems. We focus on a column-oriented NoSQL, HBase, because it is widely used by many Internet enterprises such as Facebook, Twitter, and LinkedIn. Because HBase does not support SQL, we use Apache Phoenix as an SQL layer on top of HBase. Experimental results using TPC-H show that column-level denormalization with atomicity significantly improves query performance, the use of secondary indexes on foreign keys is not as effective as in RDBMSs, and the query optimizer of Phoenix is not very sophisticated. Important open problems are supporting complex SQL queries, automatic index selection, and optimizing SQL queries for NoSQL.

Keywords: Migration · RDBMS · NoSQL · HBase · Phoenix · Denormalization · Secondary index · Query optimization

1 Introduction

NoSQL databases have become a popular alternative to traditional relational databases due to the capability of handling big data, and the demand on the migration from RDBMS to NoSQL is growing rapidly [1]. Because NoSQL has different data and query model comparing with RDBMS, the migration is a challenging research problem. For example, NoSQL does not provide sufficient support for SQL queries, join operations, and ACID transactions.

This research was supported by Basic Science Research Program through the National Research Foundation of Korea (NRF) funded by the Ministry of Science, ICT & Future Planning (NRF-2015R 1C 1A 1A02036517).

W. Lee et al. (eds.), *Proceedings of the 7th International Conference on Emerging Databases*, Lecture Notes in Electrical Engineering 461,
https://doi.org/10.1007/978-981-10-6520-0_3

In this paper, we provide a comprehensive study on important issues in the migration from RDBMS to NoSQL. We make three main contributions. First, we investigate the challenges faced in translating SQL queries for NoSQL. Second, we evaluate the effect of denormalization, secondary indexes, and join algorithms on query performance of NoSQL. Third, we identify open problems and future work. We focus on HBase because it is widely used by many Internet enterprises such as Facebook, Twitter, and LinkedIn. Because HBase does not support SQL, we use Apache Phoenix as an SQL layer on top of HBase.

Experimental results using TPC-H show that column-level denormalization with atomicity significantly improves query performance, the use of secondary indexes on foreign keys is not as effective as in RDBMSs, and the query optimizer of Phoenix is not very sophisticated. Important open problems are supporting complex SQL queries, automatic index selection, and optimizing SQL queries for NoSQL.

The remainder of this paper is organized as follows. Section 2 presents background and related work. Section 3 discusses important issues in the migration from RDBMS to column-oriented NoSQL. Section 4 presents experimental results on the issues and open problems. Section 5 provides conclusions.

2 Background and Related Work

HBase is a column-oriented NoSQL and uses Hadoop Distributed File System (HDFS) as underlying storage for providing data replication and fault tolerance. HBase does not support SQL queries and secondary indexes. Apache Phoenix works as an SQL layer for HBase by compiling SQL queries into HBase native calls and supports secondary indexes.

Reference [1] proposed a denormalization method called CLDA that avoids join operations and supports atomicity using the notions of column-level denormalization and atomic aggregates. The CLDA method improves query performance with less space compared with table-level denormalization methods [2–8], which duplicate whole tables. For a column-oriented NoSQL, [9] proposed a column partitioning algorithm. Reference [10] studied the implementation of secondary indexes for HBase.

3 Migration from RDBMS to Column-Oriented NoSQL

In this section, we provide a comprehensive study on important issues in the migration from RDBMS to HBase with Phoenix. The issues are exemplified and discussed using a case study on TPC-H.

3.1 Translating SQL Queries

Phoenix does not provide sufficient support for complex SQL queries with complex predicates, subqueries, and views. To migrate such complex queries, we need to simplify complex queries using query unnesting techniques [11–14] and temporary tables.

For example, benchmark queries of TPC-H are very complex, and Phoenix does not sufficiently support queries Q11, Q15, Q18, Q19, and Q21. For Q11, we unnest the subquery in the HAVING clause because Phoenix does not support it. For Q15, we store the result of a view into a temporary table because Phoenix supports only a view defined over a single table using a SELECT * statement. For Q18, we unnest the subquery with the GROUP BY and HAVING clauses because Phoenix produces wrong results. For Q19, Phoenix does not efficiently evaluate a complex predicate of the disjunctive normal form, which is a disjunction of multiple condition clauses. For the query, Phoenix does not push down predicates. To efficiently evaluate the query, we compute results for each condition clause and union the results using temporary tables. For Q21, we unnest the subqueries because Phoenix does not support non-equi correlated-subquery conditions.

3.2 Denormalization

Because NoSQL systems do not efficiently support join operations, we need denormalization, which duplicates data so that one can retrieve data from a single table without joining multiple tables. To denormalize relational schema, we use the method called *Column-Level Denormalization with Atomicity* (*CLDA*) [1], which is the state-of-the-art denormalization method. Although CLDA was originally proposed for a document-oriented NoSQL, it is general enough to be applied to other types of NoSQL. CLDA avoids join operations without denormalizing entire tables by duplicating only columns that are accessed in non-primary-foreign-key-join predicates. CLDA also combines tables that are modified within the same transaction into a unit of atomic updates to support atomicity.

For example, Fig. 1 shows TPC-H Q8 where non-primary-foreign-key-join predicates are shaded. If we add `r_name` to `orders` and `p_type` to `lineitem`, we can avoid "`orders` ⋈ `customer` ⋈ `nation` ⋈ `region`" and "`lineitem` ⋈ `part`." Table 1 shows the columns duplicated by CLDA for the 22 TPC-H queries. The name of each column contains the names of the foreign keys. The number of duplicated columns is small because there are common columns appearing in multiple non-primary-foreign-key-join predicates. According to the TPC-H specifications, the `lineitem` and `orders` tables should be modified within the same transaction. To support transaction-like behavior, CLDA combines the `lineitem` and `orders` tables into a single table. Thus, we can avoid "`orders` ⋈ `lineitem`" with atomicity.

```
select
    o_year,
    sum(case
            when nation = 'BRAZIL' then volume
            else 0
    end) / sum(volume) as mkt_share
from
    (
        select
            extract(year from o_orderdate) as o_year,
            l_extendedprice * (1 - l_discount) as volume,
            n2.n_name as nation
        from
            part, supplier, lineitem, orders, customer,
            nation n1,nation n2, region
        where
            p_partkey = l_partkey
            and s_suppkey = l_suppkey
            and l_orderkey = o_orderkey
            and o_custkey = c_custkey
            and c_nationkey = n1.n_nationkey
            and n1.n_regionkey = r_regionkey
            and r_name = 'AMERICA'
            and s_nationkey = n2.n_nationkey
            and o_orderdate between
                date '1995-01-01' and date '1996-12-31'
            and p_type = 'ECONOMY ANODIZED STEEL'
    ) as all_nations
group by
    o_year
order by
    o_year;
```

Fig. 1. TPC-H Q8.

Table 1. Columns duplicated by the CLDA method for the 22 TPC-H queries.

Table	Duplicated columns
supplier	s_nationkey_n_name
partsupp	ps_partkey_p_brand ps_partkey_p_type ps_partkey_p_size ps_suppkey_s_nationkey_n_name ps_suppkey_s_nationkey_n_regoinkey_r_name
orders	o_custkey_c_nationkey o_custkey_c_mktsegment o_custkey_c_nationkey_n_name o_custkey_c_nationkey_n_regoinkey_r_name
lineitem	l_partkey_p_name l_partkey_p_brand l_partkey_p_type l_partkey_p_size l_partkey_p_container l_suppkey_s_nationkey l_suppkey_s_nationkey_n_name

3.3 Secondary Indexes

Phoenix offers a secondary index on top of HBase using an index table, which consists of index columns and the primary key of the indexed data table. The query optimizer of Phoenix internally rewrites the query to use the index table if it is estimated to be beneficial. If the index table does not contain all the columns referenced in the query, Phoenix accesses the data table to retrieve the columns not in the index table. Phoenix also offers a covered index, which is an index that contains all the columns referenced in the query. Using a covered index, we can avoid the costly access to the data table, but the overhead of data synchronization and space consumption increase.

3.4 Join Algorithms

Phoenix supports a sort-merge join and a broadcast hash join. The broadcast hash join first computes the result for the expression at the right-hand side of a join condition and then broadcasts the result onto all the cluster nodes; each cluster node has a partition of the table at the left-hand side and computes the join locally. When both sides of the join are bigger than the available memory size, the sort-merge join should be used. Currently, the query optimizer of Phoenix does not make this determination by itself. We can force the optimizer to use a sort-merge join by using the USE_SORT_MERGE_JOIN hint.

4 Experimental Evaluation

4.1 Experimental Setup

For the migration from RDBMS to HBase with Phoenix, we evaluate the effect of denormalization, secondary indexes, and join algorithms on query performance. Using the TPC-H benchmark with scale factors (SFs) 1 and 10, we measure the average query execution time for the TPC-H queries. For each query, we first run the query once to warm up the cache and then measure the average execution time for two subsequent runs.

We use HBase 0.9.22, Phoenix 4.8.1, and MySQL 5.7.18. All experiments were conducted on a cluster of four PCs with an Intel Core i5-6600 CPU, 16 GB of memory, Samsung 850 PRO 256 GB SSDs, and Ubuntu 16.04. We set the JVM memory to 12 GB. One PC is a master, and the other three PCs are slaves. For MySQL, we use only one PC.

We conduct the following experiments.

Experiment 1: The effect of denormalization
To see the effect of denormalization, we compare query performance for the denormalized schema generated by the CLDA method and for the normalized schema, which has a one-to-one correspondence with the relational schema. We also compare database size. We use secondary indexes on foreign keys and the USE_SORT_MERGE_JOIN hint for all the queries.

Experiment 2: The effect of secondary indexes on foreign keys
To see the effect of secondary indexes on foreign keys, we compare query performance for databases with and without secondary indexes on foreign keys. We use the normalized schema and the USE_SORT_MERGE_JOIN hint for all the queries. We also run the same test for MySQL to see the effect of secondary indexes on RDBMS.

Experiment 3: The effect of join algorithms
To see the effect of join algorithms, we compare query performance with and without the USE_SORT_MERGE_JOIN hint. We exclude queries that are failed due to out-of-memory errors if the USE_SORT_MERGE_JOIN hint is not used. We use foreign key indexes and the normalized schema.

4.2 Experimental Results

Experiment 1: The effect of denormalization
Figure 2 shows that the CLDA method significantly improves query performance at the expense of using more space compared with the normalization method that uses the

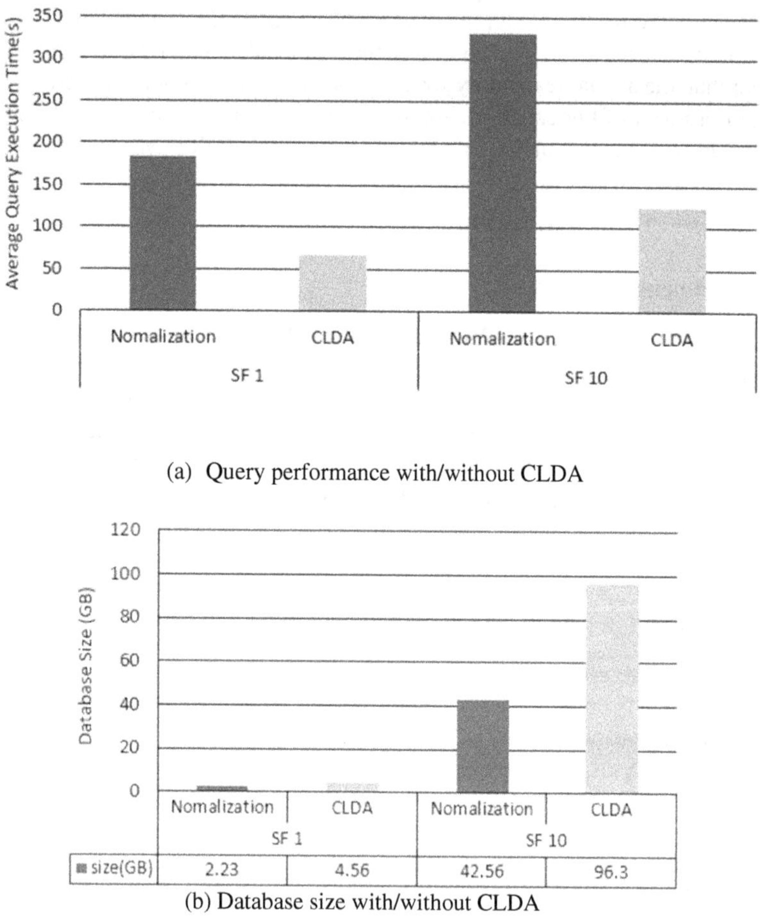

(a) Query performance with/without CLDA

(b) Database size with/without CLDA

Fig. 2. The effect of column-level denormalization with atomicity

relational schema as it is. This is because the CLDA method reduces the number of joins by duplicating columns. For SF 1, the CLDA method is 2.7 times faster, but uses 2.0 times more space. For SF 10, the CLDA method is 2.7 times faster, but uses 2.3 times more space. We note that, for SF 10, queries Q2, Q7, Q8, Q9, Q17, and Q21 failed for the normalization method; queries Q9, Q13, Q17, Q18, and Q21 failed for the CLDA method. Queries Q2, Q7, Q8, Q9, Q13, and Q18 failed due to out-of-memory errors; queries Q17 and Q21 failed due to HRegionServer failures. We exclude the failed queries.

Experiment 2: The effect of secondary indexes on foreign keys
For MySQL, secondary indexes on foreign keys are very effective. Without secondary indexes, 73% of queries (16 queries) takes more than one hour, and the average query execution time of the other 27% (6 queries) is 5.4 s even for SF 1. With secondary indexes, the average query execution time of all queries is 0.2 s for SF 1. Figure 3 shows that for HBase with Phoenix, the average query execution times with and without secondary indexes are almost the same for both SFs 1 and 10. For SF 10, we exclude the failed queries.

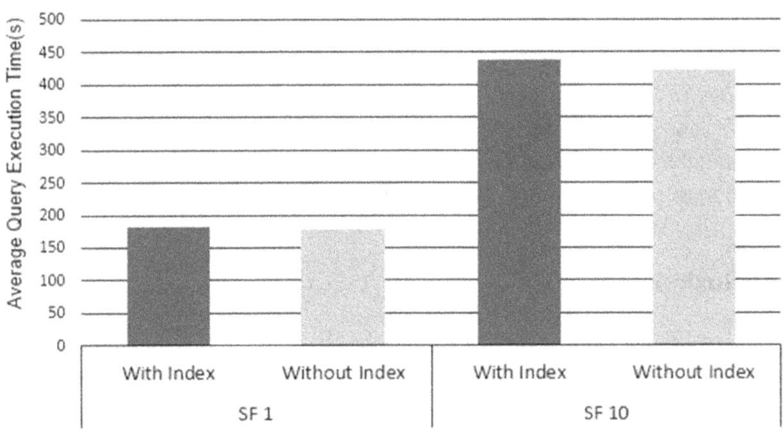

Fig. 3. Query performance with/without secondary indexes on foreign keys

Experiment 3: The effect of join algorithms
The broadcast hash join incurs out-of-memory errors for 27% of queries (6 queries) for SF 1 and 59% of queries (13 queries) for SF 10. For the other queries without any errors, the broadcast hash join improves the average query execution time by 2.0 times for both SFs 1 and 10 compared with the sort-merge join as shown in Fig. 4.

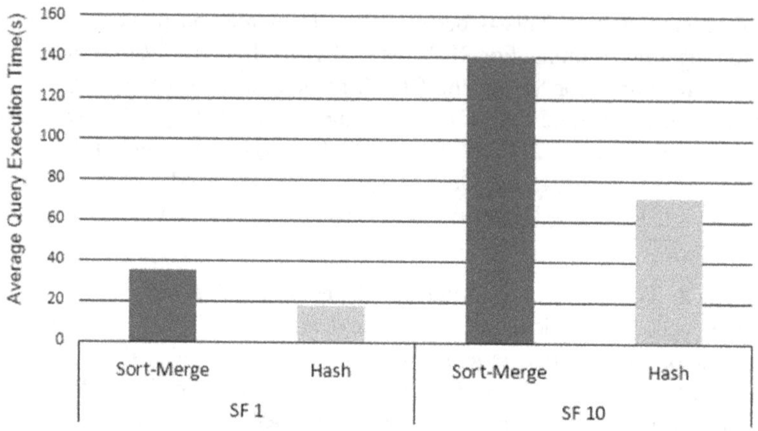

Fig. 4. Query performance with/without the USE_SORT_MERGE_JOIN hint

4.3 Discussion

Column-level denormalization with atomicity proposed for a document-oriented NoSQL is also effective for a column-oriented NoSQL. Because the secondary index is implemented outside HBase, it is not as efficient as in RDBMSs. We should use covered indexes for performance. Because the query optimizer of Phoenix does not consider the case where the broadcast hash join incurs out-of-memory errors, we often need to manually specify to use the sort-merge join. For a large database, many queries are failed due to out-of-memory errors or HRegionServer failures.

5 Conclusions

We summarized the challenges faced, lessons learned, and open problems for the migration from RDBMS to HBase with Phoenix. We addressed important issues of query translation, denormalization, secondary indexes, and join processing. Extensive experiments show that column-level denormalization with atomicity improves query performance by up to 2.7 times, the use of secondary indexes on foreign keys is not as effective as in RDBMSs, and the query optimizer of Phoenix is not very sophisticated. Important open problems for future work are supporting complex SQL queries, automatic index selection, and optimizing SQL queries for NoSQL.

References

1. Yoo, J., Lee, K.-H., Jeon, Y.-H.: Migration from RDBMS to NoSQL using column-level denormalization and atomic aggregates. J. Inf. Sci. Eng. **34**(1) (2018 to appear). http://journal.iis.sinica.edu.tw/paper/1/160464-2.pdf?cd=EF95C1D50DCDC958E
2. Karnitis, G., Arnicans, G.: Migration of relational database to document-oriented database: structure denormalization and data transformation. In: CICSyN, pp. 113–118 (2015)

3. Zhao, G., Lin, Q., Li, L., Li, Z.: Schema conversion model of SQL database to NoSQL. In: 3PGCIC, pp. 355–362 (2014)
4. Lee, C.-H., Zheng, Y.-L.: Automatic SQL-to-NoSQL schema transformation over the MySQL and HBase databases. In: IEEE ICCE-TW, pp. 426–427 (2015)
5. Zhao, G., Li, L., Li, Z., Lin, Q.: Multiple nested schema of HBase for migration from SQL. In: 3PGCIC, pp. 338–343 (2014)
6. Lee, C.-H., Zheng, Y.-L.: SQL-to-NoSQL schema denormalization and migration: a study on content management systems. In: IEEE SMC, pp. 2022–2026 (2015)
7. Vajk, T., Feher, P., Fekete, K., Charaf, H.: Denormalizing data into schema-free databases. In: IEEE CogInfoCom, pp. 747–752 (2013)
8. Vajk, T., Deak, L., Fekete, K., Mezei, G.: Automatic NoSQL schema development: a case study. In: PDCN, pp. 656–663 (2013)
9. Ho, L.-Y., Hsieh, M.-J., Wu, J.-J., Liu, P.: Data partition optimization for column-family NoSQL databases. In: IEEE Smart City, pp. 668–675 (2015)
10. Ge, W., Huang, Y., Zhao, D., Luo, S., Yuan, C., Zhou, W., Tang, Y., Zhou, J.: A secondary index with hotscore caching policy on key-value data store. In: ADMA 2014. LNCS, vol. 8933, pp. 602–615 (2014)
11. Lee, K.-H., Park, Y.-H.: Revisiting source-level XQuery normalization. IEICE Trans. Inf. Syst. **E94-D**(3), 622–631 (2011)
12. Lee, K.-H., Kim, S.-Y., Whang, E., Lee, J.-G.: A practitioner's approach to normalizing XQuery expressions. In: DASFAA 2006. LNCS, vol. 3882, pp. 437–453 (2006)
13. Kim, W.: On optimizing an SQL-like nested query. ACM Trans. Database Syst. **7**(3), 443–469 (1982)
14. Ganski, R., Wong, H.: Optimization of nested SQL queries revisited. In: ACM SIGMOD, pp. 23–33 (1987)

Personalized Social Search Based on User Context Analysis

SoYeop Yoo and OkRan Jeong

Gachon University, Seongnam 13120, Republic of Korea
bbusso90@gmail.com, orjeong@gachon.ac.kr

Abstract. As SNS is widely spreading on the development of internet and smart phones, many users are able to share various forms of social media such as texts, photos, and videos on the social network. The user context such as interest, relationship, or emotion exists in the social media contents. So it is very meaningful to analyze and use the information from the social media contents. In this paper, we propose personalized social search mechanism which is designed to provide the search results to the user based on user context analysis. We extract the user's information such as social relationship, interest, emotion, and the correlation of hashtags by user's social context. The proposed search system utilizes the analyzed results. To verify the performance of the proposed system, we have implemented to show the experiment results.

Keywords: Personalized social search · User context · Social search system · Social media mining

1 Introduction

Social Networking Service ('SNS') has started to spread widely due to many reasons such as the development of internet, spread of mobile devices, demand of users, and so on. SNS is of great research value because it possesses social data that the users make on their own and SNS is actually being actively studied at this time [1–4]. While the previous social networks were mostly text-oriented ones, the present social networks are mainly the multimedia-oriented ones that help to share various media such as photos and videos. Accordingly, the analysis of users' interest on social networks also should be focused on media such as photos or videos in addition to the texts as the subject of analysis.

We can figure out user's characteristics by analyzing contents because a large amount of social media contents on the social media sites are created by users. So many researches have been performed with social media contents analysis, analysis framework or tools.

As SNS is composed of the data that the users have created on their own, such characteristics of SNS should be taken into account in classifying the categories. For this reason, the Facebook category classification to which such characteristics seem to

W. Lee et al. (eds.), *Proceedings of the 7th International Conference on Emerging Databases*, Lecture Notes in Electrical Engineering 461,
https://doi.org/10.1007/978-981-10-6520-0_4

be applied has been reclassified by ODP (DMOZ ODP) basis which is the largest classification standard. We have defined this as a social category in [5]. In this study, in order to add category characteristics of multimedia data to the existing social category, the categories of iTunes and YouTube have been analyzed and adopted to enlarge the existing social category.

However, it is difficult to exactly figure out the user's characteristic only with interests. Lots of information is exposed or imposed on the contents through various methods. It is also important to analyze the social relationships and the interests as well as the user's emotions. The existing analysis frameworks analyze only the contents. To utilize the analyzed data in various ways, we need to make another framework or system.

It is possible to utilize the analyzed results of the social media contents in various ways such as recommendation or search. It is possible to aware user's social contexts. If we apply user's social contexts to search system, we could get personalized search results. For social search, it is important to aware about the unstructured data and apply user's social contexts on ranking algorithm.

In this study, we propose personalized social search mechanism based on user context analysis. The existing researches for the traditional search focus on ranking algorithm using the correlation between a query and data. Unlike this, we extract social contexts such as social relationship, interests, and emotions from user-generated social media contents and utilize the extracted features on ranking algorithm for the personalized search. We have validated the proposed system through implementation and experiments.

2 Social Context: Social Media Contents Analysis

The data extracted from the social media sites are used in many areas like multimedia recommendation or advertisement of books or goods by analyzing them. However, the gathered data are atypical because in the social media sites, the users are able to create contents without formality and constraint. As the sites have strengthened their privacy statement, we can only gather limited data. Under this restricted condition, we proposed a social context based personalized search system that can analyze the user's social relationships, interests, emotions, and association of hashtags on the social media contents.

2.1 Overall Structure

The social media content analysis is based on the contents that are created and shared by the users on the social media sites. Users share their interests and emotions by sharing contents with others and freely taking actions on each other. Analyzing the user's social relationships, interests, and emotions can be helpful to be utilized as a service for understanding the user's characteristics and providing the personalized search results.

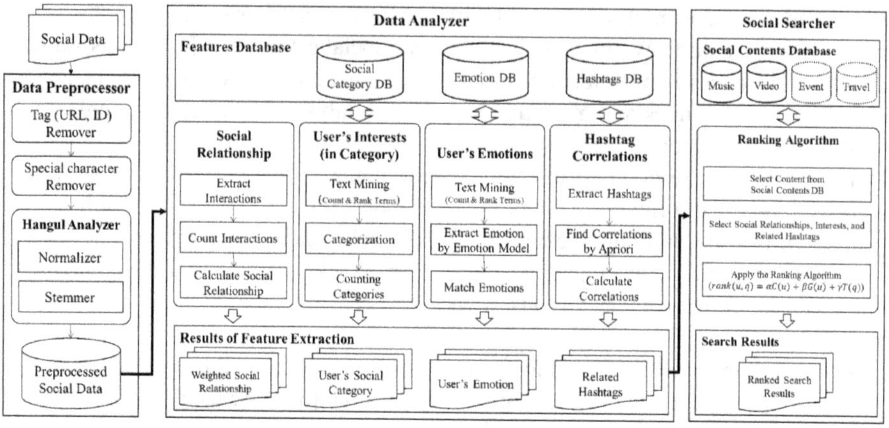

Fig. 1. Overall structure of personalized social search system based on user context analysis

Figure 1 shows the structure of personalized social search system based on user context analysis. It is composed of 3 modules: preprocessing module, analyzing module, and search module. The preprocessing module collects the social media contents from the sites and filters the gathered data. The analyzing module analyzes the user's social relationships, interests, emotions, and hashtag correlation. And the search module computes the contents score using the analyzed results.

2.2 Preprocessing Module

Users express their thoughts, interests, or social issues in various ways through the social media sites. Because the freedom of expression is guaranteed, the form of data exists as different kinds such as text, image, audio and video, etc. To utilize these atypical data, the process of handling atypical data to the fixed data is necessary.

Table 1. Abbreviations

Abbr.	Descriptions
DOCID	ID of document (Tweet)
Date	Tweeted date (YYYYMMDDHHMMSS)
TXT	Tweeted text
INRPLTOSTTSID	ID of original tweet (−1 if this is original tweet)
INRPLTOUSRID	Users' ID of original tweet (−1 if this is original tweet)
USRID	User's ID of tweet

In this study, we chose Twitter and gathered the tweet data to verify our proposed social search system which is using user's social contexts. We collected tweet data at random, and treated all data with tweet ID, date, text, and user's ID, etc. as shown in Table 1. We chose the most necessary features in a number of tweet data and used them for filtering to analyze the user's social relationships, interests, and emotions.

2.3 Data Analyzer

The most important part in the social search system in this study is the analyzing module. To grasp the user's characteristics reflected in the social media contents, it is necessary to analyze with more diversity and accuracy. For this reason, we chose the user's social relationships, interests, emotions, and hashtag correlation. The analyzing module can be divided into the parts of analyzing social relationships, interests, emotions, and related hashtags. The stages of social relationship analysis, interest analysis, emotion, and hashtag correlation analysis as key parts of analyzing user's characteristics from the social media contents will be specified.

Social Relationship Analysis. Social media sites have definitive friendships like the friendship in Facebook or the Follower/Followee in Twitter. However, these lists of friends are protected with the policy and the relationships exist without communication. For these reasons, there are some limits on analyzing the social relationship by just depending on the user's friends list. Therefore, we need to analyze new social relationships based on the social media contents and make a new list for social relationships. Although there have been studies on how to establish the social relationship network already, we made a different analysis of social relationship which is customized for our proposed system.

For the calculation of social relationship, it is necessary to calculate the weight based on the shared behavior between the user and the user's friends. Especially in case of [5], we can calculate how much weight a user assigns to a certain friend, based on the mutual behaviors like text, tag, or comment between the user and the friend using Eq. 1, where $\mathrm{u}_{\mathrm{actions}}$ indicates the number of user u, and $\mathrm{u} \cdot \mathrm{v}_{\mathrm{actions}}$ represents the number of interactions between the user u and the friend v.

$$SR(u, v) = \frac{u \cdot v_{actions}}{\sum u_{actions}} \tag{1}$$

The interaction which is used for calculating social relationships varies depending on the social media sites and exists as several different forms. If a user shares one content, other users show actions such as 'Likes', comments or sharing the content with others. Then we can regard these actions as the interactions between the user and other users. Especially in Twitter, we consider 'RT (retweet)' and 'Mention (message)' as interactions.

Interest Analysis. The user's interest is one of the most important features in the field of recommendation. On social media sites, the user's interests are reflected on the contents created by the users so that we analyze the user's interests by analyzing the data. There are many ways to analyze interests, but we will expand and use the analyzing method using the social category which is used in [5] for a simpler and easier analysis.

The category classification list as a basis of category classification is needed to be changed in compliance with the features of YouTube mentioned above. The category classification list for [5] was made by combining Facebook as a text-only based site with the ODP (Open Directory Project) of DMOZ as the largest classification site since

[5] was related to the text-based system. In this study, two more category lists (YouTube category list as a representative multimedia-based system plus iTunes category list as a representative site of Podcast enabling to share the user's self-video) have been added to the category classification list used in [5]. The same category classification list used in [13, 14] was selected for the Podcast category classification list. As a result, Facebook classification, ODP classification, Podcast classification, and YouTube classification have been integrated into a new classification list which is applicable to both texts-based and multimedia-based categories.

KOMORAN(v2.0.4), a Korean morpheme analyzer, provided by Shineware, has been used to classify the texts on the preprocessed social media contents into each word. Then the most frequently shown words from each category have been identified and ranked. The final category classification list has been created eventually. To evenly apply all properties of ODP, Facebook, YouTube, and Podcast, we made some rules for creating the new social category classification list. First, we use the category which is commonly used in more than 3 places. Then, we unite the similar categories into one category. For the example, the category 'Arts' is commonly used in ODP, Facebook, and Podcast while 'Animals' is only used in YouTube. So we select 'Arts' for our new social category list and do not choose 'Animals'. The final 15 classification categories are as follows: Arts, Business, Comedy, Education, Games, Health, Kids/Home, Movies, Music, New, Science, Society, Sports, Technology, and ETC.

Extracting the user's interests using social category is based on the data shared or generated from the users. It is the way of expressing the user's interests to generate contents or take action on other users such as comments or sharing. So it is possible to extract the category of user's interests by collecting contents and analyzing them.

Emotion Analysis. Many studies for analyzing emotions have been made since the user's emotions can change easily for many reasons such as their surroundings or mood, and these emotions are reflected in the life style. So we use the emotion analysis to analyze more accurately to be utilized in many areas. For the digitization of emotions, we use the Arousal-Valence model proposed by [15], the user's social relationship and text analysis. Since most users express emotions to their acquaintances for the most part, using this can be a meaningful analysis.

In this study, we have analyzed the user's emotions using the extraction method of MOAF proposed in [16]. To digitize and calculate the user's emotions, we have used the Arousal-Valence model. 'Arousal' represents the intensity of emotion. The higher the score of 'Arousal' is, the brighter the emotion is. And the lower the score is, the calmer the emotion is. 'Valence' represents the degree of positivity-negativity. A higher score means positivity, and a lower score means negativity. Also, we use ANEW [17] and WordNet [18] to create the emotion dictionary. With the emotion dictionary, we analyze the words, and digitize the emotions using an emotion model. Using this method, we can analyze user's emotions.

Hashtag Association Analysis. On social media sites, hashtags which are formed of '#+word' are widely used to express the topic of contents or user's interests and to provide faster retrieval. Using hashtags makes it easier to search on social media sites and to be connected with other unconnected users who have similar interests. Due to these advantages and usability of hashtags, they are used commonly.

Users could use one hashtag on one content, however, it is more general to use two or more related hashtags simultaneously because there is no limitation on the number of hashtags. Several hashtags are usually presented on one content, so it might be true that cooccurred hashtags have association. Analyzing correlation of hashtags could give an effect of related search and make it possible for users to get more various and accurate information.

In this paper, we compute hashtag association with 'Apriori'. However, it is easy to analyze and figure out the association using Apriori, it is necessarily required to filter hashtags. There are many typographical errors because of the characteristic of user-generated contents. Also there are many hashtags which are related to gambling and advertisement, so it is very important to filter these tags.

3 Social Search System

The analyzed social contexts of user's social relationships, interests, emotions, and hashtag correlation are used for searches. In this paper, we propose social search system using user's contexts. Unlike the existing traditional search engine, the proposed social search system has put goal to provide the personalized retrieved results to users by extracting imposed contexts in social media contents. We use social context which analyzes user's social relationship, interests, emotions, and hashtag association for searching.

The social search system provides the search results which are applied to ranking algorithm with analyzed social contexts such as social relationship, interests, and related hashtags. One of the most important elements in the search system is ranking algorithm. Most users of a search system want to access the necessary information faster with the minimum click. Because of these reasons, ranking algorithm which determines order of the results is very important. In this paper, we rank the search results using the analyzed social contexts, user's social relationship, interests, and the related hashtags.

$$\mathrm{rank}(u,q) = \sum (\alpha S(u) + \beta G(u) + \gamma T(q)) \tag{2}$$

Equation 2 computes the rank of each retrieved result if a user u enters a query q. α, β, and γ are the weight for social relationship, interests based group, and related hashtags of the query respectively. The sum of three weights is 1. These weights are used to emphasize the importance of each factors. If the user searches with the query, we assort users with the categories of interests and social relationships. Users could get more useful information in the shared contents which are generated by users who have similar interests. Also if there are associated hashtags of the query, users could access the related information easily.

4 Experiments

To verify the performance of the proposed mechanism, we have implemented to show the experiment results. We have implemented the system of analyzing and recommending social media contents. Then we have chosen Twitter for experiments to prove the validation of our proposed framework.

4.1 Experiment Method

In this study, we propose a social context based search system through analyzing user's social relationship, interests, emotions, and hashtag association. The social search system could not only analyze various social contexts of users, but also provide the retrieved results based on the analyzed results. In order to prove the validation on how our proposed system can be effective, Twitter has been used for experiments. We have collected 1 million tweet data for 15 days. For the experiments, we have used 1 million tweets and 53,064 users.

Also, we have used the biggest classification site called ODP as the base in addition to the categories of Facebook, YouTube, and iTunes Podcast to build a social category classification list which is used in the category extractor. About 1330 data of Facebook pages, 600 data of YouTube videos, and 4400 data of Podcasts have been collected for the social category classification list.

To implement the search system based on the gathered tweet data, we have used JAVA. First, we have preprocessed the gathered data by filtering through the preprocessing module, computed the user's social relationships using 'RT' and 'Mention' based on the preprocessed data. Then we have analyzed the user's interests and emotions after applying the text mining to tweet data. We have used the visualization toolkits to make the users check the analyzed and retrieved result more easily.

We have applied the analyzed results on the ranking algorithm of the social search system. We have used three quarters of the gathered tweet data as training data, and the rest have been used as testing data to verify the ranking algorithm of our proposed system. To confirm the accuracy of our search results, we have calculated NDCG@10. NDCG [20] is the method to evaluate the quality of ranking, and it could calculate the average performance of ranking algorithm. It is very important for a search system to provide well ranked results, so we use NDCG.

4.2 Results

As the given query, social searchers find out the matched results on Twitter, Instagram, and YouTube, then show the ranked results based on the analyzed user's social contexts. Because it reflects user's features, the user could access data which he/she wants faster and more accurate. We also provide contents on Twitter, Instagram, and YouTube, so users can get broad data.

To normalize the rank score, we set the sum of three weights as 1. The weight of α which is for user's social relationship shows that the higher the weight, the higher the accuracy is. The score of the social relationship is based on the actions what the user act directly, so it works as an important feature in ranking algorithm. Following the experiment result, we set α as 0.5 to give more portion to social relationships. We also set the weight of user's interests, β and the query related hashtags, γ as 0.3 and 0.2, respectively.

We also experimented the comparison result with the existing research. Reference [11] proposed a social search method which could show every object of document, person, and tag when users enter queries. Reference [12] proposed a social search engine called Aardvark which indexed using persons not documents. We compared these two existing researches with our proposed results of social search.

Table 2. Comparison result

Test case	NDCG@10
Amitay et al. [11]	0.703
Horowitz et al. [12]	0.715
Social search with S	0.734
Social search with S and G	0.763
Proposed social search (with **S**, **G**, and **T**)	0.746

Table 2 shows the comparison result with the existing two researches and our proposed research with variations. S refers to social relationship, and G refers to the group based on user's interests. T indicates the related hashtags. 'Social Search with S' indicates the experiment result of our proposed system which applies user's social relationship only. Our proposed social search system shows 0.746 as the experiment result. When we compared with these two existing researches, we verified that the social context based ranking algorithm has relatively higher performance.

5 Conclusion

On the social media sites, lots of information, emotions, or contents are generated and shared. So we propose a system for analyzing and retrieving the social media contents. The existing social search system had been focused on user's social relationship or frequently used words. But we have preprocessed the gathered social media contents to extract the data needed for analyzing to analyze the user's social relationships, interests based on the social categories, digitized emotions as well as the correlation of hashtags. Also, we have consolidated analyzed results to apply on the ranking algorithm of social search. We have implemented the user's social context based personalized search system and also performed an experiment to verify the possibility of more useful retrieval if social context is applied to ranking algorithm.

Acknowledgments. This research was supported by Basic Science Research Program through the NRF(National Research Foundation of Korea), and the MIFP(Ministry of Science, ICT & Future Planning), Korea, under the National Program for Excellence in SW(2015-0-00932) supervised by the IITP(Institute for Information & communications Technology Promotion (Nos. NRF- 2015R1C1A2A01051729, 2015-0-00932)

References

1. Hu, B., Jamali, M., Ester, M.: Learning the strength of the factors influencing user behavior in online social networks. In: 2012 IEEE/ACM International Conference on Advances in Social Networks Analysis and Mining, pp. 368–375. IEEE (2012)
2. Kim, Y., Moon, I.: A study on algorithm for selection priority of contents in social network service. In: The 30th KIMICS, vol. 15, pp. 753–754, KISTI (2011)
3. Lawton, G.: Knowledge management: ready for prime time? IEEE Comput. Soc. **34**(2), 12–14 (2001)
4. Musial, K., Kazienko, P., Kajdanowicz, T.: Social recommendations within the multimedia sharing systems. Emerg. Technol. Inf. Syst. Knowl. Soc. **5288**, 364–372 (2008)
5. Yoo, S.Y., Jeong, O.R.: Social category based recommendation method. JICS **15**(5), 73–82 (2014)
6. Godoy, D., Amandi, A.: Modeling user interests by conceptual clustering. Inf. Syst. **31**(4–5), 247–265 (2006)
7. Hopfgartner, F., Jose, J.M.: Semantic user profiling techniques for personalised multimedia recommendation. Multimed. Syst. **16**(4–5), 255–274 (2010)
8. Ramanathan, K., Kapoor, K.: Creating user profiles using Wikipedia. In: The 28th International Conference on Conceptual Modeling, pp. 415–427. Springer, Heidelberg (2009)
9. Tao, K., Abel, F., Gao, Q., Houben, G.-J.: TUMS: Twitter-based user modeling service. In: ESWC 2011 Workshops, pp. 269–283. Springer, Heidelberg (2011)
10. Kapanipathi, P., Jain, P., Venkataramani, C., Sheth, A.: User interests identification on twitter using a hierarchical knowledge base. Semant. Web Trends Challenges **8465**(1), 99–113 (2014)
11. Amitay, E., Carmel, D., Har'El, N., Ofek-Koifman, S., Soffer, A., Yogev, S., Golbandi, N.: Social search and discovery using a unified approach. In: Proceedings of the 20th ACM Conference on Hypertext and Hypermedia, pp. 199–208. ACM (2009)
12. Horowitz, D., Kamvar, S.D.: The anatomy of a large-scale social search engine. In: Proceedings of the 19th International Conference on World Wide Web, pp. 431–440. ACM (2010)
13. Yoo, C.J., Jeong, O.R.: Category extraction for multimedia file search. In: 2013 International Conference on Information Science and Applications (ICISA), pp. 1–3. IEEE (2013)
14. Yoo, S.Y., Jeong, O.R.: SNS based recommendation algorithm. In: 2013 International Conference on Information Science and Applications (ICISA), pp. 1–3. IEEE (2013)
15. Thayer, R.E.: The Biopsychology of Mood and Arousal. Oxford University Press, New York (1989)
16. Park, T., Yoo, S., Jeong, O.: Social network based music recommendation system. In: 2014 Korea Computer Congress Conference, pp. 1569–1571, KIISE (2015)
17. Bradley, M.M., Lang, P.J.: Affective norms for English words (ANEW): instruction manual and affective ratings, Technical report C-1, The Center for Research in Psychophysiology, University of Florida, pp. 1–45 (1999)
18. Miller, G.A.: WordNet: a lexical database for English. Commun. ACM **38**(11), 39–41 (1995)
19. Li, J., Cardie, C.: Timeline generation : tracking individuals on Twitter. In: The 23rd International Conference on World Wide Web, pp. 643–652. ACM (2014)
20. Yilmaz, E., Kanoulas, E., Aslam, J.A.: A simple and efficient sampling method for estimating AP and NDCG. In: Proceedings of the 31st Annual International ACM SIGIR Conference on Research and Development in Information Retrieval, pp. 603–610. ACM (2008)

Dynamic Partitioning of Large Scale RDF Graph in Dynamic Environments

Kyoungsoo Bok, Cheonjung Kim, Jaeyun Jeong, Jongtae Lim, and Jaesoo Yoo(✉)

School of Information and Communication Engineering, Chungbuk National University, Chungdae-ro 1, Seowon-gu, Cheongju, Chungbuk 28644, Korea
{ksbok, jtlim, yjs}@chungbuk.ac.kr, {attain94, wodbs603}@naver.com

Abstract. In this paper, we propose a dynamic partitioning scheme of RDF graph to support load balancing in the dynamic environment. It generates clusters by grouping RDF subgraphs with high use frequencies while generating subclusters with RDF subgraphs with lower use frequencies. These clusters and subclusters perform load balancing based on the mean frequency of queries for the distributed server. In addition, the number of edge-cuts connected to clusters and subclusters is minimized to minimize the cost of communication between servers.

Keywords: Dynamic partition · RDF · Cluster · Load balancing

1 Introduction

The Semantic Web can define meaningful relationships among information and generate new knowledge through reasoning [1, 4]. As metadata and ontology play an important role in the semantic web, the World Wide Web Consortium (W3C) has proposed the Resource Description Framework (RDF) to describe web data formally [4, 5, 7, 9]. Recently, RDF data are growing continuously with the advent of big data and the spread of the semantic web. When processing large-scale RDF data in a single server, it becomes difficult to provide services to many users due to the decrease in performance in terms of processing storage and speed [2, 3]. Therefore, RDF data partitioning schemes are required to store and manage data using multiple servers.

The existing RDF partitioning schemes focus on minimizing the number of edge-cuts generated by partitioning [6, 8]. However, since these schemes store data by partitioning entered data based on predetermined criteria, data may become concentrated to specific servers. In addition, a variety of query requests from users increase communication costs among certain servers. For example, when RDF graph data are partitioned in distributed servers and join queries on partitioned edges occur, the communication cost between servers increases, and the reply time for the query becomes delayed [8, 11]. To address this problem, research has been conducted on dynamic partitioning schemes [10–12]. The existing schemes are not suitable for dynamic RDF data environments in terms of data storage, load balancing, and query processing.

W. Lee et al. (eds.), *Proceedings of the 7th International Conference on Emerging Databases*, Lecture Notes in Electrical Engineering 461,
https://doi.org/10.1007/978-981-10-6520-0_5

In the existing dynamic partitioning schemes, the communication cost between servers increases when join queries are processed.

In this paper, we propose an RDF partitioning scheme to provide load balancing to a distributed server without data replication in a dynamic RDF data environment. Clustering is performed based on the query data frequently used by users. A partitioning minimizes the number of edge-cuts among partitions. In addition, partitioning by calculating the mean data use frequency for the distributed server provides load balancing.

2 Related Work

SPAR is a social portioning and replication middleware that leverages the social graph to achieve data locality while minimizing replication [11]. When a new sever is added, SPAR perform force redistribution of the master replicas from the other servers to the new one to immediately balance all servers or wait for new arrivals to fill up the server. When the exiting server is removed, the master replicas are redistributed to the other servers equally.

Sedge use two-level partition architecture with complementary partitions and on-demand partitioning [12]. Complementary partitioning generates multiple partitions such that their partition boundaries do not overlap. On-demand partitioning uses two partitioning scheme such as partition replication and dynamic partitioning to deal with internal hotspots and cross-partition hotspots. Partition replication replicates the same data in multiple servers to share the workload on these partitions. Dynamic Partitioning reconstruct new partitions to serve cross-partition queries locally.

Hermes proposed a lightweight repartitioner for distributed social graph [10]. The initial partition generated by Metis [6] is repartitioned by a lightweight repartitioner. Each server maintains auxiliary data to perform repartitioning. The repartitioning process has two phases. In the first phase, each server runs the repartitioner algorithm using the auxiliary data to indicate some vertice in its partition that should be migrated to other partitions. These vertice are logically moved to their target partitions until no further vertice are chosen for migration. In the second phase, physical data migration is performed. Vertice and relationships that were marked for migration by the repartitioner are moved to the target partitions.

3 The Proposed Dynamic Partitioning Scheme

3.1 Architecture

In this paper, we propose an RDF dynamic partitioning scheme considering load balancing by analyzing the query data frequently used by users. The proposed scheme provides load balancing by grouping considering the use frequency of the data used by user-requested queries. The query frequency is defined as the use frequency of the data used by queries. The RDF subgraphs with high query frequency are grouped to generate clusters, and the RDF subgraphs with relatively lower query frequency compared to the clusters are grouped to generate subclusters. In addition, it minimizes the communication cost between servers during join processing by minimizing the number of edge-cuts connected to partitions.

Figure 1 shows the processing process of the proposed dynamic partitioning scheme. If the queries are concentrated on a particular server and the communication cost are increased between servers during query processing, the proposed scheme repartitions RDF graphs through the following three steps. The first stage is the cluster generation stage, in which RDF subgraphs that are frequently used in queries are grouped to generate clusters. The clusters are the criteria for data partitioning for load balancing and are generated with RDF graphs with high query frequency. The second stage is the subcluster generation stage, and the subgroups are generated with data with lower query frequency than the clusters. The last stage is the graph partitioning stage, and partitions are generated by considering data size, query frequency, and the number of edges.

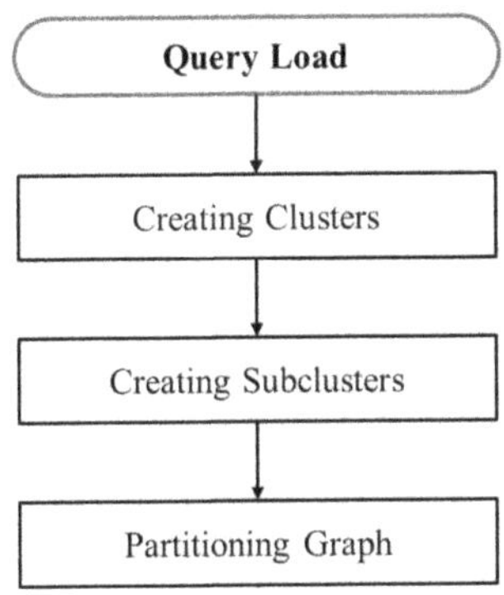

Fig. 1. The processing process of the proposed RDF dynamic partitioning scheme

3.2 Cluster Generation

To process user-requested queries quickly in a dynamic distributed environment, a scheme to minimize join processing between servers and communication cost is required. The proposed scheme stores the queries frequently requested by users in the past in their same servers, expecting them to be requested again in the future. Therefore, the data with high query frequency are grouped into clusters. The clusters are composed of RDF subgraphs that have been frequently requested by users. To meet users' constantly changing diverse needs, clusters are generated based on the queries that have been frequently used by users.

Grouping is performed in such a way that the frequency of the query data is close to the mean value of the cluster frequency. In addition, clusters are used as a reference point in each server to perform equal load balancing among servers, and as many clusters are generated as there are servers. By grouping the query data that used to be stored in multiple different servers into the same server, this scheme can minimize the communication costs that occur during query processing and provide fast query responses.

Equation (1) is the mean cluster frequency (*AvgCF*) which is the criterion used to generate a cluster in each server, where *SumofHighQF* is the sum of the high frequency values. *NumofServer* is the number of servers in the distributive server. In the cluster generation stage, the data with query frequencies higher than *AvgCF* are grouped into clusters. The data with query frequencies lower than *AvgCF* are grouped into subclusters.

$$AvgCF = \frac{SumofHighQF}{NumofServer} \tag{1}$$

3.3 Subcluster Generation

To perform equal load balancing among servers, query frequency is used based on the mean query frequency of the distributed server. The mean query frequency of the distributed server is the mean query frequency value as the criterion to perform equal load balancing during data partitioning. It is calculated based on the query frequency from statistical data. Specifically, each server uses one cluster, and subgroups with query frequencies lower than the cluster frequency perform load balancing of the distributed server by adding the mean query frequency of the clusters to the subgroups. In other words, in the subcluster generation stage, subclusters are generated by grouping data with low query frequencies, excluding the clusters generated in the cluster generation stage.

Subclusters refer to the subgroups that are grouped based on query data as in clusters but that have lower query frequencies than those of clusters. These are used to perform equal load balancing among servers, and subclusters are distributed around a cluster in each server to perform efficient load balancing. In other words, subgroups are used to improve the efficiency of load balancing. In addition, subclusters calculate hops in distance using connecting edges based on clusters generated in the cluster generation stage, and they perform distribution based on the range of threshold values established by users. Furthermore, to minimize communication cost between servers, the data that are never requested by users are not included in the grouping. This prevents the collapse of the previously partitioned graph data structure and reduces additional repartitioning cost through distribution based on the prediction of the range of queries that can be requested by users in the future.

3.4 Graph Partitioning

This is the last stage of the three-stage processing process in the proposed scheme and the stage that partitions the graph data. In this stage, partitioning is performed by considering the size of the partitions to solve the problem of data concentration to specific servers due to a dynamic RDF data environment. In addition, to solve the problem of load balancing in existing dynamic partitioning schemes, the mean query frequency of the distributed server is calculated. Finally, to minimize communication cost between servers, a scheme to minimize the number of edge-cuts is employed.

The edge-cut minimization scheme performs cuts not based on the edges connected between vertices in RDF data but based on the edges connected between clusters and subclusters. First, the candidate subclusters to be moved to another server for edge cutting are selected based on the criteria that few edges are connected to the subclusters within each server and many edges are connected to the subclusters in other servers. Edge-cut minimization is performed by relocating selected cluster candidates. By relocating the selected candidates to the servers with which they have many connections, the number of edges cuts can be minimized.

The proposed scheme considers the size of the partitions to solve the problem of data concentration to specific servers. Therefore, it offers the minimum partition size ($PSize_{min}$) and the maximum partition size ($PSize_{max}$). The partition size means the number of vertice in a RDF subgraph to be stored in each server. Equation (2) describes the minimum partition size ($PSize_{min}$) to be stored in each server, and Eq. (3) describes the maximum partition size ($PSize_{max}$) to be stored in each server. Here, $Node_{cnt}$ is the number of servers in the distributed server system, $Size_{full}$ is the total size of the data that is stored or needs to be stored in the distributed server. α and β are parameters for adjusting the minimum partition size and the maximum partition size to be stored in each distributed server, respectively.

$$PSize_{\min} = \frac{Size_{full}}{Node_{cnt}} \times \alpha \tag{2}$$

$$PSize_{\max} = \frac{Size_{full}}{Node_{cnt}} \times \beta \tag{3}$$

4 Performance Evaluation

To demonstrate the superiority of the proposed scheme, a performance evaluation was conducted by comparing it with Sedge and Hermes, which showed high performance among recently studied schemes. The evaluation was conducted in the cluster environment with an Intel® Core™ i3 CPU 540 processor with 4G memory, 1 TB hard disk, and eight distributed servers. Table 1 shows dataset used in the performance evaluation.

Table 1. Dataset

Classification	DBLP
Number of vertices	317 thousand
Number of edges	1 million

Figure 2 shows the storage ratio of data stored across distributed servers when performing RDF graph data partitioning. In the proposed scheme, the storage ratios of distributed servers are almost similar since it considers both a cluster size and the edge-cut ratios between clusters and subclusters. Figure 3 shows the results of comparison in query response time of Sedge, Hermes, and the proposed scheme using eight different queries. The proposed scheme performs grouping based on the query data with high data use frequency, and was designed to generate partitions in the size within the range of threshold values specified by users. Performance evaluation results showed the query response time of the proposed scheme was improved by approximately 10% than Sedge, and approximately 3% than Hermes.

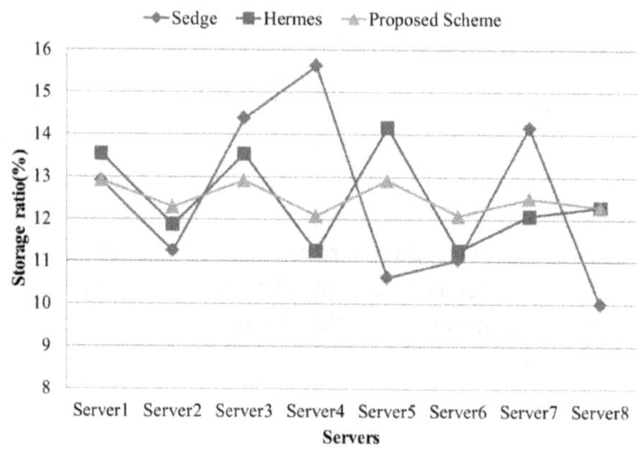

Fig. 2. Ratio of distributed storage

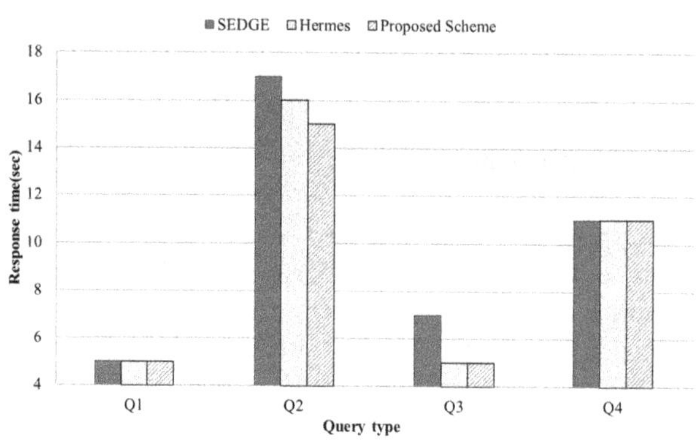

Fig. 3. Query response time

5 Conclusions

In this paper, we proposed an RDF dynamic partitioning scheme considering load balancing in the dynamic RDF data environment. The proposed scheme addressed data concentration to specific servers by performing grouping based on frequently used query data. In addition, it performed graph data partitioning by minimizing the number of edge-cuts to reduce communication cost between servers. These allowed the proposed scheme to outperform existing systems by maximizing the advantages of parallel systems and providing fast query responses. In a future work, we will apply the dynamic partitioning scheme to an actual distributed environment.

Acknowledgments. This work was supported by the National Research Foundation of Korea (NRF) grant funded by the Korea government(MSIP) (No. 2016R1A2B3007527), by the MSIP (Ministry of Science, ICT and Future Planning), Korea, under the ITRC(Information Technology Research Center) support program (IITP-2017-2013-0-00881) supervised by the IITP(Institute for Information & communication Technology Promotion), and by Basic Science Research Program through the National Research Foundation of Korea(NRF) funded by the Ministry of Education(2015R1D1A3A01015962)

References

1. Abadi, D., Marcus, A., Madden, S., Hollenbach, K.: Scalable semantic web data management using vertical partitioning. In: Proceedings of International Conference on Very Large Data Bases, pp. 411–422 (2007)
2. Ali, L., Janson, T., Schindelhauer, C.: Towards load balancing and parallelizing of RDF query processing in P2P based distributed RDF data stores. In: Proceedings of Euromicro International Conference on Parallel, Distributed, and Network-Based Processing, pp. 307–311 (2014)
3. Galarraga, L., Hose, K., Schenkel, R.: Partout: A distributed engine for efficient RDF processing. In: Proceedings of International World Wide Web Conference, pp. 267–268 (2014)
4. Gomez-Perez, A., Corcho, O.: Ontology languages for the semantic web. IEEE Intell. Syst. **17**(1), 54–60 (2002)
5. Huang, J., Abadi, D.J., Ren, K.: Scalable SPARQL querying of large RDF graphs. Proc. VLDB Endowment **4**(11), 1123–1134 (2011)
6. Karypis, G., Kumar, V.: METIS-Unstructured Graph Partitioning and Sparse Matrix Ordering System Version 2.0. Technical report, Department of Computer Science, University of Minnesota (1995)
7. Kim, K., Moon, B., Kim, H.: R3F: RDF triple filtering method for efficient SPARQL query processing. World Wide Web **18**(2), 317–357 (2015)
8. Malewicz, G., Austern, M. H., Bik, A.J.C., Dehnert, J.C., Horn, I., Leiser, N., Czajkowski, G.: Pregel: a system for large-scale graph processing. In: Proceedings of ACM SIGMOD International Conference on Management of Data, pp. 135–146 (2010)
9. Neumann, T., Weikum, G.: The RDF-3X engine for scalable management of RDF data. VLDB J. **19**(1), 91–133 (2010)
10. Nicoara, D., Kamali, S., Daudjee, K., Chen, L.: Hermes: dynamic partitioning for distributed social network graph databases. In: Proceedings of International Conference on Extending Database Technology, pp. 25–36 (2015)
11. Pujol, J.M., Erramilli, V., Siganos, G., Yang, X., Laoutaris, N., Chhabra, P., Rodriguez, P.: The little engine(s) that could: scaling online social networks. IEEE/ACM Trans. Netw. **20**(4), 1162–1175 (2012)
12. Yang, S., Yan, X., Zong, B., Khan, A.: Towards effective partition management for large graphs. In: Proceedings of ACM SIGMOD International Conference on Management of Data, pp. 517–528 (2012)

Efficient Combined Algorithm for Multiplication and Squaring for Fast Exponentiation over Finite Fields $GF(2^m)$

Kee-Won Kim[1], Hyun-Ho Lee[2], and Seung-Hoon Kim[1(✉)]

[1] Department of Applied Computer Engineering, Dankook University, Yongin, Republic of Korea
{nirkim,edian}@dankook.ac.kr
[2] Department of Computer Science, Dankook University, Yongin, Republic of Korea
leehh4016@naver.com

Abstract. For big data security, high-speed arithmetic units are needed. Finite field arithmetic has received much attention in error-control codes, cryptography, etc. The modular exponentiation over finite fields is an important and essential cryptographic operation. The modular exponentiation can be performed by a sequence of modular squaring and multiplication based on the binary method. In this paper, we propose a combined algorithm to concurrently perform multiplication and squaring over $GF(2^m)$ using the bipartite method and common operations. We expect that the architecture based on the proposed algorithm can reduce almost a half of latency compared to the existing architecture. Therefore, we expect that our algorithm can be efficiently used for various applications including big data security which demands high-speed computation.

Keywords: Finite fields · Multiplication · Squaring · Exponentiation · Cryptography

1 Introduction

As data is growing exponentially in big data environment, data security and privacy has been considered major issues for the growth of big data technology. When the volume of data has a dramatic growth, it means that the amount of data that needs to be encrypted and decrypted also increases. If the amount of data cannot be reduced, the best solution is to develop methods to speed up computation time of encryption and decryption.

The arithmetic over finite fields $GF(2^m)$ is very important for many practical applications in error-correcting codes and cryptography [1,2]. Among the arithmetic operations over finite fields, the multiplication is an important operation. This is because the time-consuming operations such as exponentiation, division, and multiplicative inversion can be performed by repeated multiplications.

W. Lee et al. (eds.), *Proceedings of the 7th International Conference on Emerging Databases*, Lecture Notes in Electrical Engineering 461,
https://doi.org/10.1007/978-981-10-6520-0_6

Thus, an efficient multiplication algorithm with low complexity and latency, high throughput rate, and short computation delay is required.

The modular exponentiation is one of the important and essential cryptographic operations. The modular exponentiation can be performed by a sequence of modular squaring and multiplication based on the binary method. There are two modular exponentiation schemes: least significant bit(LSB)-first and most significant bit(MSB)-first modular exponentiation, where LSB and MSB are the LSB and MSB of the exponent. For fast LSB-first modular exponentiation, modular squaring and multiplication can be performed in parallel [3–5]. Using the common operations from multiplication and squaring, Choi and Lee [6] proposed the combined systolic multiplier/squarer that computes modular squaring and multiplication concurrently instead of computing them separately for the LSB-first exponentiation.

In this paper, we propose an efficient combined algorithm to concurrently perform multiplication and squaring over $GF(2^m)$ using the bipartite method and deriving common operations. The proposed algorithm enables multiplication and squaring to operate in pipelined computation in parallel. Also, our algorithm can lead to a low-latency hardware architecture for the multiplication and squaring over $GF(2^m)$. The remainder of this paper is organized as follows. In Sect. 2, we review a modular exponentiation algorithm. In Sect. 3, we propose an efficient combined algorithm to concurrently perform multiplication and squaring over $GF(2^m)$. Finally, Sect. 4 gives our conclusions.

2 Modular Exponentiation

The exponentiation is a crucial part of modern cryptographic algorithms. The most commonly used algorithms for exponentiation are the binary methods (also called square-and-multiply methods) [7]. Its basic idea is to compute modular exponentiation by using the binary expression of exponent E and the exponentiation operation is broken into a series of squaring and multiplication operations. The modular exponentiation has two methods: LSB-first and MSB-first modular exponentiation, where LSB and MSB are the LSB and MSB of the exponent. The LSB-first modular exponentiation can be used to compute modular squaring and modular multiplication concurrently. The LSB binary modular exponentiation algorithm is represented as Algorithm 1 which computes the modular exponentiation starting from the LSB of the exponent and proceeding to the left.

Algorithm 1. LSB binary modular exponentiation algorithm in $GF(2^m)$

```
Input : M, E = ∑_{i=0}^{m-1} e_i 2^i (where e^i ∈ {0,1}), G
C = 1
S = M
for i = 0 to m − 1 do
    if (e_i = 1) then C = C × S mod G
    S = S × S mod G
end for
return C
```

Note that modular squaring and multiplication can be performed concurrently to improve speed of exponentiation [3–6]. Using the common-multiplicand method [4,5], Choi and Lee [6] proposed the combined systolic multiplier/squarer that computes modular squaring and multiplication concurrently instead of computing them separately for the LSB-first exponentiation.

3 Proposed Combined Algorithm for Multiplication and Squaring

In this section, we propose a new combined algorithm to compute multiplication and squaring over $GF(2^m)$ in parallel. We can adopt the bipartite concept [8–11] to decrease the latency required for calculating multiplication and squaring over finite fields. In [8–11], they have used the Montgomery multiplication algorithm [12,13] for deriving a bipartite structure. The bipartite multiplication using Montgomery multiplication algorithm requires converting the operand to Montgomery representation and final result to the original representation. Our proposed algorithm does not require additional conversion operations because we adopt the bipartite concept not using Montgomery multiplication algorithm. Also, we decrease the space complexity by deriving common operations from bipartite parts like common-multiplicand method [4–6].

Let the finite field over $GF(2^m)$ be defined, in general, by an irreducible polynomial of degree m, given by $G = x^m + \sum_{j=0}^{m-1} g_j x^j$, where $g_i \in GF(2)$. The polynomial basis $\{1, x, \cdots, x^{m-2}, x^{m-1}\}$ is used to represent the field elements, so that any two arbitrary elements A and B in $GF(2^m)$ can be represented in the form of polynomials of degree $(m-1)$ as $A = \sum_{j=0}^{m-1} a_j x^j$ and $B = \sum_{j=0}^{m-1} b_j x^j$, where a_j and $b_j \in \{0, 1\}$, for $0 \leq j \leq m-1$.

Since x is a root of $G(x)$, i.e. $G(x) = 0$, $x^m \bmod G$ and $x^{m+1} \bmod G$ are as follows:

$$x^m \bmod G = \sum_{j=0}^{m-1} g_j x^j, \tag{1}$$

$$x^{m+1} \bmod G = \sum_{j=1}^{m-1} (g_{m-1} g_j + g_{j-1}) x^j + g_{m-1} g_0 \equiv G' = \sum_{j=0}^{m-1} g'_j x^j. \tag{2}$$

Assume that $G' = x^{m+1} \bmod G$ is given in advance. Let $k = \lfloor m/2 \rfloor$ and $l = \lceil m/2 \rceil$. The multiplication and squaring of A and B, $P = AB \bmod G$, $S = AA \bmod G$ can be expressed as follows:

$$\begin{aligned} P = AB \bmod G &= \sum_{i=0}^{m-1} b_i A x^i \bmod G \\ &= \sum_{i=0}^{l-1} b_{2i} A x^{2i} \bmod G + x \sum_{i=0}^{k-1} b_{2i+1} A x^{2i} \bmod G, \end{aligned} \tag{3}$$

$$\begin{aligned} S = AA \bmod G &= \sum_{i=0}^{m-1} a_i A x^i \bmod G \\ &= \sum_{i=0}^{l-1} a_{2i} A x^{2i} \bmod G + x \sum_{i=0}^{k-1} a_{2i+1} A x^{2i} \bmod G. \end{aligned} \tag{4}$$

From (3) and (4), we can see that P and S can be divided into two parts, respectively. We define P and S as follows:

$$P = Q + xR \bmod G, \tag{5}$$

$$S = T + xU \bmod G, \tag{6}$$

where

$$Q = \sum_{i=0}^{l-1} b_{2i} A x^{2i} \bmod G, \tag{7}$$

$$R = \sum_{i=0}^{k-1} b_{2i+1} A x^{2i} \bmod G, \tag{8}$$

$$T = \sum_{i=0}^{l-1} a_{2i} A x^{2i} \bmod G, \tag{9}$$

$$U = \sum_{i=0}^{k-1} a_{2i+1} A x^{2i} \bmod G. \tag{10}$$

We can observe that the computations of Q, R, T, and U require $Ax^{2i} \bmod G$ in common. We define $A^{(i)} = Ax^{2i} \bmod G$, for $0 \le i \le l-1$. Then, the equations can be expressed as $A^{(i)} = \sum_{j=0}^{m-1} a_j^{(i)} x^j \bmod G$, where $A^{(0)} = A$. Then, based on (1) and (2), $A^{(i)}$ can be expressed as

$$\begin{aligned} A^{(i)} &= A^{(i-1)} x^2 \bmod G \\ &= \sum_{j=0}^{m-1} a_j^{(i-1)} x^{j+2} \bmod G \\ &= \sum_{j=0}^{m-3} a_j^{(i-1)} x^{j+2} + (a_{m-2}^{(i-1)} x^m + a_{m-1}^{(i-1)} x^{m+1}) \bmod G \\ &= \sum_{j=0}^{m-3} a_j^{(i-1)} x^{j+2} + \sum_{j=0}^{m-1} a_{m-2}^{(i-1)} g_j x^j + \sum_{j=0}^{m-1} a_{m-1}^{(i-1)} g'_j x^j \\ &= \sum_{j=0}^{m-1} (a_{j-2}^{(i-1)} + a_{m-2}^{(i-1)} g_j + a_{m-1}^{(i-1)} g'_j) x^j, \end{aligned} \tag{11}$$

where $A^{(0)} = A$, $a_{-2}^{(i-1)} = a_{-1}^{(i-1)} = 0$, and $1 \le i \le l-1$.

From (11), we can obtain the coefficient of $A^{(i)}$ as follows:

$$a_j^{(i)} = a_{j-2}^{(i-1)} + a_{m-2}^{(i-1)} g_j + a_{m-1}^{(i-1)} g'_j, \tag{12}$$

where $a_j^{(0)} = a_j$, $a_{-2}^{(i-1)} = a_{-1}^{(i-1)} = 0$, and $1 \le i \le l-1$.

Using $A^{(i)}$, Q, R, T, and U from (7) to (10) are represented as follows:

$$Q = \sum_{i=1}^{l} b_{2(i-1)} A^{(i-1)}, \tag{13}$$

$$R = \sum_{i=1}^{k} b_{2i-1} A^{(i-1)}, \tag{14}$$

$$T = \sum_{i=1}^{l} a_{2(i-1)} A^{(i-1)}, \tag{15}$$

$$U = \sum_{i=1}^{k} a_{2i-1} A^{(i-1)}. \tag{16}$$

From above equations, the recurrence equations of Q, R, T, and U can be formulated as

$$Q^{(i)} = Q^{(i-1)} + b_{2(i-1)} A^{(i-1)}, \tag{17}$$

$$R^{(i)} = R^{(i-1)} + b_{2i-1} A^{(i-1)}, \tag{18}$$

$$T^{(i)} = T^{(i-1)} + a_{2(i-1)} A^{(i-1)}, \tag{19}$$

$$U^{(i)} = U^{(i-1)} + a_{2i-1} A^{(i-1)}, \tag{20}$$

where $Q^{(0)} = R^{(0)} = T^{(0)} = U^{(0)} = 0$, $Q^{(i)} = \sum_{j=0}^{m-1} q_j^{(i)} x^j$, $R^{(i)} = \sum_{j=0}^{m-1} r_j^{(i)} x^j$, $T^{(i)} = \sum_{j=0}^{m-1} t_j^{(i)} x^j$, and $U^{(i)} = \sum_{j=0}^{m-1} u_j^{(i)} x^j$ are ith intermediate results.

Therefore, the coefficients of $Q^{(i)}$, $R^{(i)}$, $T^{(i)}$, and $U^{(i)}$ can be represented as follows:

$$q_j^{(i)} = q_j^{(i-1)} + b_{2(i-1)} a_j^{(i-1)}, \text{ for } 1 \le i \le l, \tag{21}$$

$$r_j^{(i)} = r_j^{(i-1)} + b_{2i-1} a_j^{(i-1)}, \text{ for } 1 \le i \le k, \tag{22}$$

$$t_j^{(i)} = t_j^{(i-1)} + a_{2(i-1)} a_j^{(i-1)}, \text{ for } 1 \le i \le l, \tag{23}$$

$$u_j^{(i)} = u_j^{(i-1)} + a_{2i-1} a_j^{(i-1)}, \text{ for } 1 \le i \le k, \tag{24}$$

where $q_j^{(0)} = r_j^{(0)} = t_j^{(0)} = u_j^{(0)} = 0$ and $0 \le j \le m-1$. The equations from (13) to (16) can be simultaneously executed because there are no data dependency between computations of Q, R, T, and U.

After computing $Q^{(l)}$, $R^{(k)}$, $T^{(l)}$, and $U^{(k)}$, we finally require computing Eqs. (5) and (6) in order to obtain the result of multiplication and squaring. Therefore, the Eqs. (5) and (6) are represented as follows:

$$\begin{aligned}
P &= Q^{(l)} + xR^{(k)} \bmod G \\
&= \sum_{j=0}^{m-1} q_j^{(l)} x^j + x \sum_{j=0}^{m-1} r_j^{(k)} x^j \bmod G \\
&= \sum_{j=0}^{m-1} q_j^{(l)} x^j + \sum_{j=0}^{m-2} r_j^{(k)} x^{j+1} + r_{m-1}^{(k)} x^m \bmod G \\
&= \sum_{j=0}^{m-1} (q_j^{(l)} + r_{j-1}^{(k)} + r_{m-1}^{(k)} g_j) x^j,
\end{aligned} \tag{25}$$

$$\begin{aligned}
S &= T^{(l)} + xU^{(k)} \bmod G \\
&= \sum_{j=0}^{m-1} t_j^{(l)} x^j + x \sum_{j=0}^{m-1} u_j^{(k)} x^j \bmod G \\
&= \sum_{j=0}^{m-1} t_j^{(l)} x^j + \sum_{j=0}^{m-2} u_j^{(k)} x^{j+1} + u_{m-1}^{(k)} x^m \bmod G \\
&= \sum_{j=0}^{m-1} (t_j^{(l)} + u_{j-1}^{(k)} + u_{m-1}^{(k)} g_j) x^j,
\end{aligned} \tag{26}$$

where $r_{-1}^{(k)} = 0$ and $u_{-1}^{(k)} = 0$.

By above equations, we can derive Algorithm 2 for the efficient combined multiplication and squaring for fast exponentiation over $GF(2^m)$. As can be seen in Algorithm 1, $a_j^{(i)}$, $q_j^{(i)}$, $r_j^{(i)}$, $t_j^{(i)}$, and, $u_j^{(i)}$ can be executed simultaneously and can be induced very similarly. The proposed algorithm is regular and simple and is well-suited to implementing systolic architecture by fully exploiting inherent parallelism of the input datum. Based on the proposed Algorithm 2, the hardware architecture can be efficiently composed.

The systolic array for multiplication and squaring of Choi and Lee [6] has the latency of $3m$ clock cycles. We expect that the architecture based on the proposed algorithm can reduce almost a half of latency compared to the Choi and Lee's architecture.

Algorithm 2. Proposed combined algorithm for multiplication and squaring

Input : $A, B \in GF(2^m)$, G, G'
Output : $P = AB \bmod G$, $S = AA \bmod G$

$A^{(0)} = A$, $Q^{(0)} = 0$, $R^{(0)} = 0$, $T^{(0)} = 0$, $U^{(0)} = 0$
for $i = 0$ to $\lceil m/2 \rceil$ do
    $a_{-2}^{(i-1)} = 0, a_{-1}^{(i-1)} = 0$
    for $j = 0$ to $m - 1$ do
        in parallel do:
            $a_j^{(i)} = a_{j-2}^{(i-1)} + a_{m-2}^{(i-1)} g_j + a_{m-1}^{(i-1)} g'_j$
            $q_j^{(i)} = q_j^{(i-1)} + b_{2(i-1)} a_j^{(i-1)}$
            $r_j^{(i)} = r_j^{(i-1)} + b_{2i-1} a_j^{(i-1)}$
            $t_j^{(i)} = t_j^{(i-1)} + a_{2(i-1)} a_j^{(i-1)}$
            $u_j^{(i)} = u_j^{(i-1)} + a_{2i-1} a_j^{(i-1)}$
        end do
    end for
end for
$r_{-1}^{(k)} = 0$, $u_{-1}^{(k)} = 0$
for $j = 0$ to $m - 1$ do
    in parallel do:
        $p_j = q_j^{(l)} + r_{j-1}^{(k)} + r_{m-1}^{(k)} g_j$
        $s_j = t_j^{(l)} + u_{j-1}^{(k)} + u_{m-1}^{(k)} g_j$
    end do
end for
return P, S

4 Conclusion

In this paper, we have proposed a combined algorithm for multiplication and squaring for fast exponentiation over finite fields. We have induced an efficient algorithm which is highly suitable for the design of pipelined structures. The proposed algorithm enables finite field multiplication and squaring to operate in pipelined computation. Also, our algorithm has low latency. Moreover, our algorithm enables the computation to share the hardware architecture so that we expect that it reduces not only time complexity but also hardware complexity compared to the recent study. Therefore, we expect that our algorithm can be efficiently used for various applications including big data security which demands high-speed computation.

Acknowledgments. This research was supported by Basic Science Research Program through the National Research Foundation of Korea(NRF) funded by the Ministry of Education (NRF-2015R1D1A1A01059739).

References

1. Blahut, R.E.: Theory and Practice of Error Control Codes. Addison-Wesley, Reading (1983)
2. Menezes, A.J., Van Oorschot, P.C., Vanstone, S.A.: Handbook of Applied Cryptography. CRC Press, Boca Raton (1996)
3. Scott, P.A., Simmons, S.J., Tavares, S.E., Peppard, L.E.: Architectures for exponentiation in $GF(2^m)$. IEEE J. Sel. Areas Commun. **6**, 578–585 (1988)
4. Lee, K.J., Yoo, K.Y.: Linear systolic multiplier/squarer for fast exponentiation. Inf. Process. Lett. **76**, 105–111 (2000)
5. Ha, J.C., Moon, S.J.: A common-multiplicand method to the Montgomery algorithm for speeding up exponentiation. Inf. Process. Lett. **66**, 105–107 (1998)
6. Choi, S.H., Lee, K.J.: Efficient systolic modular multiplier/squarer for fast exponentiation over $GF(2^m)$. IEICE Electron. Express **12**, 20150222 (2015)
7. Knuth, D.E.: The Art of Computer Programming. Seminumerical Algorithms, vol. II. Addison-Wesley, Reading, MA (1997)
8. Kim, K.W., Jeon, J.C.: Polynomial basis multiplier using cellular systolic architecture. IETE J. Res. **60**, 194–199 (2014)
9. Kim, K.W., Jeon, J.C.: A semi-systolic Montgomery multiplier over $GF(2^m)$. IEICE Electron. Express **12**, 20150769 (2015)
10. Kaihara, M.E., Takagi, N.: Bipartite modular multiplication. In: CHES 2005, vol. 3659, pp. 201–210 (2005)
11. Kaihara, M.E., Takagi, N.: Bipartite modular multiplication method. IEEE Trans. Comput. **57**, 157–164 (2008)
12. Montgomery, P.: Modular multiplication without trial division. Math. Comput. **44**, 519–521 (1985)
13. Koc, C., Acar, T.: Montgomery multiplication in $GF(2^k)$. Des. Codes Cryptogr. **14**, 57–69 (1998)

Efficient Processing of Alternating Least Squares on a Single Machine

Yong-Yeon Jo, Myung-Hwan Jang, and Sang-Wook Kim(✉)

Hanyang University, Seoul, South Korea
{jyy0430,sugichiin,wook}@hanyang.ac.kr

Abstract. Alternating least squares (ALS) is one of the algorithms widely used in recommendation systems. In this paper, we propose a method to perform ALS on a graph engine on a single machine. We employ our graph engine, RealGraph, to handle big graphs and develop ALS efficiently performed on top of it. Real-world graphs in performing ALS follow *the power-law degree distribution*, specifying that a few nodes have a lot of edges while a lot of nodes do only a few edges. Prior graph engines do not consider this important characteristic, which slows down their performance. According to our extensive performance evaluation, our ALS running on RealGraph significantly outperforms those on other engines up to 2.5 times.

Keywords: Alternating least squares · Graph engine · Performance

1 Introduction

Collaborative filtering (CF) is the underlying technique for recommendation systems that recommend items to a target user by referring to a group of users whose preferences are similar to that of him/her. Many companies such as Amazon, Netflix, and TiVo have adopted this technique to their business [1,2]. In order to realize this technique, the algorithms based on *the matrix factorization* (MF) have been widely used [3,4]. MF approximates large sparse matrix R representing users and items by two low-rank matrices U and V ($R \approx U \times V$). Here, the matrices U and V represent the users with latent factors and the items with latent factors, respectively. *Alternating least squares* (ALS) is a typical algorithm based on MF. The recommendation system based on this algorithm showed the best-performance in Netflix's movie rating prediction [3,5].

Matrix R can be defined as the bipartite graph for the users with the items. In this graph, each edge indicates an element in matrix R. A row in matrix U and a column in matrix V can be defined as a set of latent factors of a user and an item, respectively. Thus, ALS can be interpreted as a graph algorithm [6–8].

As recommendation systems have been widely used, the number of users and items gets larger and larger in the online world so that the graphs representing

W. Lee et al. (eds.), *Proceedings of the 7th International Conference on Emerging Databases*, Lecture Notes in Electrical Engineering 461,
https://doi.org/10.1007/978-981-10-6520-0_7

them have also increased dramatically. In order to efficiently perform the recommendation algorithms with such large-scale graphs (i.e., big graphs), various graph engines on a single machine have emerged [7,8]. In this context, we have also developed *RealGraph* [9], a graph engine that can handle big graphs with the limited computing resources. RealGraph bases its design on the principles adopted in well-known database storage systems such as WiSS [10], Exodus [11], and Shore [12] that efficiently store/retrieve big data. RealGraph consists of four layers: a storage management layer, a buffer management layer, an object management layer, and a thread management layer.

We aim at the development of ALS running on RealGraph and demonstrate its superior performance. The graphs generally used in ALS are *real-world ones* following *the power-law degree distribution* [13], which specifies that a few nodes have a lot of edges while a lot of nodes have a few edges as shown in Fig. 1. Existing graph engines use *the node- or edge-based workload assignment* to threads without taking this important characteristic into consideration [7,8]. This leads to *load imbalance* across threads or frequent thread confliction which slows down the performance of graph engines.

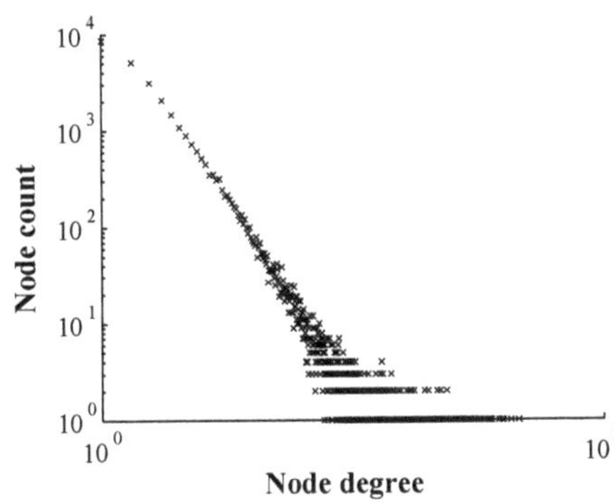

Fig. 1. Power-law degree distribution.

Considering this characteristic, we use *a segment-based workload distribution* that assigns each thread a storage segment whose size is fixed and also matched to a standard I/O unit. A storage segment could have multiple nodes with different degrees or a single node whose degree is so large. We note, However, the sum of total degrees in a storage segment is almost the same. Thus, this approach leads to the uniform distribution of workloads to threads and also infrequent thread confliction, which help achieve higher performance of ALS.

We compare the performance of ALS on RealGraph with those on existing graph engines such as GraphChi [7] and X-Stream [8] whose codes are available. They also employ the external-memory model like RealGraph. The experimental results show that the performance of ALS on RealGraph outperforms substantially those on other graph engines up to 2.5 times.

The contributions of this paper are as follows. We develop ALS to be efficiently performed on RealGraph by adopting the characteristic of real-world graphs. We evaluate the performance of ALS in various graph engines through extensive experiments with changing parameters of graph engines and ALS to demonstrate the superiority of our ALS on RealGraph.

The organization of the paper is as follows. Section 2 overviews RealGraph. Section 3 introduces the procedures of MF and ALS, and also presents how to implement the ALS of RealGraph. Section 4 shows and analyzes experimental results. Finally, Sect. 5 summarizes and concludes the paper.

2 RealGraph

This section introduces our graph engine, RealGraph [9], running on a single machine. It adopts *the vertex-centric model* as the programming model and *the external-memory model* as the memory model [7]. RealGraph aims at efficiently handling big graphs with the limited computing resources as well as applying itself to various graph algorithms.

2.1 Overview of Architecture

The underlying architecture of RealGraph is based on well-known database storage systems, such as WiSS [10], EXODUS [11], and Shore [12] to efficiently store/access the data. We modify this architecture and develop the layers for processing graphs on top of it to efficiently handle big graphs. We describe the role and structure of each layer.

(1) Storage management layer: as the lowest layer, it manages the space in the secondary storage which is partitioned into the fixed-size storage segments matching a standard I/O unit defined by the graph engine. In RealGraph, we set the size of storage segment as 1MiB. The role of this layer is to allocate and release the space for each storage segment in the secondary storage, and also perform their I/Os.

(2) Buffer management layer: it manages the space of the main memory. Its role is to allocate/load/release storage segments in the main memory. There exist a buffer and a buffer index. The buffer is space for maintaining the storage segments in the main memory. The buffer index is to index the storage segments currently loaded in the buffer to quickly access to them.

(3) Object management layer: it manages the objects in storage segments. Each storage segment is composed of multiple objects. If the length of an object exceeds the size of a storage segment, it is stored across multiple storage segments. In RealGraph, we represent a graph by adjacency lists, each of which is composed of a node and its adjacent nodes with their weights. Then, we construct an adjacency list as an object in a storage segment. There are two types of an object such as an incoming object regarding to incoming edges of a node and an outgoing object regarding to outgoing edges of a node. This layer has

the object index which is to index the objects in all storage segments to identify the object corresponding to a specific node.

(4) Thread management layer: it manages threads and accesses/processes the graph data via multi-threading. In RealGraph, there are two types of threads: a main thread and a worker thread. The main thread controls a number of worker threads, each of which executes a certain job. This layer provides an attribute vector containing final (or intermediate) computation results obtained while running graph algorithms. It also has indicators to identify those nodes to be processed in each iteration.

2.2 Processing Procedure

RealGraph performs graph algorithms as follows:

- Step 1: the main thread searches for the nodes to be processed in the current iteration by scanning *the indicator*
- Step 2: the main thread identifies the objects corresponding to the required nodes from the object index in the object management layer
- Step 3: the main thread verifies whether the storage segment containing the objects exists in the buffer by using the buffer index in the buffer management layer
- Step 4: the main thread assigns the objects thus found to worker threads
- Step 5-1: if the objects exist in the buffer, each worker thread performs the job for the objects
- Step 5-2: otherwise, each worker thread requests the storage segment containing the objects to the buffer management layer which requires the storage segment to the storage management layer. It processes the job after the storage segment being loaded

3 Development of ALS on RealGraph

This section introduces the matrix factorization (MF) and alternating least squares (ALS) with their procedures, and explains how to implement ALS to be efficiently performed on RealGraph.

3.1 Alternating Least Squares

Figure 2 shows *the matrix factorization* (MF). It is a matrix operation that large-sparse matrix R is approximated by the matrix multiplication of two low-rank matrices U and V ($R \approx U \times V$). In matrix R, a row indicates a user, a column does an item, and each element means a rating of the item given by the user. In matrices U and V, a row/column indicates a user/item, respectively, a column/row means a latent factor, and each element has a latent value.

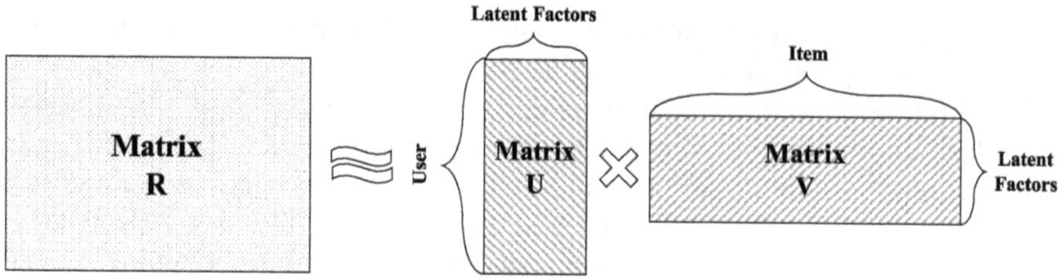

Fig. 2. Matrix factorization ($R \approx U \times V$).

ALS is one of the algorithms for MF whose steps are as follows:

- Step 1: initialize matrix V with random values between 0 and 1
- Step 2: hold matrix V constant, and solve matrix U by minimizing Eq. (1)
- Step 3: hold matrix U constant, and solve matrix V by minimizing Eq. (1)
- Step 4: repeat steps 2 and 3 until the objective function converges

$$objectivefunction = \sqrt{\frac{1}{n}\sum |p(u,v) - r(u,v)|^2} \tag{1}$$

where $p(u, v)$ and $r(u, v)$ are predicted and observed ratings for user u and item v, respectively [5].

Matrix R can be defined as the bipartite graph for users with items. Each edge indicates a rating that each user gives to an item. A row in matrix U and a column in matrix V can be defined as a set of latent factors of a user and an item, respectively. ALS can be seen as a graph algorithm by computing a set of latent factors using graph R, thus it is performed in graph engines.

3.2 Realization of ALS on RealGraph

In this section, we address how to implement ALS to be efficiently performed on RealGraph. We first convert bipartite graph R to a set of objects in the secondary storage. Each user is represented as an outgoing object which is composed of a user with his or her ratings to items. Each item is represented as an incoming object which consists of an item with its ratings from users.

The values of the elements of matrices U and V continuously change during performing ALS. In order to store these changing values, we represent matrices U and V as attribute vectors. That is, matrices U and V consist of attribute vectors as many as the number of latent factors, whose size is the number of users and items, respectively.

The real-world graphs follow *the power-law degree distribution* [13] that a few nodes (i.e., hub nodes) have a lot of edges while a lot of nodes have only a few edges as shown in Fig. 1. In other words, it means that a few items are rated by a lot of users while most items are rated by only a few users.

The existing graph engines use the node- or edge-based workload distribution across threads which do not leverage this characteristic. Here, the node-based approach means that a thread processes a single node and its adjacent nodes [7]. It incurs the skewed distribution across threads by the power-law degree distribution. The edge-based approach means that a thread processes a single edge [8]. In a graph engine using this approach, the result of a hub node is updated from its incoming edges processed by multiple threads. This leads to thread confliction. In conclusion, such workload distributions used in existing graph engines cause to slow down their performance.

We employ *the segment-based workload distribution* to solve these observed problems. The segment-based approach has each thread assigned with a storage segment as a workload. A storage segment, as mentioned in Sect. 2.1, has multiple objects in general or a partial object rarely, but the total size of object(s) in a storage segment is almost constant. Thus, threads can be assigned with similar workloads.

This distribution can also incur the thread confliction as in the edge-based approach, in the case that the result of a (hub) node is updated from its adjacent nodes stored across storage segments (e.g., updating the rank value in PageRank) by multiple threads concurrently. However, this case happens *very rare*. In real-world datasets, such objects stored across multiple storage segments (i.e., hub objects) exist only about 0.0027% of the total objects in the case of 1MiB of storage segment in RealGraph[1]. That is, the occurrence of the thread confliction is negligible.

4 Evaluation

This section compares and analyzes the performance of ALS running on graph engines through extensive experiments.

4.1 Experimental Setup

We conducted the experiments on the system equipped by Intel Core i7-4770 3.40Ghz, Samsung SSD 850 PRO, 32GiB main memory and installed on Ubuntu 16.4 64bit. We used three graph engines such as RealGraph [9], GraphChi [7], and X-Stream [8]. Although there are more graph engines on a single machine, GraphChi and X-Stream are the graph engines whose source codes and ALS implementation are available in public. The three graph engines adopt an external-memory model [7] as the memory model to handle big graphs. RealGraph and GraphChi adopt the vertex-centric model while X-Stream does the edge-centric model as the programming model [8].

We used the Netflix dataset which consists of about 2,649 K users and 17 K items. The number of ratings between users and items is 100,480K. It is widely used to evaluate recommendation systems [7,8,14].

We set the default parameters of graph engines as follows: threads to 4 and memory size to 4GiB. Other than these, there are two more parameters in terms

[1] Existing graph engines use more than 1MiB for a storage segment.

of ALS: (1) D, the number of latent factors, and (2) the number of iterations. We set the number of latent factors as 10 and the number of iterations as 10 as default settings.

4.2 Experimental Results

We measured the overall execution time including I/O and processing times and verified the performance of ALS on the three graph engines with changing parameters. In Fig. 3, the y-axis indicates the execution time and the x-axis does parameters in terms of computing resources in graph engines. In Fig. 4, the y-axis also denotes the execution time and the x-axis does parameters related to ALS.

Varying the Number of Threads. This experiment verifies the performance of ALS running on three graph engines, with changing the number of threads from 2 to 8 with other parameters fixed as mentioned above. Figure 3a shows the results that ALS on RealGraph outperforms those on other graph engines regardless of the number of threads.

We note that the performance of ALS on RealGraph and GraphChi gets improved as the number of threads increases while that on X-Stream has no change. RealGraph and GraphChi use the segment- or node-based workload distribution, which causes extremely rare or no thread confliction. However, a lot of thread confliction occurs in X-Stream using the edge-based approach due to the computations of some hub nodes. This is a main bottleneck of the performance. Although the number of threads increases to 8, the performance gain is achieved with up to four threads because our system has only four physical cores inside.

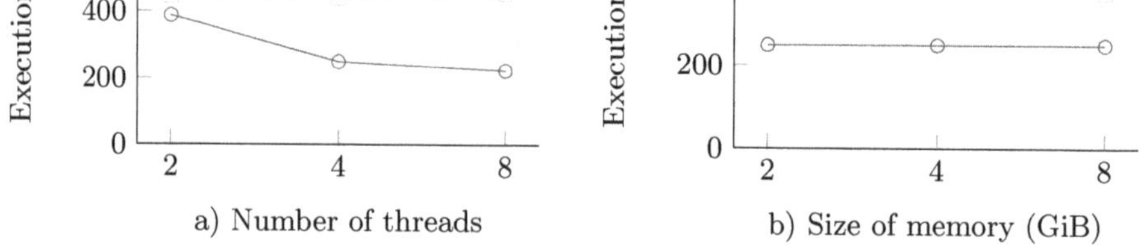

Fig. 3. The execution time of ALS running on graph engines with changing parameters related to graph engines

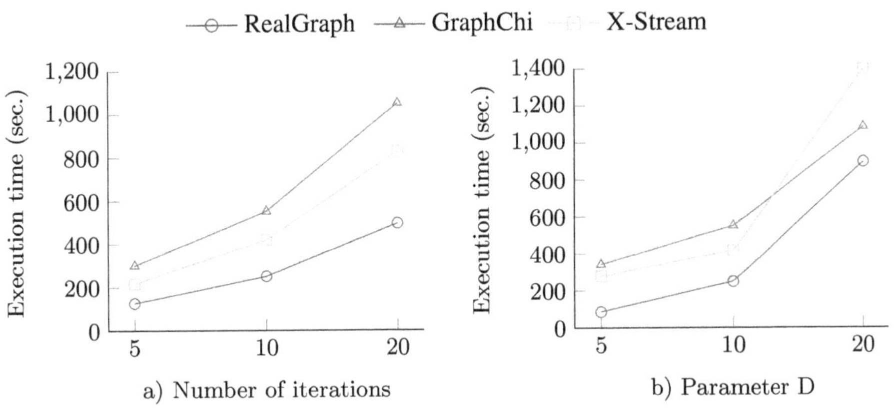

Fig. 4. The execution time of ALS running on graph engines with changing parameters related to ALS

Varying the Size of Memory. As the second experiment, we measured the performance of ALS, changing the given memory size from 2GiB to 8GiB with other parameters fixed. In Fig. 3b, the performance of RealGraph is also better than that of others, but its execution time is not changed. This is due to the graph representation, the way to store the graph into the storage. RealGraph requires the smallest space to store the graph by employing efficient graph representation while GraphChi does the largest space.

For this reason, there is no change with the execution time in RealGraph because ALS is performed at the situation that the whole graph can reside in main memory. Other graph engines, however, require frequent I/Os from the storage because they can not load the graph into the memory at once. As the memory size increases to 8GiB, the execution time of X-Stream decreases because the I/O requirement reduces. Interestingly, the execution time of GraphChi even slightly grows up due to write operation to store *the intermediate results* by its graph representation although the given main memory gets larger.

Varying the Number of Iterations. In the third experiment, we measured the performance of ALS, varying the number of iterations from 5 to 20 with parameters related to graph engines fixed. Figure 4a shows that the execution times of ALS on all graph engines linearly grow up with the increasing number of iterations. ALS on RealGraph still outperforms significantly those on the other engines as in the previous experiments.

Varying Parameter D. As the last experiment, we measured the performance of ALS, varying parameter D (i.e., the number of latent factors) from 5 to 20 with other parameters fixed. Figure 4b shows that the execution times of ALS on the three graph engines are all grown up rapidly as parameter D increases. This is because the number of latent factors affects the amount of computations of ALS considerably. In the case of parameter D set to 20, the execution time of ALS on X-Stream exceeds that on GraphChi because more frequent thread confliction occurs by rapidly increasing the amount of computations. Of course, our proposal also performs better than the others in this experiment.

5 Conclusion

ALS is a representative algorithm widely used in recommendation systems [3]. We developed an efficient method to perform ALS on top of RealGraph. We first observed and analyzed ALS running on the existing graph engines. Existing graph engines cause the load imbalance and frequent thread confliction because they use node- or edge-based workload distribution among threads without taking the power-law degree distribution of real-world graphs into consideration. After all, their performances are degraded. We used segment-based workload distribution leveraging this important characteristic. It helps distribute the workloads evenly to threads and achieve higher performance of ALS.

Through extensive experiments, we demonstrated that ALS on RealGraph outperforms ALS significantly on the other graph engines up to 2.5 times. By virtue of the results, we expect that our proposal improves the performance of various recommendation systems. As our future work, we consider to develop additional algorithms on RealGraph to contribute the recommendation systems domain.

Acknowledgment. This research was supported by (1) Semiconductor Industry Collaborative Project between Hanyang University and Samsung Electronics Co. Ltd. and (2) the MSIP (Ministry of Science, ICT and Future Planning), Korea, under the ITRC (Information Technology Research Center) support program (IITP-2017-2013-0-00881) supervised by the IITP (Institute for Information & communication Technology Promotion).

References

1. Hwang, W.-S., et al.: Told you I didn't like it: exploiting uninteresting items for effective collaborative filtering. In: Proceedings of the 32nd IEEE International Conference on Data Engineering (ICDE), pp. 349–360 (2016)
2. Lee, J., et al.: Improving the accuracy of top-n recommendation using a preference model. Inf. Sci. **348**, 290–304 (2016). doi:10.1016/j.ins.2016.02.005
3. Koren, Y., Bell, R., Volinsky, C.: Matrix factorization techniques for recommender systems. IEEE Comput. **42**(8), 30–37 (2009)
4. Koren, Y.: A multifaceted collaborative filtering model. In: Proceedings of the ACM International Conference on Knowledge Discovery and Data Mining (SIGKDD), pp. 426–434 (2008)
5. Zhou, Y., et al.: Large-scale parallel collaborative filtering for the Netflix prize. In: Proceedings of International Conference on Algorithmic Applications in Management (AAIM), pp. 337–348 (2008)

6. Low, Y., et al.: Distributed GraphLab: a framework for machine learning and data mining in the cloud. Proc. Very Large Database Endow. (PVLDB) **5**(8), 716–727 (2012)
7. Kyrola, A., Blelloch, G.E., Buestrub, C.: Large-scale graph computation on just a PC. In: Proceedings of USENIX Symposium on Operating Systems Design and Implementation (OSDI), pp. 31–46 (2012)
8. Roy, A., Mihailovic, I., Zwaenepoel, W.: Edge-centric graph processing using streaming partitions. In: Proceedings of the ACM Symposium on Operating Systems Principles (SOSP), pp. 472–488 (2013)
9. Jo, Y.-Y., et al.: RealGraph: a graph engine leveraging the power-law distribution of real-world graphs. Technical report of Dake laboratory
10. Chou, H.-T., et al.: Design, implementation of the Wisconsin storage system. Softw. Pract. Exp. **15**, 943–962 (1985). doi:10.1002/spe.4380151003
11. Carey, M.J., et al.: Object and File Management in the EXODUS Extensible Database System. University of Wisconsin-Madison, Computer Sciences Department, Madison (1986)
12. Dewitt, D.J., et al.: Shoring up persistent applications. In: Proceeding ACM SIGMOD International Conference on Management of Data (SIGMOD), pp. 383–394 (1994)
13. Leskovec, J., Kleinberg, J., Faloutsos, C.: Graph evolution: densification and shrinking diameters. ACM Trans. Knowl. Discov. Data (TKDD) **1**(2), 1–41 (2007). doi:10.1145/1217299.1217301
14. Cremonesi, P., Koren, Y., Turrin, R.: Performance of recommender algorithms on top-N recommendation tasks. In: Proceedings of the ACM Conference on Recommender Systems (RecSys), pp. 39–46 (2010)

Parallel Compression of Weighted Graphs

Elena En, Aftab Alam, Kifayat Ullah Khan, and Young-Koo Lee(✉)

Department of Computer Science and Engineering, Kyung Hee University,
Yongin-si 446-701, Republic of Korea
yklee@khu.ac.kr
http://www.khu.ac.kr

Abstract. Large graphs such as social network, web graph, biological network, are complex and facing the challenges of processing and visualization. Motivated by such issues, Taivonen et al. [1] proposed models and sequential algorithms for weighted graph with the intentions to generate a candidate compress graph. The proposed compression algorithm is expensive in terms of computation time because of the sequential process. The weighted graph compression algorithms can be made faster while adopting parallel processing technique. In this paper, we adopt parallel processing technique for weighted graph compression problem while using multi-selection nodes to perform merge-able technique with various graph clustering algorithms to avoid overlapping between nodes from different threads. For the performance evaluation purposes of the proposed method, we carry out series of tests on the real networks. We perform extensive experiments on parallel graph summarization while using different graph clustering algorithms. Our results demonstrate their effectiveness for parallel graph compression on real networks.

Keywords: Weighted graph · Network · Graph compression · Graph clustering · Parallel processing · Graph mining

1 Introduction

Real-world data are transformed to graph structure with the intention to extract hidden knowledge while exploiting the linked structure. This linked structure is playing a vital role in various domains ranging from Web structure to the data generated by scientists. In the graph structure, the node represents the entities e.g. person, Web page, module etc., and the link is the relationship between entities such as a friend link, hyperlink etc. Below, we look at some of these application domains:

World Wide Web. The web-graph describes the directed links between webpages of the World Wide Web. The link structure of Web has been playing a significant role in the search engine such as a Google, where quality of search has been improved. Web-graph is a prime example of large-scale graph. Google estimates the total number of Webpages exceeds one trillion; experimental graphs of the World Wide Web contain more than twenty billion nodes (pages) and one hundred and sixty billion edges (hyperlinks) [2].

Social Networking Graph. In the context of the internet, social graph signifies relations of internet users where node and edge represent users and relations among them respectively.

W. Lee et al. (eds.), *Proceedings of the 7th International Conference on Emerging Databases*, Lecture Notes in Electrical Engineering 461,
https://doi.org/10.1007/978-981-10-6520-0_8

Best known examples of social graphs are the Twitter graph and the Facebook graph, with millions of users. The social networks give birth to various groups and communities of interests, for example, music, car, school, work and much more. Communications between users of such associations are quite strong that allows identifying them easily [3].

Biological Network. In the biological domain, the nodes represent entities such as genes, proteins, and enzymes, whereas the edges encode the relationships, such as control of the reaction and the correlation, between these objects [4]. The graph generated in such domains consists of million and billions of nodes and edges. Performing various data mining operations on such huge graph are quite challenging and expensive in terms of memory, processing and time.

Contributing in the area of graph summarization Toivonen et al. [1] proposed the randomized semi-greedy method [1] that computes possibility of all pairwise merges and then recursively performs the best merge until the required compression ratio is reached. To understand this approach, we remind that graph is contained many pairs of nodes that gives reduction in cost by merging operation. However, this algorithm is expansive in terms of execution time, because of sequential selection of nodes during graph compression. Therefore, the algorithm [1] can be made more efficient in terms of execution of time adopting the concept of parallelism. To achieve better performance, we use different clustering algorithms to cluster a large graph in to small independent graphs so that to be executed independently.

In this paper, we exploit different clustering algorithms to create independent small graph and then we apply the algorithm of [1] for compression. Then, we compare the performance of each clustering algorithms in the context of graph compression.

2 Related Work

Graph Compression. The problem of graph compression is relevant in the study of graphs. Most of the works have been done based on common ideas and theories of graph analysis. In fact, those researchers are proposing to utilize similar connections to merge graph nodes. Navlakva et al. [5] and Tian et al. [6] made a fundamental contribution to the graph compression research. They have proposed a compact graph as a major summary model, where nodes represents a set of nodes and edges are the relationships among of the nodes. Moreover, the original graph can be recreated from summary graph by employing the set of edge corrections. The sum of size of the edge correction set and size of the summary graph is minimized by the principle of the Minimum Description Length (MDL) [7]. The most related methods to our work have been proposed by Toivonen et al. [1]. They apply several methods to determine the efficiency of compressing large graphs such as co-authorship graphs. According to the experiments, the randomized semi-greedy method, that is a generalized version of the randomized algorithm of [5] can efficiently compress a weighted graph with a little effect on the node clustering. The authors present notions of supernodes and superedges by 2-hops optimization. The selected node merges a suitable nodes with minimized mean weight in its 2-hop neighborhood to produce a smaller graph [1].

The above-mentioned works have been focused on sequential data processing for graph representation, therefore we have explored parallel data processing and graph partitioning for better performance.

Parallelization. Parallel algorithms on graphs [8] are used in various scientific and practical applications. Multi-core systems, provide good localization of data and provide good acceleration with an increase in the number of cores. Acceleration of application processing strongly correlates with the used memory bandwidth. Therefore, parallel processing plays a significant role in many areas of scientific computing. The graph represents the vertex-centric view in the case of parallel processing. Parallel processing of graphs occurs by partitioning the graph data through processing resources and resolving dependencies along edges through iterative computation and communication.

Graph Clustering Algorithms. We mainly reviewed the clustering algorithms such as MCODE, Graph entropy, IPCA, and COACH that have performed well in previous surveys [9].

Molecular complex detection (MCODE) is well studied method for finding dense regions of the network. Dense regions are determined based on the connectivity data in the network. This algorithm works in three ways: (1) node scoring; (2) cluster finding; and (3) post processing is repeated and clusters that do not contain the maximum connected subgraph are filtered out in the resulting subgraphs.

Proposed by Li et al., Identification of protein complexes (IPCA) algorithm is focused on vertex distance and density subgraphs. The algorithm can be managed by two equations the shortest path between a pair of vertices, and the vertex interaction probability, both in subgraph level. There are four dominant divisions in the algorithm: (1) weighting vertex and computing the weight for every edge and vertex; (2) selecting seed, where the highest weighted vertex is selected as the seed; (3) extending cluster, working on extension of some cluster K; and (4) extending judgment. The IPCA algorithm requires more time to process the data, since it is related to the number, as well as the size of the created clusters.

Core-attachment based method (COACH) operates in two phases; detection a strongly connected area, which calls the "preliminary cores" of the network, and the expansion of regions by highly connected neighbors. The COACH algorithm begins from discovering cores and end with formation complexes from the graph. If the core graph is not dense enough, core nodes will be removed from the core graph. The extended graph can contain a number of connected elements and recursively perform this process to obtain strongly connected sub-graphs in each connected elements. The deleted cores are recursively added back into resulting sub-graph. Then, applying the post-processing of the discovered cores, the maximal dense form can be obtained.

Kenley and Cho presented a novel theoretical term, "*Graph entropy*" that measures the structural complexity of the given graph. The algorithm includes the method of seed increment. Moreover, this method not requires threshold, due to it searches through examination, representing advantageous solution by reducing graph entropy. This technique based on searching locally the most favorable clusters, which have a minimal value of a graph entropy. Parallelization and clustering algorithms is related to each other because they have the same properties of data partitioning for solving the time complexities.

However, the task of clustering is expensive, since many of algorithms require iterative or recursive procedure, and most of real-world networks are huge. Consequently, the parallelization of clustering algorithms should not be avoid.

3 Preliminary

We denote some basic graph notations, which can be useful for understanding this paper. Let $G = (V, E, w)$ be a directed weighted graph, where V is the set of vertices, $E \subset V \times V$ is the set of edges, and w: $E \rightarrow R^+$ assigns positive weight to each edge $e \in E$. $S = (V', E', w')$ is defined as a compressed weighted graph of G if V' is a partition of V, where nodes $v' \in V'$ are grouped to supernodes, and edges $e' \in E'$ are grouped to superedges. A supernode is a node $v \in VS$, correspond to a set of Av nodes in G, and each edge $(u, v) \in ES$ is called a superedge, which represents the edges between all pairs of nodes in Au and Av [5].

4 Parallel Compression of Graphs Using Clustering Algorithms

In this section, we introduce a parallel graph compression approach to improve the existing work [1].

4.1 Using Graph Clustering Towards Summarization

Graph clustering is the clustering of data in the form of graphs. The graph clustering algorithms are focused on summarization graph nodes by their similarities. The main purpose of clustering is the division of the graph into two or more partitions based on the similarity of objects. Clustering affects the simplification of extracting the meaningful information from the graph. The clustering algorithm generates clustering for any input data taking into account the fact that not all graphs have natural clusters. The discovery of clusters makes it possible to identify certain hidden structures in the graph.

4.2 Proposed System Overview

The task of our parallel algorithm is to divide the problem into sub-problems and executed in parallel to obtain outputs independently. For our implementation, we use more than one processor cores to run a program for processing computational intensive jobs. The program is split and executed by several CPUs at the same time, where processed results are then recombined.

The parallel compression on a single system can be achieved by using two different approaches. In the first approach, parallel processing can be performed on a single large graph. The idea is to select multiple nodes according to the available CPU cores and then merge it with a suitable node in its 2-hop neighborhood. In the second approach, the large graph can be partitioned into independent subgraphs and then can be executed in parallel by compressing each sub-graph independently. In this paper, we use the graph partitioning technique for parallel graph compression while leaving the former one for future work.

In order, to avoid the node's overlap among the nodes in the procedure of 2-hop optimization in parallel processing and generating independent sub-graphs, we use available graph clustering algorithms which we mentioned in above section. Once we generate independent sub-graphs, then we randomly choose subgraphs accordingly to the number of CPU cores and process them in parallel.

The architecture of the proposed parallel graph compression is shown in Fig. 1. In the first stage, we partation a large graph into small independent subgraphs. Then we apply above algorithms. There are well-known methods attempt to detect highly interconnected subgraphs within the datasets. It can be applied to various datasets such as social networks, literature and other interaction network.

After this step, we select randomly different number of subgraph according to the number of available CPU cores and execute for compression independently.

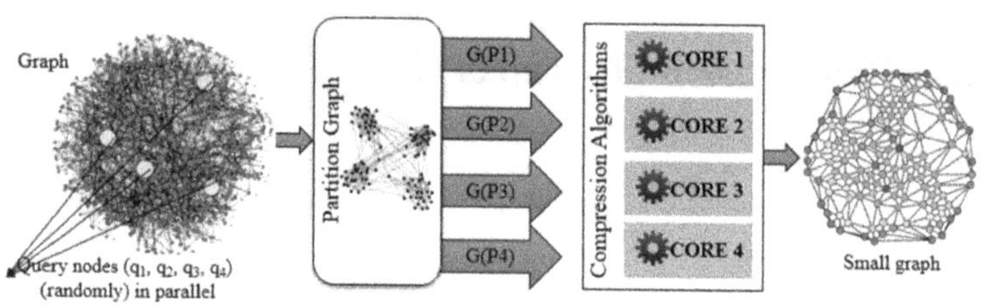

Fig. 1. Architecture of the proposed solution.

4.3 The Proposed Algorithm

Algorithm 1 is the proposed algorithm for parallel compression of the weighted graph. The input to the algorithm is G where the output is S. The proposed algorithm consists of four main parts. In the part first part, we set the main required variables i.e. *graphPartitioner* (select the algorithm to partition the graph), *cpuCount* (get the numbers of processes), and *executeService* (create a thread pool according to the available CPUs). In the second part of the algorithm, we call a function called *graphPartition* which is responsible to partition the graph according to the *graphPartitioner* parameter (being set in step 1) while returning the numbers of partitions *numOfPartitions*.

One of the challenges in parallel processing is load balancing. In our algorithm, step 5 solves the problem of load balancing with the aim to utilize the resources efficiently. Step 1 gives us the number of processors (*cpuCount*) while step 4 returns the number of graph partitions (*numOfPartations*). We divide the *numOfPartations* by *cpuCount* and get the *partationSlab* which means that how many numbers of partitions should be executed per CPU. Finally, the submit function is used to assign the number of partitioned graphs to each processor and execute in parallel.

Algorithm 1 Parallel Compression of Weighted Graph

S (V', E', w') ← G (V, E, w)

Requires:

1: graphPartitioner = {CoAch, IPCA, MCODE, GraphEntropy}

2: cpuCount ← Runtime.getNumOfProcessors ()

3: executorService ← Executors.newFixedThreadPool (cpuCount)

Partition Selector:

4: numOfPartitions ← graphPartition (graphG, graphPartitioner)

Load Balancing:

5: partitionSlab ← numOfSubGraphs/ cpuCount

Parallel Execution:

6: **for** *i=0* **to** *cpuCount*

  executorService.submit (start, partitionSlab,randSemiGreedyAlgo)

  start ← start + slab

end for

5 Experiments

In this section, we describe the experimental setup being used for the evaluation of the proposed algorithm. Furthermore, we present results showing favorable performance for weighted graph compression in parallel processing.

5.1 Experimental Setup

The purpose of our experimental study is to show the effectiveness and compare the scalability of various methods. For the evaluation purpose we have implemented our approach as an extension to the compression graph. We used four different clustering algorithms that are intended for partitioning the huge amount of data that is needed for parallel processing.

All the algorithms have been implemented framework program by using Java (JDK 1.7) by using NetBeans IDE 8.2 and Python 27, and all experiments were run on a standard PC with 8 GB of main memory and Intel Core i3-4130 3.40 GHz CPU processor.

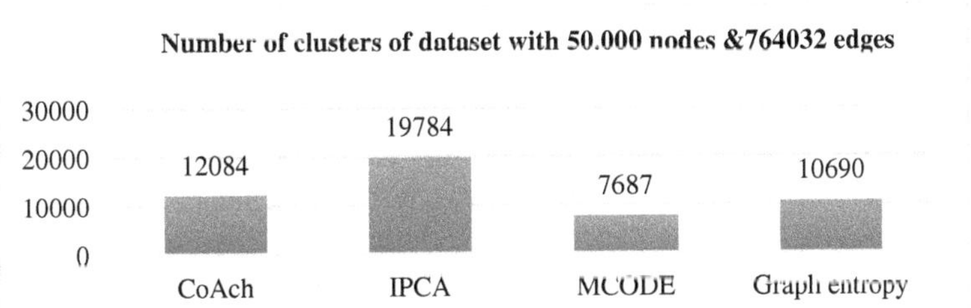

Fig. 2. The number of clusters generated by different algorithms

5.2 Datasets

The following Table 1 introduces the datasets for our experiments. We used four real-world networks i.e., Web graph, Social Networking graph and Biological graphs [10]. The size of graph increases from 1000 to 50000 nodes. The dataset of 1 K consist of 1000 nodes and 7692 edges. The dataset of 10 K has 10000 nodes and 153308 edges, whereas 20 K dataset has 20000 nodes and 164586 edges. The biggest dataset consist of 50000 nodes with 764032 edges. Each edge of graphs has an associated numerical value as a weight. The weighted graphs are used to represent structures, where pairwise connections have some numerical value.

Table 1. Datasets.

S#	Name	Nodes	Edges
1	1 K	1000	7692
2	10 K	10000	153308
3	20 K	20000	164586
4	50 K	50000	764032

5.3 Results

In this section, we present experimental results of our proposed data cauterization techniques for parallel processing of compression of weighted graphs with real data sets.

Depending on algorithms properties number of generated clusters might be different. In our proposed algorithm, each thread store almost equal number of clusters. The Fig. 2 illustrates the number of clusters from datasets with 50000 nodes and 764032 edges generated by various algorithms. It can be seen that MCODE algorithm produced only 7687 number of clusters, whereas IPCA has comparable number of clusters with 19784. By contrast, the algorithms of CoAch and Graph entropy produce approximately the same amount of clusters 12084 and 10690 respectively.

In Fig. 3, we compared the results of supernodes obtained by the sequential method and our parallel cluster approach for compressing the weighted graph. It can be observed that the outperforming results for graph compression are from MCODE and Graph entropy clustering algorithms. There are less number of supernodes than existing work. The minimum number of supernodes leads to the fact that the compressed graph will be small. A small size of graph allows users to better understand and analyze the graph. Whereas, the MCODE algorithm has a directed node that allows fine-tuning of clusters of interest without considering the rest of the network and allows examination of cluster interconnectivity. The graph entropy produce a high quality clusters that can be evaluated by connectivity. A cluster has a higher quality if it has denser intra-connections within the cluster and sparser interconnections to vertices outside of the cluster. The graph entropy definition is formulated to measure the cluster quality effectively. A graph with lower entropy indicates that the vertices in the cluster have more inner links and less outer links. The IPCA and CoAch methods detect the clusters from each node and it has high probability of overlapping. In comparison with the previous listed algorithms, they produce more number of clusters due to the fact that some predicted clusters are similar and match the same benchmark clusters. However, MCODE and CoAch algorithms lost many nodes by reducing nodes, which do not have any relations. IPCA and CoAch clustering algorithm have higher number of supernodes from existing work in every dataset. Based on the given results, we can conclude, that MCODE and Graph entropy algorithms is more suitable for our approach of parallel graph partitioning and show the improvement for graph compression method of existing work.

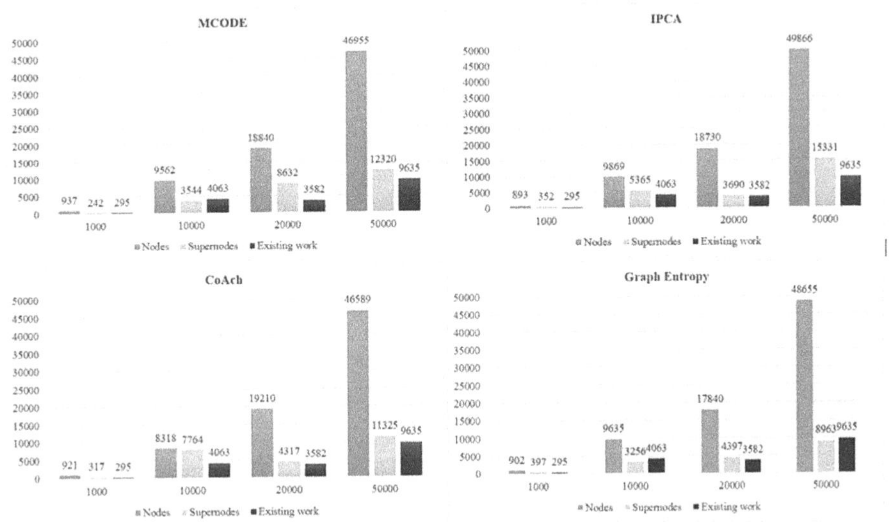

Fig. 3. Comparison of existing results of supernodes with our approach parallel graph clustering method

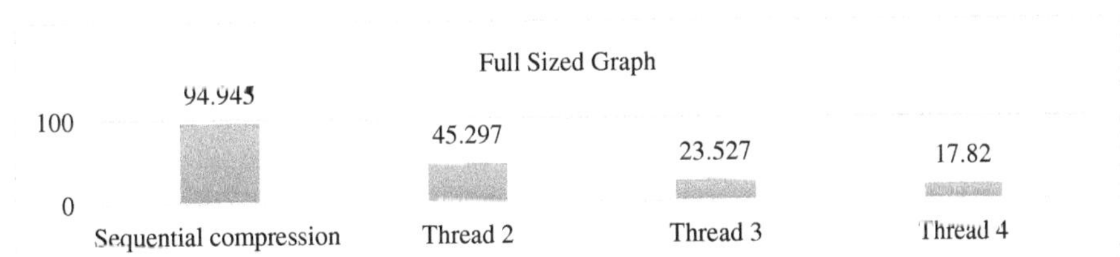

Fig. 4. Comparison of running time (in sec) between sequential compression and parallel compression

In Fig. 4, we have been illustrated the comparison of running time (sec) of existing work with sequential compression and parallel compression. There is significant difference of computational time between sequential and four threads in parallel processing, so it is clearly seen the efficiency of parallel compression method.

Additional scalability experiments were run with different dataset size and different number of threads from one to four. As our system has four threads for each clustering algorithm, it have been a maximum number of threads in our experiments in Fig. 5.

A comparative experiment conducted between a parallel processes with optimization of cluster algorithms shows that the running time decreases by more than two times compared to a sequential process with the same clustering algorithms.

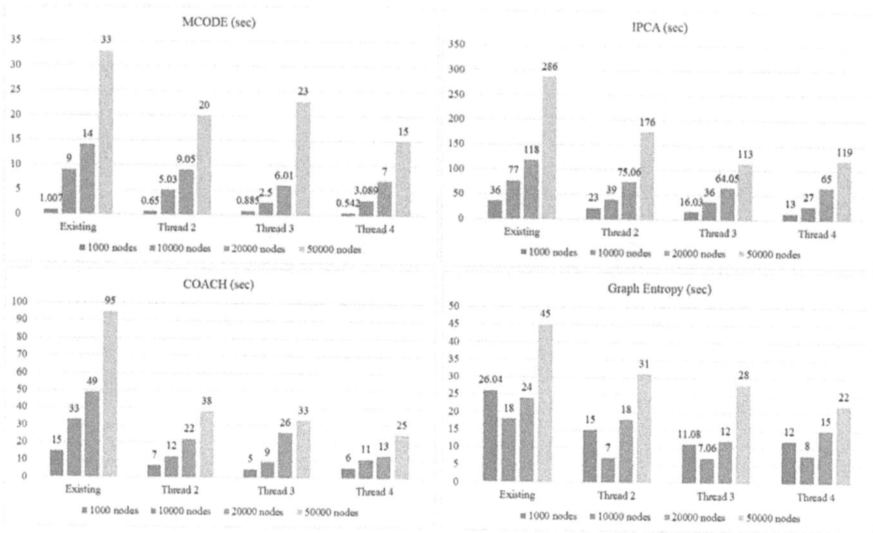

Fig. 5. Running time of parallel graph clustering algorithm for graph compression

6 Conclusion

In this paper, we compared the methods of sequential data processing and parallel data processing, as well as comparison of four clustering methods in a parallel process, to analyze a large-scale of real graph dataset. We have presented the problem of parallel compression graph and gave the solution by means of graph clustering algorithms and experimental results on multiple graph datasets. In this work, we have expanded the existing work by parallel processing, which shows significant difference between sequential and parallel processing for graph compression. Our experiments illustrates that MCODE clustering method processed 764032 edges graph in 15 s in parallel, and Graph entropy method processed same size of data in 22 s, which are suitable for our approach.

Our proposed method has several advantages such as, notable results in saving the memory space and reducing the computational time. However, during multiple selection nodes in parallel, it is possible, that same nodes can be selected and for some of the query node, all CPU cycles will be wasted, but, the probability of this scenario is very low.

Acknowledgment. This work was supported by the National Research Foundation of Korea (NRF) grant funded by the Korea government (MEST) (No. 2015R1A2A2A01008209).

References

1. Toivonen, H., et al.: Compression of weighted graphs. In: Proceedings of the 17th ACM SIGKDD International Conference on Knowledge Discovery and Data Mining. ACM (2011)
2. Sakr, S.: Processing large-scale graph data: a guide to current technology. IBM Developerworks, p. 15 (2013)
3. Baruah, T.D.: Effectiveness of Social Media as a tool of communication and its potential for technology enabled connections: a micro-level study. Int. J. Sci. Res. Publ. **2**(5), 1–10 (2012)
4. Cline, M.S., et al.: Integration of biological networks and gene expression data using Cytoscape. Nat. Protoc. **2**(10), 2366 (2007)
5. Navlakha, S., Rastogi, R., Shrivastava, N.: Graph summarization with bounded error. In: Proceedings of the 2008 ACM SIGMOD International Conference on Management of Data. ACM (2008)
6. Tian, Y., Hankins, R.A., Patel, J.M.: Efficient aggregation for graph summarization. In: Proceedings of the 2008 ACM SIGMOD International Conference on Management of Data. ACM (2008)
7. Rissanen, J.: Modeling by shortest data description. Automatica **14**(5), 465–471 (1978)
8. Nielsen, F.: Introduction to HPC with MPI for Data Science. Springer, Switzerland (2016)
9. Price, T., Peña, F.I., Cho, Y.-R.: Survey: Enhancing protein complex prediction in PPI networks with GO similarity weighting. Interdisc. Sci. Comput. Life Sci. **5**(3), 196–210 (2013)
10. Stanford Large Network Dataset Collection.html. http://snap.stanford.edu/data/

An Efficient Subgraph Compression-Based Technique for Reducing the I/O Cost of Join-Based Graph Mining Algorithms

Mostofa Kamal Rasel and Young-Koo Lee(✉)

Department of Computer Science and Engineering,
Kyung Hee University, Seoul, South Korea
{rasel,yklee}@khu.ac.kr

Abstract. Many join-based graph mining algorithms such as triangle listing and clique enumeration output a large size of intermediate or final data that sometimes dominates the mining cost. A few researches highlighted on the size of output data. However, those techniques have limitation that they are highly specific to their corresponding graph mining algorithms. In this paper, through the careful observations of the output patterns, we propose a general compression solution that can be applied to any join-based graph algorithm. It first categorizes the overlapping and non-overlapping vertices in a resultant subgraph set of a join-based graph mining algorithm. Then it compresses the output data by removing the redundancy from the overlapping vertices and by encoding the non-overlapping vertices using a non-aligned hybrid bit vector compression technique. Our proposed technique performs the compression on-the-fly and can easily be adopted by the join-based graph mining algorithms. Experiments on the real datasets show that our proposed technique, which is adopted in a triangle listing algorithm, reduces the size of the output data and the running time by three times and more than two times, respectively. The proposed technique also reduces the I/O cost for a maximal clique listing algorithm.

Keywords: Graph I/O compression · Bitmap compression · Graph mining algorithms

1 Introduction

The join-based graph mining algorithms, which perform join among the list of vertices in iterations, usually focus on the efficient accessibility of the input graph to reduce the I/O cost. However, the intermediate or the final output data of these graph mining algorithms can be too large compared to the input graph. For example, the size of the output of a triangle listing algorithm for the WebUK dataset is over 52 times higher than the dataset size, which dominates the cost of triangle listing algorithm.

W. Lee et al. (eds.), *Proceedings of the 7th International Conference on Emerging Databases*, Lecture Notes in Electrical Engineering 461,
https://doi.org/10.1007/978-981-10-6520-0_9

The output of a join-based graph mining algorithm contains many overlapping vertices. The I/O cost of such an algorithm can be reduced by removing the redundancy in the output. Some of the existing join-based graph mining algorithms reduce the I/O cost by merging the redundant patterns or vertices in the output. For example, the clique listing algorithms [8,10] proposed to merge some smaller structural patterns into a larger structural pattern to reduce the I/O cost for the intermediate output. However, their compression approach is dependent on the technique of their proposed clique listing algorithm.

We observe that the join-based graph mining algorithms output a set of subgraphs from a join of multiple adjacency lists and/or subgraphs. We also observe that the overlapping vertices share and connect to a significant number of non-overlapping vertices in each set of subgraphs. For example, a join of a triangle listing algorithm finds the common neighbors between the adjacency lists of vertices 0 and 1 of a graph in the Fig. 1 and outputs the triangles $\triangle_{012}$, $\triangle_{013}$, and $\triangle_{014}$. Note that the three triangles have the overlapping vertices 0 and 1, which share and connect to the non-overlapping vertices 2, 3, and 4. The given example indicates that the number of required words to represent each set of subgraphs, which are produced from each join, can be reduced by removing redundancy from the overlapping vertices. The I/O cost of a join-based graph mining algorithm can be further reduced by compressing the non-overlapping vertices with a less number of encoded words. However, the compression must be performed on-the-fly to overcome the possible delay caused from the encoding cost. We note that a general compression solution for reducing the I/O cost should easily be adopted to compress the set of resultant subgraphs of each join. The solution also needs to categorize and list the overlapping and non-overlapping vertices among each set of subgraphs and then encode the non-overlapping vertices on-the-fly with less number of encoded words.

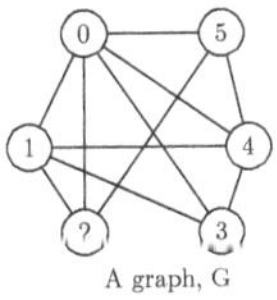

Fig. 1. A graph G that contains the triangles $\triangle_{012}$, $\triangle_{013}$, $\triangle_{014}$, $\triangle_{043}$, $\triangle_{045}$, and $\triangle_{025}$.

	WAH encoded words
1) Run-length of 0-fill words	→ 10000000000000000000000000000001
2) A literal word for 35 and 45	→ 00010000000001000000000000000000
3) Run-length of 0-fill words	→ 10000000000000000000000000000001
4) A literal word for 112	→ 00000000000000001000000000000000
5) Run-length of 0-fill words	→ 10000000000000000000000000000010
6) A literal word for 201	→ 00000000010000000000000000000000

Fig. 2. A list of WAH encoded words which are encoded for the vertices 35, 45, 112, and 201.

The traditional graph compression algorithms such as [1,4,5] usually reorders the vertices for a better compression ratio. However, the vertex reordering is expensive and applied on all vertices of a graph. Therefore, we cannot compress on-the-fly the non-overlapping vertices of a set of subgraphs using a traditional graph compression technique. We observe that the non-overlapping vertices can be represented using a *bit vector*[1] and then compressed on-the-fly by the bitmap compression algorithms such as [3,9] since they do not reorder vertices. Note that the bitmap-based compression algorithms compress a bit vector in the unit of word by encoding the *0-fill*, *1-fill*, and *literal*[2] words. However, since the bitmap-based compression algorithms contains the run length of the 0-fill words, the sparsity in the non-overlapping vertices sometimes increases the size of the compressed data. For example, the Fig. 2 shows that if a list of four non-overlapping vertices 35, 45, 112, and 201 is represented by a bit vector, the WAH [9] bitmap compression algorithm requires six words to encode them.

Based on our observation, we propose an effective compression technique that represents a set of subgraphs by the overlapping and non-overlapping vertices, where the non-overlapping vertices are encoded using the non-aligned hybrid words. The proposed technique does not encode the 0-fill words. In the compressed data, a word represents either a single vertex or a group of vertices. We introduce the *S-word*, *M-word*, and *F-word* to encode the non-overlapping vertices. Since the proposed technique does not require any preprocessing such as vertex reordering, the join-based graph mining algorithms can easily adopt our compression technique and perform the compression on-the-fly.

The rest of the paper is organized as follows. Section 2 discusses the related studies. We present the proposed compression technique in Sect. 3. Section 4 analyzes the results of the experiments. Finally, Sect. 5 concludes the paper.

2 Related Works

Only a handful number of graph mining algorithms focused on reducing the size of output data. These researches looked for the patterns or redundancy among the data to reduce the number of items. Xie et al. [10] compressed the intermediate candidates while discovering the large sized cliques in the graphs. To reduce the size of the candidate item patterns, they merged several small item patterns to a large one. However, the item patterns still contain redundant vertices. Wang et al. [8] focused on the redundancy among the group of cliques and represented each group using a maximal clique to reduce the space. These techniques consider the redundancy among the different structural patterns in the graph and are applicable only for those specific graph mining algorithms. However, our proposed technique is a general solution that categorizes the overlapping and non-overlapping vertices the resultant subgraphs and encodes the non-overlapping vertices to reduce the required space.

The bitmap compression algorithms, such as WAH [9], and some of the variants of WAH, such as PLWAH [3], can represent the neighbors of a vertex in a bit vector and compress them on-the-fly. We note that the encoded data of these techniques includes the literal word and the run-length of the 0-fill or 1-bit

[1] A bit vector represents the neighbors of a vertex in a graph using the bits 0 or 1.

[2] A 0-fill or 1-fill or literal word refers to a word that contains all 0-bits or all 1-bits or both 0 and 1-bits, respectively.

words. Because of the sparsity of the non-overlapping vertices, the bitmap compression algorithms have to include several words to encode the run-length of 0-fill words which increases the size of the encoded data. The encoding technique used in [7] prunes all 0-fill words and used word ids that increases the size of encoded data for some cases. However, our proposed technique neither encodes the 0-fill words nor uses the word ids.

3 Proposed Non-aligned Hybrid Compression Technique

Our proposed technique is adopted in a join-based graph mining algorithm and compresses the resultant subgraphs of each join on-the-fly. The proposed technique performs the compression in two different steps: (i) *Vertex Categorization* prepares the lists for overlapping and non-overlapping vertices from each join of a graph mining algorithm, (ii) *Non-aligned Hybrid Encoding* compresses the non-overlapping vertices by a list of hybrid encoded words. We define the overlapping and the non-overlapping vertices of join result as follows.

Definition 1. *Overlapping Vertices. We define the vertices as overlapping that are common in the resultant subgraphs of a join.*

Definition 2. *Non-overlapping Vertices. We define the vertices as non-overlapping that are not common in the resultant subgraphs of a join, but connected to all overlapping vertices.*

We denote the lists of the overlapping and non-overlapping vertices as L^o and L^n, respectively.

3.1 Vertex Categorization

Each join of the join-based graph mining algorithms outputs a set of subgraphs by joining multiple neighbor lists. For example, a join-based triangle listing algorithm performs a join between the lists of neighbors of an edge. We consider the vertices of that edge as L^o and the common neighbors of that L^o as L^n. Notice that the triangle listing algorithm creates a triangle for each non-overlapping vertex, which is connected with both of the overlapping vertices.

However, a join-based clique listing algorithm finds the common neighbors in the neighbor lists of either an edge or a smaller clique and outputs a set of subgraphs, which are also cliques. We consider the vertices of that edge or clique as L^o and the common neighbors of that L^o as L^n. Notice that we can represent a resultant clique by L^o and a vertex of L^n, since each vertex of L^n is connected to all vertices of L^o. We note that if the output cliques from a join do not have same number of vertices, we use a *(key, value)* pair to represent L^o and L^n for a set of subgraphs. In this approach, we consider the first $m-1$ number of vertices of each m sized subgraph as the key and the mth vertex as the value. If a key already exists, then the new value is appended with the previous value.

Algorithm 1: Joining Process

Input: A graph $G = (V, E)$, an edge (u, v)

1 $adj(u) \leftarrow$ Get the neighbors of u
2 $adj(v) \leftarrow$ Get the neighbors of v
3 $C \leftarrow adj(u) \cap adj(v)$
4 **for** *each vertex id* $w \in C$ **do**
5 | output $\triangle_{uvw}$
6 **end**

Triangles	Overlapping, L^o	Non-overlapping, L^n
$\triangle_{012}$, $\triangle_{013}$, $\triangle_{014}$	0, 1	2, 3, 4
$\triangle_{025}$	0, 2	5
$\triangle_{043}$, $\triangle_{045}$	0, 4	3, 5

Fig. 3. The list of overlapping and non-overlapping vertices to represent the triangles $\triangle_{012}$, $\triangle_{013}$, $\triangle_{014}$, $\triangle_{043}$, $\triangle_{045}$, and $\triangle_{025}$ of the graph G in Fig. 1.

We illustrate the Algorithm 1 to show the joining process of a triangle listing algorithm. The algorithm takes as input a graph $G(V, E)$ and an edge $(u, v) \in E$, where V and E denotes the vertices and edges, respectively. In the line 3, the triangle listing algorithm performs join to find the common neighbors from the neighbor lists $adj(u)$ and $adj(v)$ of u and v, respectively. The triangles are output in the lines 4 to 6 as the joining results. A clique listing algorithm performs similar joins that takes as inputs a graph G and a clique Q, which finds the common neighbors from the neighbor lists of Q. The Fig. 3 shows an example of the vertex categorization process in a triangle listing algorithm for a graph G given in Fig. 1. The algorithm finds the triangles $\triangle_{012}$, $\triangle_{013}$, and $\triangle_{014}$ from the first join between $adj(0)$ and $adj(1)$. Therefore, we categorize the overlapping vertices 0 and 1 in L^o and the non-overlapping vertices 2, 3, and 4 in L^n. The second join is performed between $adj(0)$ and $adj(2)$ to produce $\triangle_{025}$. Finally, the triangle listing algorithm can produce the triangles $\triangle_{043}$ and $\triangle_{045}$ by performing the join between $adj(0)$ and $adj(4)$. Note that a triangle listing algorithm outputs 18 words for the six triangles for the graph G of the Fig. 1. However, the list of L^o and L^n in the Fig. 3 shows that the algorithm can reduce the number of words to 12 by applying vertex categorization.

The total size of the join results can be too large to affect the efficiency of a join-based graph mining algorithm. To reduce the size of this output, we categorize and merge the vertices attended in each join as overlapping and non-overlapping. The join-based graph mining algorithms can adopt our proposed merging concept in Algorithm 1 to categorize the vertices as the overlapping and non-overlapping for each join and reduce the size of output.

3.2 Non-aligned Hybrid Encoding (NHE)

The proposed compression algorithm performs a non-aligned hybrid bit vector compression technique to encode each L^n. The encoded data contains a list of words, where each word represents either a non-overlapping vertex or a set of non-overlapping vertices in bits. We define the *S-word*, *M-word*, and *F-word* to encode the non-verlapping vertices.

Definition 3. *S-word. We define a word as S-word that represents a single vertex. The first bit of a S-word is set to 0.*

Definition 4. *M-word. We define a word as M-word that represents multiple vertices, where each high bit (1-bit) of that word is corresponded to a vertex.*

Definition 5. *F-word. We define a word as F-word that represents a single vertex and indicates that the next word is an M-word. The first bit of a F-word is set to 1.*

The proposed NHE starts by considering the first vertex c of L^n as a candidate of F-word. Then every $(v-c)^{th}$ bit of a l-length word w is set to high, where $\{v \in T | c < v \leq c+l\}$ and $T \subset L^n$. If $|T| > 1$, then we convert c to a F-word and consider w as a M-word. The next $(c+l+1)^{th}$ vertex is again considered as the new candidate of F-word. However, if $|T| == 1$, then c is encoded as S-word and the single vertex in T is considered as the new candidate for F-word. We illustrate the hybrid encoding technique in the Algorithm 2.

```
Algorithm 2: Non-aligned hybrid compression
Input  : A list of vertices L^n, word length l
Output: A list of words W that encodes L^n
1  A flag vertex, c ← −1
2  A temporary list of vertices, T ← empty
3  for each vertex id v ∈ L^n do
4      if c = −1 then
5          c ← v
6      else if v < c ≤ c + l then
7          Insert v into T
8      else
9          if T has more than one vertices then
10             Set the first bit of c to 1
11             w ← Represent T in bits
12             Insert c & w to W
13         else
14             Insert c to W
15             Insert all vertices of T to W
16         end
17         c ← v
18         T ← empty
19     end
20 end
21 Repeat lines 9 − 17
22 Return W
```

The algorithm takes as input the non-overlapping vertices L^n and the word length l and returns a list of words W that encodes L^n. For each vertex $v \in L^n$, the algorithm decides at lines $4-7$ whether v should be set as a candidate of F-word c or be inserted into a temporary list T. If c is already set and v does not satisfy $c < v \leq c+l$, the algorithm applies the hybrid encoding at lines $9-16$ and insert the encoded c and T into W. After encoding, the new vertex v is set as the new candidate for F-word at line 17 and T is set to *empty* at the line 18. Before returning W at the line 22, the algorithm repeats the lines $9-16$ at the line 21 to encode the last c and T. We note that the length of a word l must be longer enough to represent a vertex id using $(l-1)$ number of bits since a F-word in the encoded data holds a vertex id. The Fig. 4 shows an example of hybrid encoding of a list of non-overlapping vertices.

Lemma 1. *The two steps of the non-aligned hybrid vector compression always reduces the size of the output data.*

Encoded words, W =

	Encoded words
37	→ **1**0000000000000000000000000100101
45, 46, 65	→ 00000001100000000000000000010000
112	→ **0**0000000000000000000000001110000
117	→ **0**0000000000000000000000001110101
201	→ **1**0000000000000000000000011001001
207, 211, 225, 228	→ 00001000100000000000001001000000

Fig. 4. The figure shows the NHE encoding of a list of non-overlapping vertices, L^n to the 32-bit words. The words containing the first bit bold encode only a single vertex. If the first bit of a word is 1, then the next word represents multiple vertices.

Proof. The vertex categorization merges the L^o of multiple resultant subgraphs of a join into a single list, which requires less number of words compared to represent the subgraphs separately. The hybrid encoding ensures that each word represents either a non-overlapping vertex or a list of multiple non-overlapping vertices, which also requires less number of words. Since both of the steps reduce the number of words to represent the L^o and L^n, the proposed technique always reduces the size of output data.

Our proposed hybrid approach is an incremental compression technique that compresses on-the-fly for every join of a join-based graph mining algorithm. Therefore, the proposed technique can easily be adopted by any join-based graph mining algorithm by replacing the lines 4 to 6 of the Algorithm 1 with the Algorithm 2.

3.3 Decoding the Compressed Data

We describe the decoding technique to get the list of non-overlapping vertices L^n from the encoded data W as follows,

- For each word $w_i \in W$
  - If the first bit of w_i is 0, then insert w_i into L^n
  - Otherwise,
    - Set the first bit of w_i to 0 and insert w_i into L^n
    - Calculate $v = w_i + p$ for each 1-bit of word w_{i+1} and insert v into L^n, where p denotes the position of a bit in w.

We note that the output of each join of a clique listing algorithm may contain the candidates to find larger cliques. Therefore, a clique listing algorithm has to revisit the previous outputs. Since our proposed technique reduces the size of the output data, the input cost of a join-based clique listing algorithm is also reduced.

3.4 Complexity Analysis

The vertex categorization can categorize the vertices directly from each join of the neighbor lists in a join-based graph mining algorithm. Therefore, the cost for the categorization comes only from the sorting of the output cliques. The cost for the NHE comes from the probing of each non-overlapping vertex v into the range of $c < v \leq c + l$. Therefore, the cost is equal to L^n for each join. Since an encoded word does not surpass the position of a corresponding vertex in L^n, we reuse the words in L^n to represent the encoded words. Therefore, we do not need the extra memory to store the encoded data.

4 Experimental Studies

The experimental environment comprised a 64-bit Intel Core i5 3.3 GHz CPU with eight cores and 8 GB RAM. We implemented the algorithms in C language on a Linux (Ubuntu 16.04) operating system.

We used three real world datasets, Road Network (RN), Live Journal (LJ), and Web links of UK (WebUK), which have different characteristics. The RN dataset represents the road network information for California in the USA, where each vertex and edge represent a junction and a road connected to two junctions, respectively. LJ is a social network (http://www.live-journal.com) where vertices and edges represent the members and the friendships between members, respectively. In WebUK, vertices represent the pages and edges represent the hyperlinks. The statistical information of the datasets, which are used in out experiments, are given in Table 1.

Table 1. Dataset Information

Datasets	# of Nodes	# of Edges	Avg. degree	Graph size
RN	1,965,207	5,533,217	2.8	37.9 MB
LJ	4,847,571	86,220,867	17.8	383.7 MB
WebUK	18,747,771	810,803,733	43.25	3.4 GB

We implemented the proposed vertex categorization and NHE technique in a triangle listing algorithm, iTri [6] and in a maximal clique listing algorithm, FMC [2] and performed the categorization and NHE to compress the output triangles and cliques, respectively. We compared the running time and the size of the output data of the iTri algorithm with and without applying the proposed technique for all datasets. We also compared compression ratio and time with the WAH [9] algorithm. In all cases, the output triangles or cliques are represented by the lists of 32-bit words, where the first word of a list represents the number of words in that list.

We plot the experimental results for the output size () of iTri in the Fig. 5 in the percentage compression ratio. The figure includes the output data size without compression (Original), after categorization (CAT), after encoded by WAH, after encoding by NHE. The figure shows that our proposed compression technique reduces the size of output data by three times in case of WebUK dataset and more than two times in case of LiveJournal dataset. We observed that the number of non-overlapping vertices among the triangles in WebUK dataset is higher than the LiveJournal dataset. Therefore, the compression ratio is higher for WebUK dataset. However, there is no overlapping among the triangles in RN dataset. Therefore, the proposed algorithm does not gain any compression for this dataset. We observe that the average degree of the dataset affects the number of triangles as well as the occurance of number of overlapping. We also observe that WAH algorithm compresses better than the proposed NHE for LJ and WebUK dataset since WAH encodes the run length of 1-fill words. However, in case of RN dataset, WAH takes larger space compared to NHE since WAH requires extra 0-fill word to encode the single vertex of each L^n.

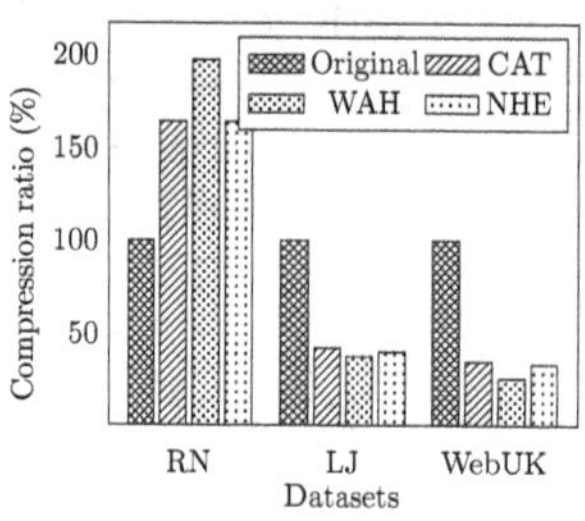

Fig. 5. The sizes of the output data in different format are plotted in the percentage of compression ratio.

Datasets	Original	CAT	WAH		NHE	
			Overall	Com.	Overall	Com.
RN	1.129	1.1	1.12	0.006	1.11	0.0022
LJ	44.08	35.42	39.46	5.24	37	1.32
WebUK	1980.37	972.61	975.25	226.89	947	39.25

Fig. 6. The running time of the iTri algorithm for different types of output format.

The overall times of the iTri algorithm with and without applying the compression technique are plotted in the Fig. 6. We also included the encoding cost for both WAH and NHE. The figure shows that the running time of iTri is reduced to $\frac{1}{2}$ times when the compression technique is applied. We observed that the reduced number of I/O requests reduces the running time. However, overall running time for WAH is higher than the proposed NHE since the compression cost for WAH is higher.

We performed the experiment for the FMC algorithm on the LiveJournal dataset. In the experiment, the memory bound and the upper bound for the number of B-vertices were set to 64 MB and 20, respectively. We categorized the overlapping and non-overlapping vertices from the maximal cliques for the first 57 in memory seed sets and measured the size of the output data. The sizes of the output data are 18.8 GB and 33.3 GB with and without applying our proposed technique on the FMC, respectively. We observed that our proposed categorization technique reduces the data size by removing the redundant vertices among the output cliques.

We observed that NHE outputs larger sized of compressed data for the 64-bit words compared to the 32-bit words, however the algorithm does not affect much on the compression time. We also observed that larger size of buffer does not affect the proposed algorithm since the vertex categorization and non-aligned compression are performed on each join result, which is assumed to be fit in the available memory, rather than the combination of multiple join-results. Due to the space limit, we did not include the resultant data for the experiments on word and buffer size.

The experimental results approve that our proposed technique can significantly reduce the output data size of the graph mining algorithms by applying our proposed compression. The compressed data also improve the running time of the mining algorithms by reducing I/O requests.

5 Conclusion

The join-based graph mining algorithms have to handle a large size of intermediate or final output data. The cost to handle the large output sometimes dominates the cost of the mining algorithms. Existing solutions for reducing the data size are specific to the graph mining algorithm. Moreover, we can not apply the traditional graph compression techniques to compress the output data of a graph mining algorithm on-the-fly. In this paper, we proposed an on-the-fly compression technique, which can be easily adopted by the join-based graph mining algorithms. The proposed technique first categorizes the vertices in the resultant subgraphs of a join into overlapping and non-overlapping. Then the non-overlapping vertices are encoded using a non-aligned hybrid compression technique. The proposed technique significantly reduces the required space and I/O cost. In future, we aim to generalize our compression technique to be adopted in any graph mining algorithms.

Acknowledgments. This research was supported by the National Research Foundation of Korea (NRF) grant funded by the Korea government (MEST; No. 2015R1A2A2A01008209).

References

1. Boldi, P., Vigna, S.: The webgraph framework i: compression techniques. In: Proceedings of the 13th International Conference on World Wide Web, WWW 2004, pp. 595–602. ACM, New York (2004)
2. Cheng, J., Ke, Y., Fu, A.W.C., Yu, J.X., Zhu, L.: Finding maximal cliques in massive networks. ACM Trans. Database Syst. (TODS) **36**(4), 21 (2011)
3. Deliège, F., Pedersen, T.B.: Position list word aligned hybrid: optimizing space and performance for compressed bitmaps. In: Proceedings of the 13th International Conference on Extending Database Technology, pp. 228–239. ACM (2010)

4. Hernández, C., Navarro, G.: Compressed representations for web and social graphs. Knowl. Inf. Syst. **40**(2), 279–313 (2014)
5. Lim, Y., Kang, U., Faloutsos, C.: Slashburn: graph compression and mining beyond caveman communities. IEEE Trans. Knowl. Data Eng. **26**(12), 3077–3089 (2014)
6. Rasel, M.K., Han, Y., Kim, J., Park, K., Tu, N.A., Lee, Y.K.: iTri: index-based triangle listing in massive graphs. Inf. Sci. **336**, 1–20 (2016)
7. Rasel, M.K., Lee, Y.K.: Exploiting CPU parallelism for triangle listing using hybrid summarized bit batch vector. In: 2016 International Conference on Big Data and Smart Computing (BigComp), pp. 183–190. IEEE (2016)
8. Wang, J., Cheng, J., Fu, A.W.C.: Redundancy-aware maximal cliques. In: Proceedings of the 19th ACM SIGKDD International Conference on Knowledge Discovery and Data Mining, pp. 122–130. ACM (2013)
9. Wu, K., Otoo, E.J., Shoshani, A.: Compressing bitmap indexes for faster search operations. In: Proceedings of the 14th International Conference on Scientific and Statistical Database Management, SSDBM 2002, pp. 99–108. IEEE Computer Society, Washington, DC (2002)
10. Xie, Y., Philip, S.Y.: Max-clique: a top-down graph-based approach to frequent pattern mining. In: 2010 IEEE 10th International Conference on Data Mining (ICDM), pp. 1139–1144. IEEE (2010)

Smoothing of Trajectory Data Recorded in Harsh Environments and Detection of Outlying Trajectories

Iq Reviessay Pulshashi[1], Hyerim Bae[1(✉)], Hyunsuk Choi[2], and Seunghwan Mun[2]

[1] Pusan National University, Busan, South Korea
{pulshashi,hrbae}@pusan.ac.kr
[2] Samsung Heavy Industry, Geoje, South Korea
{hyun_s.choi,adams91.mun}@samsung.com

Abstract. The existence of an outlying trajectory, which is a trajectory that contains significant noise, can affect the quality of the movement analysis result. However, some outlying trajectories are repairable by detection and removal of noise. We propose trajectory-smoothing algorithms for automatic outlying trajectory repair, followed by application of various outlier detection algorithms to redetect outlying trajectories from among the smoothed trajectories. Our proposed approach was validated in a case study using real-life trajectories from a shipyard in South Korea.

Keywords: Smoothing trajectory · Outlier detection · GPS · Movement analysis · Shipyard

1 Introduction

The rapid growth of emerging technologies such as Global Positioning System (GPS) has increased awareness of the necessity for mining and analysis of trajectory data [1,2] by means such as outlier detection [3]. During data acquisition, noise, a random error, can occur under any of several circumstances, one of which is environment interference [4]. The existence of noise creates an outlying trajectory that affects the quality of data analytics results [5]. Some of the outlying trajectories are repairable, which is to say that they can be rendered normal, and thus useful for analytics by detection and removal of noise. The key problems remain how to repair the trajectories and detects outlier among them.

The objectives of this paper are as follows: (1) to propose an outlying trajectory-detection framework that entails preparatory smoothing of trajectory; (2) to propose t-fixed partition (TFP) and k-ahead artificial arc (KAA) algorithms for smoothing of trajectory data, and (3) to benchmark statistical, distance and density-based outlier detection algorithms for detection of outlying trajectories on the basis of a real-life case study of a shipyard in South Korea.

W. Lee et al. (eds.), *Proceedings of the 7th International Conference on Emerging Databases*, Lecture Notes in Electrical Engineering 461,
https://doi.org/10.1007/978-981-10-6520-0_10

The remainder of this paper is organized as follows. Section 2 illustrate the problem using a running example; Sect. 3 provices a problem statement; Sect. 4 presents the proposed trajectory-smoothing approach; Sect. 5 reports an experimental evaluation based on a real-life case study; Sect. 6 discusses several related studies; Finally, Sect. 7 concludes the paper.

2 Running Example

In the shipbuilding industry, GPS technology has been adopted for monitoring movement of block transporters. Blocks are properly sized parts of ships that are worked on in any of several workareas called factories. Each block require a sequence of work including cutting and forming, block assembly, pre-outfitting and painting, pre-erection, erection, and quay in a specific factory. Because a factory is designed for a fixed-position layout, a block has to be moved from one location to another location to change a type of work. According to safety regulation, every block transporter can move at a maximum speed of 30 km per hour. Nonetheless, since many shipbuilding projects run simultaneously, a block transporter must operate on a tight schedule. Thus, monitoring and analysis of block transporter movement patterns become necessary. Figure 1 shows how one company chooses to adopt GPS technology for tracking of such patterns.

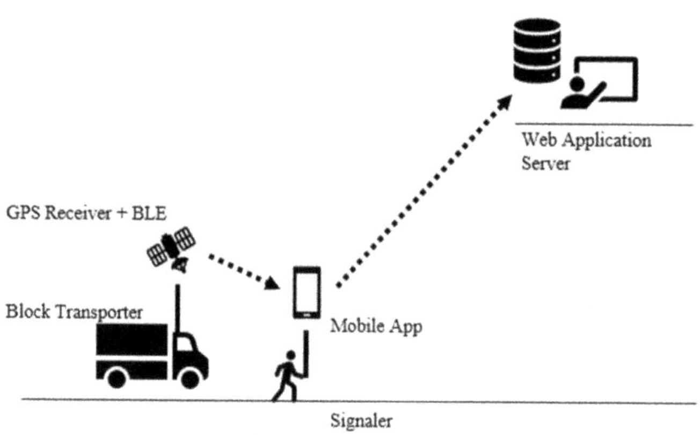

Fig. 1. GPS framework for monitoring of block transporter movement

Every block transporter is equipped with a GPS receiver device. A signaler, the person who will move along with the block transporter to guide the route, uses a mobile application that is connected wirelessly to the GPS receiver device. The application will send position data periodically to the application server in the company headquarters. The sequence of the position data collected during the movement of a block from the start location to the end location forms a trajectory. However, due to the many steel works and high buildings in the environment, the GPS signal often is deflected to another location, thereby leaving

(a) Normal trajectory (b) Outlying trajectory

Fig. 2. Examples of normal and outlying trajectories

noise in the trajectory [6]. A trajectory that contains a significant amount of noise rendering it useless for movement analysis is called an outlying trajectory. Figure 2a provides a visualization of a trajectory, while Fig. 2b shows an example of an outlying trajectory, respectively. An outlying trajectory is repairable to the extent that the noise within it can be detected and removed. For those purposes, a trajectory-smoothing algorithm must be applied before initiation of movement analysis. The issues to be resolved prior to smoothing however, are how, exactly smoothing is to be accomplished and the choice of algorithm for most efficient detection of outlying trajectories.

3 Problem Statement

We developed trajectory-smoothing algorithms for automatically detection and removal of noise from trajectories. Given a set of trajectories $T = \{t_1, \ldots, t_n\}$, our algorithm discovers a corresponding set of smoothed trajectories $S = \{s_1, \ldots, s_n\}$ then, an outlier algorithm can discover the set of outlying trajectories $O = \{o_1, \ldots, o_w\}$ from among the smoothed trajectories S such that $O \subseteq S$ and $w \leq n$.

We adopted the definition of *trajectory* from [3]; a sequence of multi-dimensional points denoted as $t_i = p_1, \ldots, p_m$. In our case, *point* p_j is a tuple $p_j = \langle lat, lng, ts \rangle$ is location data presented as a latitude (lat) and longitude (lng) pair with its corresponding timestamp (ts). *Noise* is a set of points $q_j \in t_i$ classified as random error, which is significantly unhelpful to analytics. An outlier among trajectories, or an *outlying trajectory* o_k, is a trajectory that contains a significant amount of noise.

For detection of outlying trajectories; a number of researches in the field of data mining has been conducted. Figure 3 schematizes the three popular approaches to outlier detection. (1) Statistical-based outlier detection [7], utilizes the statistical properties of data for outlier detection. This approach can work on a single object based on the threshold parameter outlier factor of and the extreme value factor ef; however, it cannot check whether it was significantly different from the other objects. (2) Distance-based outlier detection [8], detects an outlier by calculating its distance relative to the other objects. This approach can detect a significant global outlier among all data based on the parameter distance threshold r and the outlier fraction threshold; but it cannot detect a local outlier from a cluster, (i.e., a set of object that is closer to the others). (3) Density-based outlier detection [9], detects a local outlier by detecting a significant object that is far from the others among a set of closely related objects based on the parameter number of the closest-neighbor kn (we use the label kn instead of the usual k or n in order to avoid confusion with our k parameter approach).

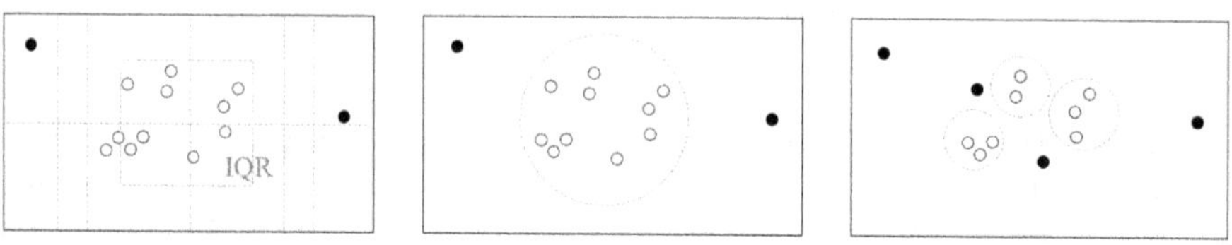

(a) Statistical-based (b) Distance-based (c) Density-based

Fig. 3. Three outlier detection algorithms

Instead of using raw data, an outlier detection algorithm usually derives a feature vector from data. A *feature vector* is a set of values that can represent the characteristics of data denoted as $\boldsymbol{f_{ki}} = \{v_l, \ldots, v_x\}$. Here, we use three kinds of feature vectors: (1) distance between two points, (2) delta time between two points, and (3) bearing or heading angle calculated from two points. Figure 4 demonstrates feature vector $\boldsymbol{f_{ki}}$ extraction from trajectory t_i.

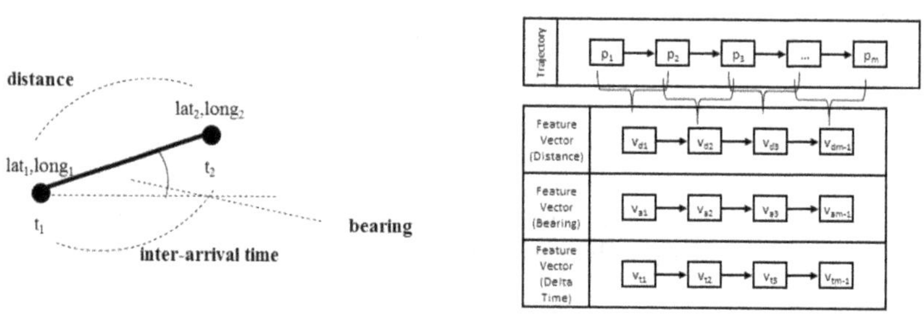

(a) Features between two pointsa (b) Trajectory-features extraction

Fig. 4. Extraction feature vector from trajectory

4 Proposed Approach

If noise can be removed from an outlying trajectory, it can be a normal trajectory that is useful for analysis. We propose an approach for smoothing of trajectory data such that only 'real' outliers are detected during subsequent outlier detection. Figure 5 illustrates that our framework entails two steps: (1) smoothing of trajectories using a trajectory-smoothing algorithm; (2) application of the outlier-detection algorithm for detection of outlying trajectories.

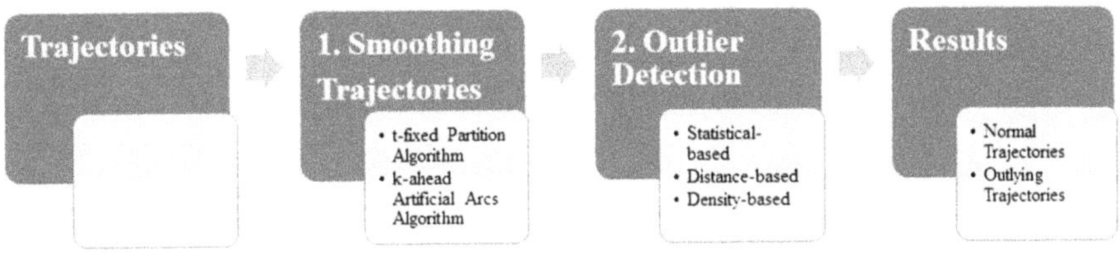

Fig. 5. Proposed outlier detection framework

4.1 t-fixed Partition (TFP) Trajectory-Smoothing Algorithm

Inspired by the work of [3] for detection of outliers from sub-trajectories, We herein propose the t-fixed Partition (TFP) smoothing algorithm for smoothing of a trajectory by partitioning every trajectory into a fixed t number of partitions. The algorithm works by reducing the number of points in a trajectory to $t+1$ number of points, these being the first point and the last one from each partition. If the length of trajectory n is less than t, we duplicate the last point from the last partition until the length of smoothed trajectory n_s is equal to t. The advantage of this approach is the fact that the length of a smoothed trajectory will be equal to t; this means that the calculation of the distance between its corresponding feature vectors will be more objective (since the lengths of the trajectories are equal). However, parameter t should be tuned in respect of the average trajectory length. If the value of t is too low, there are too many points that are removed from the trajectory; and if t is too high, most of the smoothed trajectories will be equal in length to the trajectory. Figure 6 illustrates how TFP smoothing algorithm works to smooth a trajectory.

4.2 k-ahead Artificial Arcs (KAA) Trajectory-Smoothing Algorithm

We additionally propose a trajectory-smoothing algorithm called k-ahead Artificial Arcs (KAA). It wo works as follows: first, we convert the trajectory into a graph that contains only a single path from the starting point to the end point; second, for every point in the trajectory, we create artificial arcs to the k number of points ahead from the current point; third, we apply a path-finding algorithm such as developed by Dijkstra [10] to find the shortest path from the starting point to the end point; finally, the smoothed trajectory s is a the sequence of

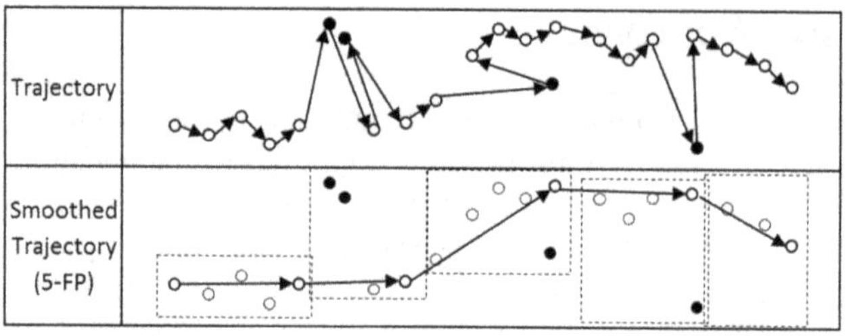

Fig. 6. t-fixed Partition (TFP) Trajectory-Smoothing Algorithm

points that is included in the shortest path. Like the t parameter in TFP, the k parameter here needs to be tuned carefully with respect to the average trajectory length. Figure 7 illustrates how the KAA smoothing algorithm works to smooth a trajectory.

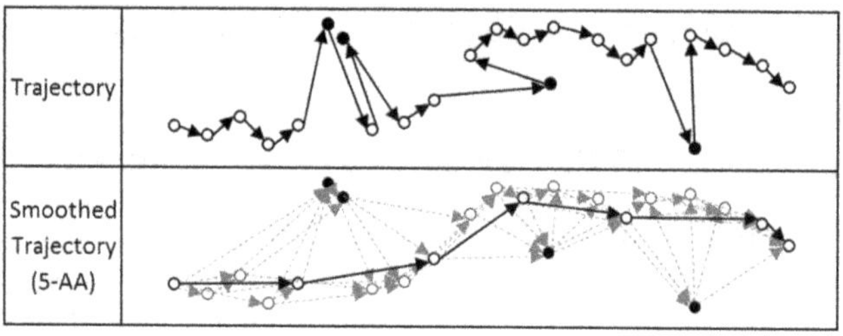

Fig. 7. k-ahead Artificial Arcs (KAA) Trajectory-Smoothing Algorithm

5 Experimental Results

We verified the proposed approach by performing experiments using a real-life case study of a shipyard in South Korea. The dataset employed contains 331 trajectories with 21,376 points in total recorded from five days of observations of block transporter movement. For every trajectory, the domain expert from a shipbuilding company manually identified 274 outlying trajectories from the data. However, only 106 outlying trajectories were unrepairable which is to say 'true' outlying trajectories that are useless for movement analysis. Figure 8 shows the outlying trajectories identified by the domain expert.

The experiments were conducted in order to benchmark the outlier-detection algorithms after trajectory smoothing using the proposed approach. The following three outlier detection algorithms were tested against the configurations respectively specified: (1) statistical-based outlier detection, with outlier factor $of = 1.5$ and extreme value factor $ef = 3$; (2) distance-based outlier detection, with distance radius $r = 1000$ and non-outlier fraction $\pi = 0.75$, and (3) density-based outlier detection, with $kn = 10$ for selection of kn-NN nearest neighbor objects. Then, the accuracy for each experiment wes calculated by comparing the outlier-detection result with the domain expert's manual identification result.

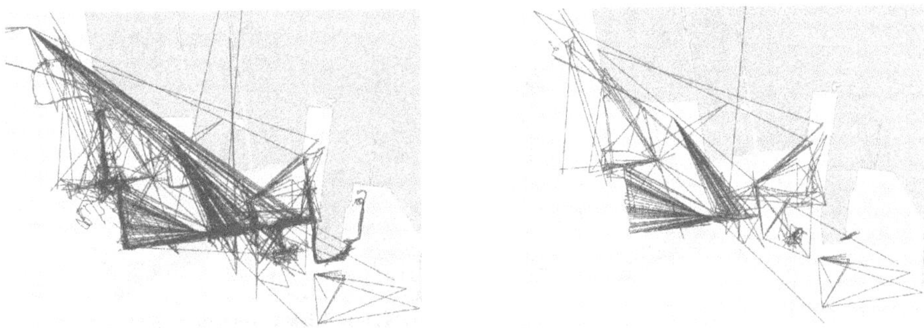

(a) 274 outlying trajectories

(b) 106 unrepairable outlying trajectories

Fig. 8. Outlying trajectories identified by domain expert

5.1 Testing t Parameter of TFP

The first experiments aimed to test the sensitivity of the t parameter used in the TFP trajectory-smoothing algorithm. Figure 9 plots the performance results for the various outlier-detection algorithms after trajectory preprocessing with the TFP smoothing algorithm.

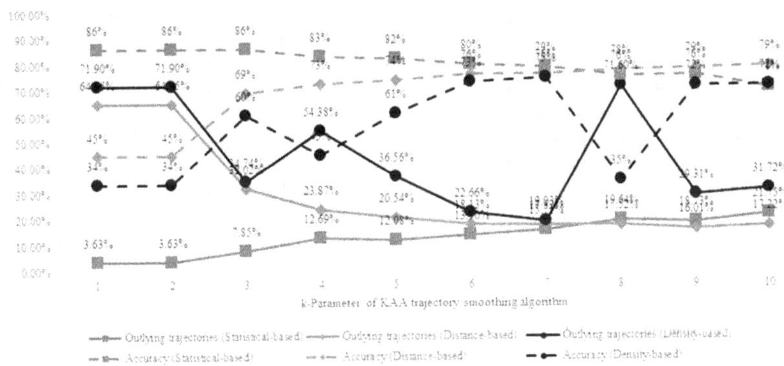

Fig. 9. TFP trajectory-smoothing algorithm: experimental results

Figure 10 shows the outliers detected with the t-parameter in experiments to determine give the best accuracy for each outlier detection algorithm after trajectory preprocessing with the TFP trajectory-smoothing algorithm.

The results after application of the TFP trajectory-smoothing algorithm for statistical-based outlier detection indicated poorer outlying-trajectory-recognition performance compared with the other outlier-detection algorithms. Whereas the density-based algorithm yielded similar results to those of our

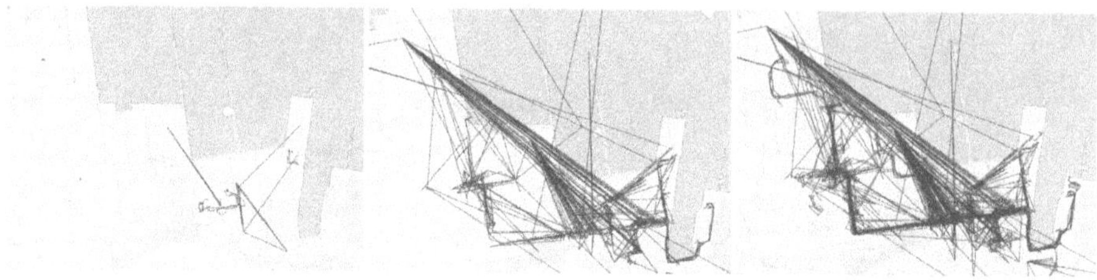

(a) Statistical-based (b) Distance-based (c) Density-based

Fig. 10. Outlier detection experimental results after smoothing using TFP algorithm

domain expert's identification, the distance-based outlier-detection algorithm delivered the best overall results, since it maintained both accuracy and a lower number of recognized outlying trajectories.

5.2 Testing *k* Parameter of KAA

The second experiments aimed to test the sensitivity of the k parameter used in the KAA trajectory-smoothing algorithm. Figure 11 plots the performance results from the various outlier detection algorithms after trajectory preprocessing with the KAA smoothing algorithm.

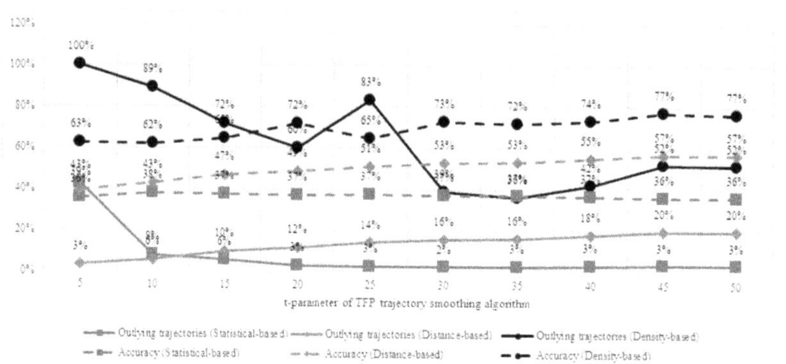

Fig. 11. KAA trajectory-smoothing algorithm: experimental results

Figure 12 shows the outliers detected with the k-parameter is configured in experiments to determine the best accuracy for each outlier-detection algorithm after trajectory preprocessing with the KAA trajectory-smoothing algorithm.

The results after application of the KAA trajectory-smoothing algorithm are quite interesting; all of the outlier-detection algorithms yielded almost the same results. In terms of accuracy, the statistical-based outlier-detection performance tended to degrade when we increased the k-parameter, while the number

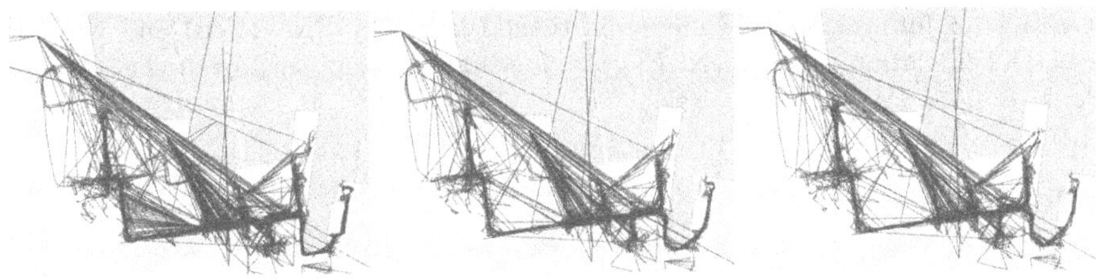

(a) Statistical-based (b) Distance-based (c) Density-based

Fig. 12. Outlier detection experimental results after smoothing using KAA algorithm

of recognized outlying trajectories was increased. However, the distance-based outlier detection algorithm showed results opposite to those of statistical-based outlier detection. Density-based outlier detection, meanwhile, yielded the worst performance, since the accuracy and detection results became unstable when we tried to increase the k-parameter of the KAA algorithm. The distance-based outlier-detection algorithm was the clear winner in terms of accuracy and low number of recognized outlying trajectories.

6 Related Work

Some excellent works on smoothing of trajectories has already been done. Zhou et al. [11] proposed a novel method for smoothing of trajectories using linear regression to approximate traces under system errors, and utilizing road information to map points to the closest adjacent roads. However, it requires prior knowledge of road information that might not be available in some cases, (including our present case study). Knapen et al. [12] matched the GPS route to a map and decomposed the path into several sub-paths. By joining the a sub-paths, the shortcut route could be found. In fact, we employed the same approach for our trajectory-smoothing algorithms.

7 Conclusions

According to our experimental results, smoothing of trajectory data and subsequent repairing of outlying trajectories can improve outlier detection. The KAA trajectory smoothing yields better results than does the TP algorithm due to it tendency to recognize more points. Subsequently, the distance based outlier detection algorithm provides stabler performance than can the other outlier detection algorithms in terms of the number of outlying trajectories detected and accuracy of detection.

Acknowledgements. This research was partially supported by the MSIP (Ministry of Science, ICT and Future Planning), Korea, under the Grand Information Technology Research Center Support Program (IITP-2017-2016-0-00318) supervised by the IITP

(Institute for Information & Communications Technology Promotion) and National Research Foundation of Korea (NRF) grant funded by the Korean government (MEST) (No. NRF-2015R1D1A1A09061331).

References

1. Zheng, Y., Li, Q., Chen, Y., Xie, X., Ma, W.-Y.: Understanding mobility based on GPS data. In: Ubicomp 2008 (2008)
2. Shi, W., Liu, Y.: Real-time urban traffic monitoring with global positioning system-equipped vehicles. IET Intel. Transport Syst. **4**(2), 113–120 (2010)
3. Lee, J.G., Han, J., Li, X.: Trajectory outlier detection: a partition-and-detect framework. In: 2008 IEEE 24th International Conference on Data Engineering, pp. 140–149 (2008)
4. Wang, J., Rui, X., Song, X., Wang, C., Tang, L., Li, C., Raghvan, V.: A weighted clustering algorithm for clarifying vehicle GPS traces. In: 2011 IEEE International Geoscience and Remote Sensing Symposium, pp. 2949–2952 (2011)
5. Chen, C., Zhang, D., Castro, P.S., Li, N., Sun, L., Li, S., Wang, Z.: iBOAT: Isolation-based online anomalous trajectory detection. IEEE Trans. Intell. Transp. Syst. **14**(2), 806–818 (2013)
6. Miura, S., Hsu, L.T., Chen, F., Kamijo, S.: GPS error correction with pseudorange evaluation using three-dimensional maps. IEEE Trans. Intell. Transp. Syst. **16**(6), 3104–3115 (2015)
7. Han, J., Kamber, M., Pei, J.: Data Mining: Concepts and Techniques, 3rd edn. Morgan Kaufmann Publishers Inc., San Francisco (2011)
8. Knorr, E.M., Ng, R.T., Tucakov, V.: Distance-based outliers: algorithms and applications. VLDB J. **8**(3), 237–253 (2000)
9. Jin, W., Tung, A.K.H., Han, J.: Mining top-n local outliers in large databases. In: Proceedings of the Seventh ACM SIGKDD International Conference on Knowledge Discovery and Data Mining, KDD 2001, pp. 293–298. ACM, New York (2001)
10. Misa, T.J., Frana, P.L.: An interview with Edsger W. Dijkstra. Commun. ACM **53**(8), 41–47 (2010)
11. Zhou, X., Li, C., Yuan, X., Xia, B., Mao, G., Xiong, L.: A novel method for smoothing raw GPS data with low cost and high reliability. In: 2016 IEEE 84th Vehicular Technology Conference (VTC-Fall), pp. 1–5 (2016)
12. Knapen, L., Hartman, I.B.-A., Schulz, D., Bellemans, T., Janssens, D., Wets, G.: Determining structural route components from GPS traces. Transp. Res. Part B Methodol. **90**, 156–171 (2016)

SSDMiner: A Scalable and Fast Disk-Based Frequent Pattern Miner

Kang-Wook Chon and Min-Soo Kim(✉)

DGIST (Daegu Gyeongbuk Institute of Science and Technology), Daegu, Republic of Korea
{kw.chon,mskim}@dgist.ac.kr
http://infolab.dgist.ac.kr

Abstract. Frequent itemset mining is widely used as a fundamental data mining technique. Recently, there have been proposed a number of disk-based methods. However, the existing methods still do not have a good scalability due to large-scale intermediate data and non-trivial disk I/Os. We propose SSDMiner, a new fast and scalable disk-based method for frequent itemset mining that is based on Apriori-like method and has no intermediate data and small disk I/O overheads by exploiting SSD. We propose a concept of bitmap chunks for storing transactional database in disks and a fast support counting based on bitmap chunks. Through experiments, we demonstrate that SSDMiner has the enhanced scalability and the good performance similar to that in memory-based methods with robustness.

Keywords: Frequent pattern mining · Scalable algorithm · SSD · Disk-based algorithm

1 Introduction

Frequent itemset mining has been widely used as a fundamental data mining tasks in a wide range of applications such as recommendation systems, social network analysis, web usage mining, bioinformatics, and market basket analysis. For example, a number of commercial recommendation systems exploit frequent itemset mining as a key module including retail stores [9,12].

However, deluge of data in order of terabytes generated by automated systems for analysis purpose makes it difficult or impossible to apply the frequent itemset mining to real world applications. The existing methods easily fail to find frequent patterns from such big data due to lack of memory, since most of the existing methods are memory-based. Therefore, developing scalable frequent itemset mining methods is still challenging problem that has not been completely solved.

W. Lee et al. (eds.), *Proceedings of the 7th International Conference on Emerging Databases*, Lecture Notes in Electrical Engineering 461,
https://doi.org/10.1007/978-981-10-6520-0_11

There have been a number of disk-based frequent itemset mining methods. To the best our knowledge, the disk-based methods are categorized by two groups: (1) partitioning-based methods [10,13] and (2) projection-based methods [2,3,6,10]. The partitioning-based methods first divide an input database into a number of partitions, which are stored in a secondary memory, perform in-memory based methods, *e.g.*, FP-Growth [7], Apriori [1], Eclat [15], on a partition at a time, and then aggregate the partial results. The projection-based methods project an input database, build independent conditional databases, *e.g.*, conditional FP-Trees in FP-Growth, equivalence classes in Eclat, and store them to a secondary memory. Then, they perform frequent itemset mining on the conditional databases one by one without aggregating the partial results.

Although the existing disk-based methods solve the limit on scalability partially, almost methods still have the following two problems. First, they do not have a good scalability due to a tremendous amount of intermediate data generated during mining tasks. The existing methods generally store the intermediate data with respect to short length of itemsets for computing the supports of their supersets. For example, in the AprioriTid [1] method, it generates and keeps the transactions in bitmaps or lists of (k-1)-itemsets for finding frequent k-itemsets. However, this approach easily fails to perform mining operations even though an input database fits into main memory, since the size of intermediate data increases more rapidly. The whole mining tasks fail when the size of intermediate data on any partition is larger than the amount of memory. Second, they do not have good performance due to non-trivial I/O cost. Especially, the projection-based methods generate conditional databases for each 1-itemset and so require lots of disk space. As a result of the large size of conditional databases, such methods cause non-trivial I/O costs. In addition, since the sizes of conditional databases are significantly different, it requires a number of random accesses to disks and degrades the performance more and more. The partitioning-based methods also suffer from I/O costs, and considering such a performance tendency becomes important when handling large-scale datasets, since the number of disk I/Os becomes large as the size of datasets becomes large within a limited size of memory.

For solving the above problems, we propose a fast and scalable disk-based frequent itemset mining method, called SSDMiner. SSDMiner solves the above two problems, and so, can find frequent patterns on much larger input and intermediate data compared with the existing memory-based methods. For solving the problem of large-scale intermediate data, SSDMiner generates equal-sized sub-databases called *bitmap chunks* in the bitmap format and performs the entire mining tasks only using bitmap chunks and bitwise `AND` operations instead of generating intermediate data during mining tasks. Therefore, SSDMiner avoids the failure of mining tasks caused by the lack of memory. For solving the problem of non-trivial I/O costs, SSDMiner avoids random I/Os by storing bitmap chunks in a contiguous disk space and fully utilizes SSD by asynchronously performing disk read and mining tasks. In addition, we use the SSD storage for storing large-scale transactional database (in a vertical format in bitmaps). Nowadays, SSDs are widely exploited from servers, *e.g.*, Intel's SSD, to laptops, *e.g.*, Apple's SSD, Samsung's SSD, and so on, as a secondary storage, since it supports faster and larger capacity SSDs based on the PCI-E interface. As a result, SSDMiner shows the much better scalability than the existing memory-based methods. In addition to the good scalability, SSDMiner significantly improves the speed of support counting by exploiting bitwise AND operations on bitmap chunk and shows the similar performance comparing to memory-based methods.

The main contributions of this paper are the following:

- We propose SSDMiner, a new fast and scalable disk-based method for frequent itemset mining that is based on Apriori-like methods and has no intermediate data and small disk I/O overheads.
- We propose a concept of bitmap chunks for storing transactional database in disks and a fast support counting based on bitmap chunks.
- Through experiments, we demonstrate that SSDMiner has the enhanced scalability and the similar performance compared with memory-based methods with robustness.

The rest of this paper is organized as the following. Section 2 reviews the related work. Section 3 proposes the SSDMiner method. In Sect. 4, we present the detailed algorithm of SSDMiner. Then, we present the experimental results in Sect. 5 and conclude this paper in Sect. 6.

2 Related Work

The problem definition of the frequent itemset mining problem is to determine all itemsets $\mathcal{F}$ that occur at least a pre-defined *minsup* as a subset of transactions in a given transaction database $D = \{t_1, t_2, ..., t_n\}$, where t_i is a subset of distinct items from D. In this paper, we mainly consider the number of occurrences as the meaning of support of an itemset.

Since SSDMiner is based on the Apriori-like methods, we briefly give a review mainly focusing on the Apriori-like methods. The Apriori method is based on the anti-monotone property which if a k-itemset is not frequent, then its supersets never become frequent itemsets. Based on such a property, the Apriori method generates a smaller number of candidate itemsets than those of the previous methods. The Apriori method is performed as the following. In the first process, it counts the supports of 1-itemsets in the database and then discards infrequent 1-itemsets. Then, in the subsequent processes, it repeatedly generates the candidate k-itemsets C_k by joining frequent (k-1)-itemsets F_{k-1} and computes the supports of C_k over the database D. However, the Apriori method suffers from two performance bottlenecks. First, it requires non-trivial overheads with respect to repeatedly scanning the whole database. It means that if there is any frequent k-itemset, this method scans the database k times. Second, its support counting is quite inefficient, and so its overall performance is significantly degraded.

In order to overcome the performance bottleneck, Agrawal et al. propose AprioriTid [1] that exploits vertical format in bitmaps or lists for representing transactions and does not require to repeatedly scan database. AprioriTid finds frequent itemsets by repeatedly performing candidate generation and testing steps as in the Apriori method. In the k-th iteration, it generates candidate k-itemsets and efficiently performs their supports by set intersections between lists or bit vectors loaded in main memory. It outperforms the original Apriori method. However, since AprioriTid generates and keeps the transactions of a large number of itemsets in bitmaps or lists, the overall mining operation could fail when the size of intermediate data is larger than the amount of memory. This performance tendency is more marked for mining large-scale datasets, since the amount of memory usage is proportional to the number of transactions in the database.

3 SSDMiner Method

SSDMiner performs two phases. The first phase of SSDMiner finds a set of itemsets up to Q levels in an enumeration tree, generates their transactions in bitmap format, and stores the itemsets with their transactions in disks (SSD). It stores the transactions of itemsets as a vertical layout in a bitmap format for fast support counting. For handling large-scale datasets, we propose a new format of transaction bit vector, called *bitmap chunk*, in Sect. 3.1 The second phase of SSDMiner iteratively performs two steps, *i.e.*, candidate generation and testing steps, as in the Apriori method. For efficiently mining large-scale database, we propose a new disk-based itemset mining technique, called *streaming bitmap chunks*, in Sect. 3.2.

3.1 Pre-computing Bitmap Chunks

For reducing computational overheads in finding frequent patterns, most of the existing methods generate the transactions with respect to intermediate levels while using more memory usage. For example, AprioriTid generates the transactions in bitmaps or lists. It generates the transactions for F_{k-1} and exploits them for finding F_k. However, this approach easily fails the entire mining tasks as the size of an input database increases, or *minsup* decreases, since keeping the intermediate data is able to require a huge amount of memory usage in such cases.

For overcoming the above problem, we propose the *pre-computing bitmap chunks*, which generates the transactions in bitmaps for the itemsets up to the length Q, where Q is a user-specified parameter. Then, it finds all other long frequent itemsets only using the bitmap chunks instead of generating the transactions for the itemsets longer than Q in an online fashion. However, it is not practical to generate all the transactions of the itemsets shorter than Q, since this approach requires a huge amount of space. For instance, we assume that the number of transactions in an input database is 10 millions, $|F_1| = 500$, and $Q = 3$.

Then, the size of transactions in bitmap is $(\binom{500}{1} + \binom{500}{2} + \binom{500}{3}) \times 10{,}000{,}000 \approx$ 23 TB. We note that there are superset and subset relationships between all the itemsets, and such relationships are the form of a lattice structure. Therefore, we could perform the entire mining tasks by partially generating transactions of the itemsets shorter than Q and exploiting them in subsequent mining processes. Our idea is based on the shell fragment approach for Online Analytical Processing (OLAP) [8]. Our technique generates transactions in a fragmented manner and exhausted manner. The fragmented manner means that it divides F_1 into multiple fragments, each of which includes Q 1-itemsets, and generates transaction bitmaps for only an itemset x within a fragment. The exhausted manner means it generates transactions for an itemset x although x is not determined to frequent. Hereafter, we refer such itemsets as false-positives. The reason why we generate transactions in the exhausted manner is due to the high cost of removing the transaction bitmap of false positives and a relatively small space cost of keeping transactions of false positives. Our technique creates a total S fragments, $S = \frac{|F_1|}{Q}$, and, for each fragment, creates the transaction bitmaps with respect to all possible itemsets, *i.e.*, $2^{|Q|} - 1$ bit vectors. Hereafter, we denote the union of all the possible itemsets within fragments as a set of pre-computed itemsets P. As a result, it requires a small amount of space. For instance, suppose that the number of transactions in an input database is 10 millions, $|F_1| = 500$, $Q = 3$. Then, our technique requires the space of up to $(7 \times 167) \times 10{,}000{,}000 = 1.46$ GB, while fully generating the transactions of the itemsets shorter than Q requires 23 TB. In order to store the result of our technique, we propose a data format called *bitmap chunk.* Suppose that the input database D is divided to $\{D_1, D_2, ..., D_n\}$. Then, it loads a partitioned database D_i to memory at a time, generates the transaction bitmaps regarding to D_i, which is regarded as a bitmap chunk BC_i, and stores BC_i in disks. Here, we note that all the bitmap chunks $BC_{1:n}$ are stored to a contiguous space in disks. As a result, our method could avoid random disk I/Os in the subsequent mining tasks and get the performance benefits compared with other disk-based methods, which might require lots of random disk I/Os.

Figure 1 shows an example of generating bitmap chunks. Here, it performs four steps. We suppose that there are two partitions D_1 and D_2, $F_1 = \{A, B, D, E\}$, and $Q = 2$. Step 1 generates fragments by dividing F_1. Here, there are two fragments $\{A, B\}$ and $\{D, E\}$, and then it enumerates all possible itemsets for each fragment. Step 2 copies a partition D_i to $DataBuf$ in main memory, and Step 3 generates bitmap chunks by scanning D_i. That is, for each transaction, if an itemset x is contained in the transaction, it sets the corresponding bit in the bit vector of x to 1. After processing all the itemsets in P, it copies $BCBuf$ to disks. A bitmap chunk BC_i includes the bit vectors of the length $|D_i|$, where bit vectors are corresponding to the pre-computed itemsets P. We denote a bit vector of x within BC_i by $BC_i[x]$. We define a concept of a set of physical pointers to the bit vectors for an itemset x in BC_i in Definition 1.

Definition 1 (Relative address). *We define a relative address of an itemset i, denoted by $RA(i)$, as the difference of offsets in bytes between the starting*

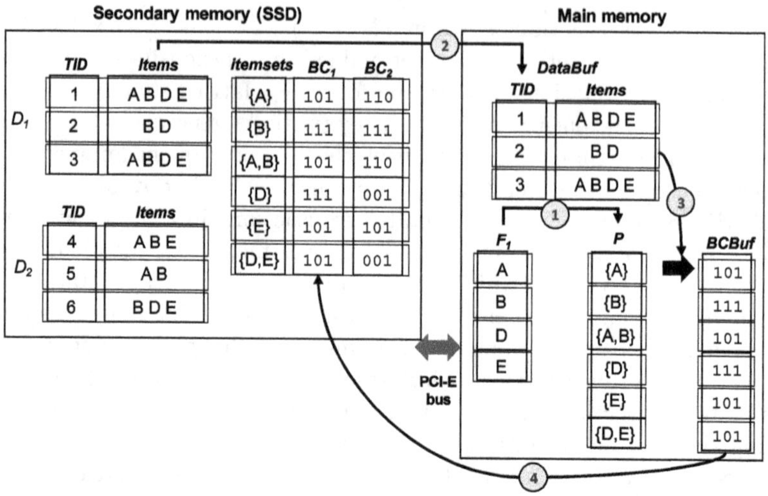

Fig. 1. Example of generating bitmap chunks.

address of BC_k *and that of* $BC_k[i]$*, for a bitmap chunk* BC_k*. Then, we define a set of relative addresses of the itemset* x*, referred as* $RA(x)$*, as* $\{RA(i)|i \in P\}$*, where* P *is a set of pre-computed itemsets.*

This concept is able to quickly access to a memory location of bit vectors corresponding to an itemset within a single bitmap chunk on main memory. We regard $RA(x)$ as an ID for an itemset x. Let the number of 1-itemsets of x be $|x|$, and the number of distinct addresses of $RA(x)$ be $|RA(x)|$. We note that in all the bitmap chunks $BC_{1:n}$, an itemset x has the same relative address $RA(x)$, because our method generates the bitmap chunks of the same size.

3.2 Streaming Bitmap Chunks

SSDMiner finds frequent patterns using the candidate-generation-and-testing approach with the BSF traversal as in the Apriori method. We refer a single series of candidate generation and testing steps as an *iteration*. The existing methods following the Apriori method with vertical format hinder from mining large-scale datasets due to a huge amount of intermediate data.

Our proposed technique solves the above issue, *i.e.*, finding frequent itemsets from large-scale datasets without degrading the performance within limited main memory. The proposed technique tests all candidate itemsets longer than Q only using the bitmap chunks, which include the transaction bit vectors of the itemsets shorter than Q+1 and so does not generate any intermediate data during the mining tasks.

We also propose a new disk-based itemset mining technique, called *streaming bitmap chunks*, for reducing the disk I/O overheads. At the L-th iteration, this technique copies the bitmap chunks from secondary memory (SSD) to main memory via the PCI-E bus asynchronously in a streaming way. SSDMiner exploits two CPU threads, *i.e.*, read thread and process thread, for streaming bitmap chunks. Here, the read thread continuously copies bitmap chunks in secondary memory (SSD) to the bitmap chunk buffers in main memory if the buffers are available. The process thread performs two operations, (1) computing the partial supports on a bitmap chunk BC_i and (2) aggregating the computed partial supports on BC_i and the partial supports previously computed. Figure 2 shows the timeline of SSDMiner. In the figure, BC_i and $Count_i$ indicate the time for copying BC_i to main memory and the time for support counting on BC_i, respectively. In general, since frequent itemset mining is computation intensive, the time to compute the partial supports is longer than that to copy bitmap chunks. Therefore, copying bitmap chunks and processing bitmap chunks are overlapped without copying the first bitmap chunk.

Definition 2 (Partial support). *We define* $\sigma_x(BC_k)$ *as the partial support of an itemset* x *within a given bitmap chunk* BC_k*. The support of* x *on the whole bitmap chunks* $BC_{1:n}$ *becomes* $\sigma(x) = \sum_{k=1}^{n} \sigma_x(BC_k)$.

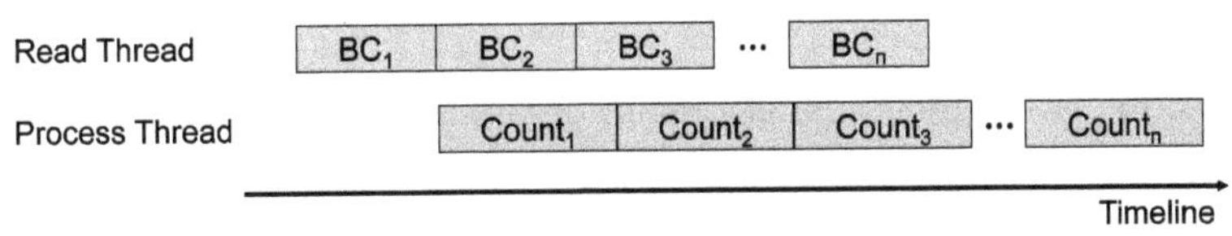

Fig. 2. Asynchronous processing of SSDMiner.

When SSDMiner calculates partial supports, our method does it very efficiently by exploiting bitmap chunks. Given a candidate itemset x, we easily partition x into multiple non-overlapping sub-itemsets $P(x)$, where each sub-itemset is within in a fragment completely and then compute the relative addresses of $P(x)$. For instance, when a candidate itemset $x = \{A, B, E\}$, $P(x) = \{\{A, B\}, \{E\}\}$ and then $RA(P(x)) = \{2, 6\}$ in Fig. 3. Then, it could access the transaction bit vectors for $P(x)$, *i.e.*, $BC_i[\{A, B\}]$ and $BC_i[\{E\}]$. Afterwards, it could calculate the partial support for x by performing bitwise AND between $BC_i[\{A, B\}]$ and $BC_i[\{E\}]$ $|RA(P(x)) - 1|$ times and counting the number of 1 s in the resultant bit vector, *i.e.*, $BC_i[\{A, B, E\}]$.

Here, it is obvious that there is a trade-off between disk (memory) usage and performance (*i.e.*, support counting time). It means that as Q increases, the number of pre-computed itemsets and the space usage become large. As a result, the amount of computations, *i.e.*, bitwise AND operations, and the elapsed time decrease. In Sect. 5, we present such a performance tendency by showing experiment results.

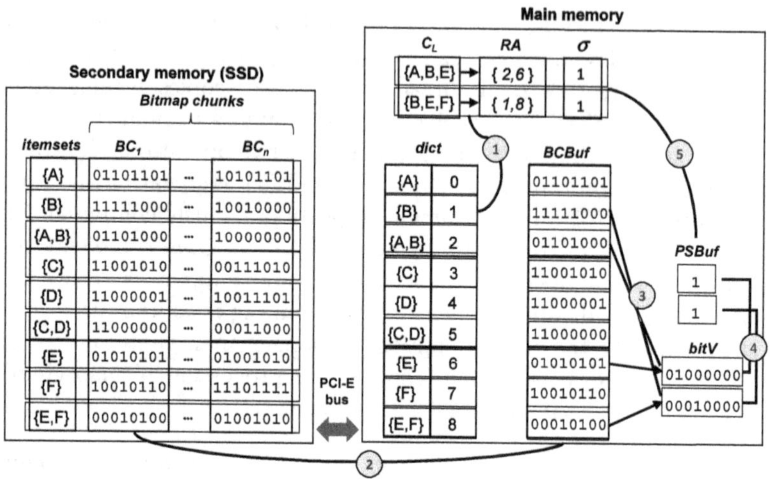

Fig. 3. Example of SSDMiner method.

4 SSDMiner Algorithm

In this section, we present the algorithm of the SSDMiner framework. We first describe the overall procedure of the algorithm using an example in Fig. 3. Then, we explain the details of SSDMiner with the pseudo code.

Our method performs a total of five steps. We refer a set of candidate itemsets at the current level L as C_L. In Step 1, it converts C_L to RA by mapping each itemset x in C_L to its relative address $RA(x)$ with dict, which is a dictionary for mapping an itemset $x \in P$ to $RA(P(x))$. In Step 2, it copies a bitmap chunk BC_i to $BCBuf$ on main memory in a streaming fashion. Step 3 and Step 4 calculate the partial supports of C_L by bitwise `AND` operations and store them $PSBuf$, respectively. In Step 5, it aggregates partial supports to those computed with other bitmap chunks. After processing all the bitmap chunks, SSDMiner finds frequent L-itemsets F_L whose supports are larger than or equal to a given *minsup*.

Algorithm 1 shows the pseudo code of the algorithm. As an initialization step, it divides a transaction database D into non-overlapping partitions D_1, D_2,...,D_n, allocates $DataBuf$, $BCBuf$, and $PSBuf$ on MM (Line 1–2). Then, it finds a set of pre-computed itemsets P by using F_1 (Line 3). Afterwards, it builds the dictionary *dict* (Line 4). We note that it is not necessary to rebuild *dict* after building *dict* once, since it is not changed during the subsequent mining tasks. Then, for each partition D_i of an input transaction database, it generates the corresponding bitmap chunk BC_i (Line 5). The main loop is composed of candidate generation step (Line 9–10) and testing step (Line 11–16). Especially, Line 11–13 efficiently calculate the partial supports by asynchronously copying bitmap chunks and computing partial supports.

Algorithm 1. The framework of SSDMiner

Input : D; /* transaction database */
Input : F_1; /* a set of frequent 1-itemsets*/
Input : *minsup*; /* minimum support */
Output : $\mathcal{F}$; /* frequent itemsets */

Divide D into D_1, D_2,...,D_n;
Allocate $\{DataBuf, BCBuf, PSBuf\}$ on MM;
$P \leftarrow$ find a set of pre-computed itemsets using F_1;
dict $\leftarrow$ dictionary mapping x to $RA(x) (x \in P)$;
Build $BC_{1:n}$ using D_1, D_2,...,D_n on MM;
$L \leftarrow 1$;
while $|F_L| > 0$ **do**
  $L \leftarrow L + 1$;
  `/* Candidate generation */`
  $C_L \leftarrow$ generate candidate itemsets using F_{L-1};
  Convert C_L to $RA(C_L)$ using *dict*;
  `/* Testing */`
  for $j \leftarrow 1$ *to* n **do**
    Copy BC_j into $BCBuf$ of MM;
    $PSBuf[j] \leftarrow$ compute the partial supports of C_L on BC_j;
  $\sigma(c) \leftarrow \sum_{k=1}^{n} PSBuf[c][k]$, for $\forall c \in C_L$;
  $F_L \leftarrow \{c | c \in C_L \bigwedge \sigma(c) \geq minsup\}$;
$\mathcal{F} \leftarrow \bigcup F_L$;
Return $\mathcal{F}$;

5 Performance Evaluation

In this section, we perform the experimental evaluation of SSDMiner compared with memory-based Apriori-like methods. We present experimental results in two categories. First, we show that SSDMiner significantly outperforms the existing memory-based methods using a real dataset while varying *minsups*. Second, we present various kinds of characteristics of SSDMiner. In detail, we present the effect of pre-computation technique while varying Q.

5.1 Experimental Setup

Datasets For experiments, we use the largest real dataset of FIMI Repository [4], called Webdocs [11]. It has been widely used for performance evaluation among the frequent itemset mining methods. Webdocs includes 1,692,082 transactions and 5,267,656 distinct items.

Environments We compare SSDMiner with two memory-based Apriori-like methods. One uses a horizontal format for representing transactions, and its implementation is from [5]. We refer this method as *M_horizontal*.

Another Apriori-like method uses a vertical format in bitmap for representing transactions, and we implement this method as it materializes intermediate data in an online fashion. We refer this method as *M_vertical*. For SSDMiner, we set a fragment size Q to ten as a default and the number of transactions in each partition D_i to 131,072. In addition, for evaluating the effect of the streaming bitmap chunks, we compare the performance between SSDMiner with the streaming bitmap chunks denoted as SSDMiner (W streaming) and SSDMiner without the streaming bitmap chunks denoted as SSDMiner (W/O streaming). We conduct all the experiments on the same workstation equipped with two Intel 8-core CPUs of 2.90GHz, 128 GB main memory, and one Intel SSD. All the experiments are performed on the same OS, Ubuntu 14.04.3 LTS. All the methods are implemented in C++ and are copiled with the same optimized option of -O3.

5.2 Experimental Results

Performance Comparison Fig. 4 presents the elapsed times of four methods, *M_horizontal*, *M_vertical*, SSDMiner (W streaming), and SSDMiner (W/O streaming) for Webdocs while varying *minsups*. We use the same range of X-axis, *i.e.*, *minsups*, with the previous work [14]. In the figure, O.O.M. indicates out of memory error.

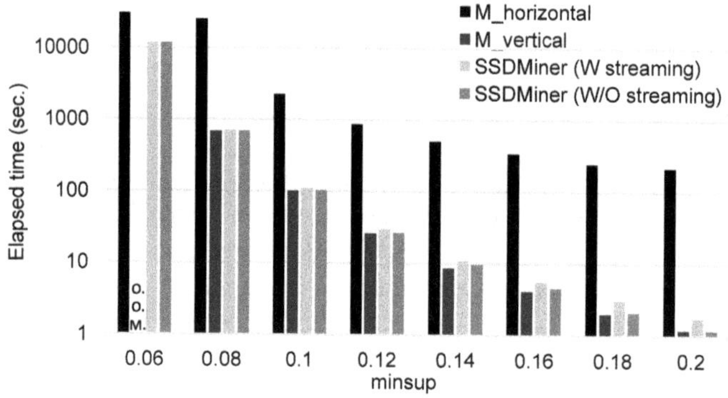

Fig. 4. Performance comparison using Webdocs.

In the figures, SSDMiner consistently and significantly outperforms *M_horizontal* for the entire range of *minsups* by a factor of 21.5–80 times due to the fast support counting. With respect to *M_vertical*, our method shows the similar performance in a range of *minsups* 0.08-0.2, even though *M_vertical* does not include the disk I/O overheads. This performance tendency is mainly due to fast disk read by exploiting SSD and asynchronous processes of copying the data from SSD to main memory and processing mining tasks. In addition to the performance, SSDMiner shows the better scalability than that of *M_vertical*.

That is, SSDMiner could find frequent patterns at the *minsup* 0.06, while *M_vertical* shows O.O.M. in the same *minsup* due to the large size of intermediate data.

With respect to the effect of the streaming bitmap chunks, the performance gaps between SSDMiner (W streaming) and SSDMiner (W/O streaming) get larger as the *minsup* decreases. This is because the number of iterations becomes large as the *minsup* decreases. As a result, the amount of data read from the disks and the number of disk I/Os become large.

Characteristics of SSDMiner Fig. 5(a)-(c) present the performance tendency while varying the size of bitmap chunks. In this experiment, we adjust the size of bitmap chunks by varying the fragment size Q. Here, we use Webdocs with $minsup = 0.08$. First, Fig. 5(a) shows that the space usage increases as Q increases as explained in Sect. 3. As Q increases by 1, the space usage exponentially increases. Second, Fig. 5(b) shows the elapsed time of generating bitmap chunks as Q increases. It is obvious that this time is proportional to the fragment size. Third, Fig. 5(c) shows that the elapsed time of mining time including the candidate generation and testing steps decreases as Q increases. The number of bit vectors in bitmap chunks becomes large as Q increases. Due to decreasing in the number of bitwise AND operations, the mining time decreases as well. Therefore, there is a trade-off between space usage and performance (*i.e.*, mining time).

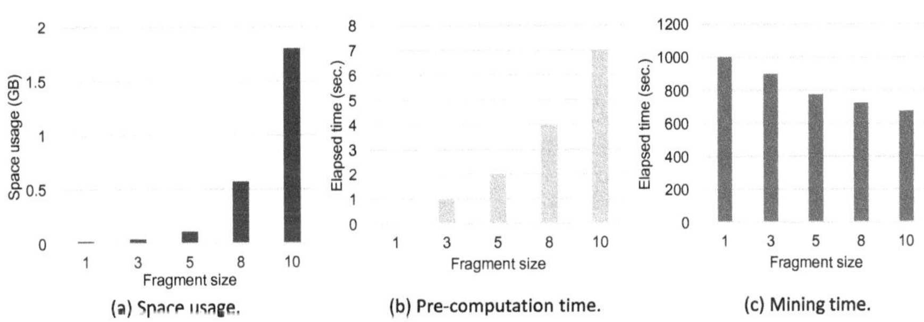

Fig. 5. Performance tendency while varying fragment size Q.

6 Conclusions

In this paper, we propose SSDMiner, a scalable disk-based method for frequent itemset mining. We carefully design SSDMiner so that it scales well with respect to the size of intermediate data and input data with the small disk I/O overheads by exploiting SSD as a secondary storage, which would be useful for big data mining. Through experiments, we demonstrate that SSDMiner has the enhanced scalability and the similar performance compared with memory-based methods with robustness. Future research directions include extending the work to more accelerate the computation intensive parts using the computing power of GPUs.

Acknowledgments. This work was partly supported by Institute for Information & communications Technology Promotion(IITP) grant funded by the Korea government(MSIP) (No. R0190-15-2012, High Performance Big Data Analytics Platform Performance Acceleration Technologies Development) and the DGIST R&D Program of the DGIST R&D Program of the Ministry of Science, ICT and Future Planning (16-BD-0404).

References

1. Agrawal, R., Srikant, R.: Fast algorithms for mining association rules in large databases. In: VLDB, pp. 487–499 (1994). http://www.vldb.org/conf/1994/P487.PDF
2. Baralis, E., Cerquitelli, T., Chiusano, S., Grand, A.: Scalable out-of-core itemset mining. Inf. Sci. **293**, 146–162 (2015)
3. Buehrer, G., Parthasarathy, S., Ghoting, A.: Out-of-core frequent pattern mining on a commodity pc. In: SIGKDD, pp. 86–95. ACM (2006)
4. FIMI Repository (2005). http://fimi.ua.ac.be
5. Goethals, B.: Survey on frequent pattern mining (2003)
6. Grahne, G., Zhu, J.: Mining frequent itemsets from secondary memory. In: ICDM, pp. 91–98. IEEE (2004)
7. Han, J., Pei, J., Yin, Y.: Mining frequent patterns without candidate generation. In: ACM SIGMOD Record, vol. 29, pp. 1–12. ACM (2000)
8. Li, X., Han, J., Gonzalez, H.: High-dimensional olap: a minimal cubing approach. In: PVLDB, pp. 528–539. VLDB Endowment (2004)
9. Lin, W., Alvarez, S.A., Ruiz, C.: Efficient adaptive-support association rule mining for recommender systems. Data Min. Knowl. Disc. **6**(1), 83–105 (2002)
10. Lucchese, C., Orlando, S., Perego, R.: Mining frequent closed itemsets out of core. In: SIAM, pp. 419–429. SIAM (2006)
11. Lucchese, C., Orlando, S., Perego, R., Silvestri, F.: Webdocs: a real-life huge transactional dataset. In: FIMI, vol. 126 (2004)
12. Sandvig, J.J., Mobasher, B., Burke, R.: Robustness of collaborative recommendation based on association rule mining. In: Recsys, pp. 105–112. ACM (2007)
13. Savasere, A., Omiecinski, E.R., Navathe, S.B.: An efficient algorithm for mining association rules in large databases. Technical report, Georgia Institute of Technology (1995)
14. Schlegel, B.: Frequent Itemset Mining on Multiprocessor Systems. Dissertation, Technischen Universitat Dresden (2013)
15. Zaki, M.J., Parthasarathy, S., Ogihara, M., Li, W., et al.: New algorithms for fast discovery of association rules. In: KDD, vol. 97, pp. 283–286 (1997)

A Study on Adjustable Dissimilarity Measure for Efficient Piano Learning

So-Hyun Park[1], Sun-Young Ihm[2], and Young-Ho Park[1(✉)]

[1] Department of IT Engineering, Sookmyung Women's University, Seoul, South Korea
{shpark, yhpark}@sm.ac.kr

[2] Department of Big Data Using Research Center, Sookmyung Women's University, Seoul, South Korea
sunnyihm@sm.ac.kr

Abstract. Recently, Kinect sensors have been used in many applications. Among them, there have been numerous studies that used Kinect sensors to improve efficiency of piano education. However, one of the main disadvantages of existing methods is that they did not take into consideration a dissimilarity measure that can occur when comparing the students' and educator's data. In addition, manually adjusting these settings can be cumbersome as for each case, there must be a set of optimal settings that match learner's and educator's data. In this paper, we propose a method to automatically set the piano learning stage by automatically adjusting the dissimilarity measure. Automatically adjusting the dissimilarity measure enables a gradual piano education and ultimately enhances the educational effect of the piano.

Keywords: Kinect sensor · Piano education · Dissimilarity measure

1 Introduction

Recently, Kinect sensors have been used in many applications, rehabilitation [1], robotics [2] and education [3–6]. Among them, there have been numerous studies that used Kinect sensors to improve efficiency of piano education. The implementation of Kinect-enabled systems can be efficient in piano education, as these devices are inexpensive and accurate comparing to other wearable and multi-camera motion capture systems. However, one of the main disadvantages of existing methods is that they did not take into consideration a dissimilarity measure that can occur when comparing the students' and educator's data. More specifically, dissimilarity measure is a numerical measure of how different two data objects are, is fixed to determine whether the educator and the learner data match.

Park et al. [3] proposed a method to solve a dissimilarity measure problem of Kinect sensors. The authors manually adjusting every student's and educator's data. However, manually adjusting these settings can be cumbersome as for each case, there must be a set of optimal settings that match learner's and educator's data. In this paper, we propose a method to automatically set the piano learning stage by automatically adjusting the dissimilarity measure. Automatically adjusting the dissimilarity measure

W. Lee et al. (eds.), *Proceedings of the 7th International Conference on Emerging Databases*, Lecture Notes in Electrical Engineering 461,
https://doi.org/10.1007/978-981-10-6520-0_12

enables a gradual piano education and ultimately enhances the educational effect of the piano. More precisely, we make the following contributions in this paper:

- We define the level of learning, the dissimilarity measure by level, and the timing of level up.
- We introduce a method that automatically sets the piano learning level when the program is first accessed.
- We propose a method that automatically sets the piano learning level by automatically adjusting the dissimilarity measure.

The paper proceeds as follow. Section 2 introduces the related study that used Kinect sensors to provide efficiency for piano education. In Sect. 3, we introduce preliminaries for our research. Section 4 describes an adjustable dissimilarity measure for efficient piano learning. Here, we first describe a method for calculating an average dissimilarity measure for joints, adjusting initial and automatic piano learning level. Section 5 discusses conclusion and future work.

2 Related Study

Kinect sensors have been used in many educational applications, such as golf [7], ballet [8] and piano [3–6]. Among them, there have been numerous studies that used Kinect sensors to improve efficiency of piano education. In this section, we briefly describe them.

Payeur et al. [4] studied feasibility of using a Kinect sensor for a piano education, and performance evaluation when somatic training methods are used. Beacon [5] conducted a study on efficiency of using a motion-tracking tracking technologies, such as Dartfish and Kinect sensor for pianist posture. Hasdjakos [6] proposed a method of capturing the motion of a pianist using a depth camera of a Kinect sensor, and detecting body landmarks. The experiments conducted by [4–6] demonstrate that Kinect sensors can be efficient in piano education, as these are inexpensive and accurate comparing to other wearable and multi-camera motion capture systems. However, one of the main disadvantages of existing methods is that they did not take into consideration a dissimilarity measure that can occur when comparing the students' and educator's data.

Park et al. [3] proposed a method to solve a dissimilarity measure problem of Kinect sensors. Specifically, the authors proposed a piano posture training system that compares data of learners and educators and provides feedback based on the comparison results. In addition, research has been conducted on whether Kinect sensor can be used as a tool to measure the effectiveness of piano education. Here, the dissimilarity measure problem is solved by manually adjusting every student's and educator's data. However, manually adjusting these settings can be cumbersome as for each case, there must be a set of optimal settings that match learner's and educator's data. Thus, comparing to [3], in this paper, we propose a method for automatic adjustment of dissimilarity measure. Automatically adjusting the dissimilarity measure enables gradual piano education and ultimately enhances the educational effect of the piano.

3 Preliminaries

This section describes the proposed method for adjustable dissimilarity measure. For this, we first describe preliminaries, including notations and definition on piano learning level in Sect. 3.1. We further describe a method for calculating average dissimilarity measure for joints in Sect. 4.1. In Sect. 4.2, we describe a procedure to initially adjust piano learning level, and in Sect. 4.3, we describe how to set piano learning level by automatically adjusting dissimilarity measure.

3.1 Notations

The notations that are used throughout this paper are summarized in Table 1. Here, *J* is a joint and *All_J* is all joints. *JM* is a joint measure, which indicates coordinate or angle of a joint. *LJM* is a learner's joint measure, such as angle or coordinate. *Av_LJM* is an average of learner's joints measure which is calculated in units of one second. *EJM* is an educator's joints measure. *Av_EJM* is an average of educator's joints measure which is also calculated in units of one second. *DM* is a dissimilarity measure i.e., the allowable error range, is fixed to determine whether the educator and the learner data match. *Av_DM* is an average dissimilarity measure. *LL* is learner's level which exists from level 1 to level 10. *PLL* is a piano learning level. *Total_PT* is a total performance time.

Table 1. The notations.

Symbols	Definitions
J	Joint
All_J	All joints
JM	Joint measure
LJM	Learner's joint measure
Av_LJM	Average of learner's joint measure
EJM	Educator's joint measure
Av_EJM	Average of educator's joint measure
DM	Dissimilarity measure
Av_DM	Average dissimilarity measure
LL	Learner's level
PLL	Piano learning level
Total_PT	Total performance time

3.2 Definition of PLL and Scope of DM

Effective piano education requires gradual learning. In the piano textbook called the piano adventure, they divided the learning stage according to the difficulty of the song [11, 12]. In addition, a composer named Carl Czerny provided the students with a textbook that they could practice in sequence according to the difficulty of the song for technical practice [13–15]. The previous methods divided the level according to the difficulty of the song. In this paper, we divided the learning level according to the accuracy of the learners' movement recognized by the Kinect sensor. Before explaining in detail, we will explain the basic concept of the dissimilarity measure.

The Dissimilarity Measure measures how A and B are different. The concept of measuring how different between two models is used in many papers [9, 10]. In this paper, we compare educator data with learner data and how educator data and learner data differ. The final goal is to build a system that provides learners with feedback for correct piano education by using a system that automatically adjusts the error rate.

We define piano learning level by its range of *Av_DM* and practice repeat time. More specifically, Table 2 describes the followings. At level 1, if *Av_DM* is more than 0.4, less than 0.5 during 5times, it goes up to level 2. Levels 2 to 8 are performed in the same way as in the previous step. At level 9, if *Av_DM* is more than 0.0, less than 0.1 during 10times, this stage is passed. If the user plays at two or more higher levels than the current level, it is raised up two levels above the current level.

Table 2. Definition of *PPL* and scope of *DM*.

PLL	Scope of *Av_DM*	Repeat times
1	0.4 <= *Av_DM* < 0.5	3
2	0.4 <= *Av_DM* < 0.5	5
3	0.3 <= *Av_DM* < 0.4	3
4	0.3 <= *Av_DM* < 0.4	5
5	0.2 <= *Av_DM* < 0.3	3
6	0.2 <= *Av_DM* < 0.3	5
7	0.1 <= *Av_DM* < 0.2	5
8	0.1 <= *Av_DM* < 0.2	10
9	0.0 <= *Av_DM* < 0.1	5
10	0.0 <= *Av_DM* < 0.1	10

4 Adjustable Dissimilarity Measure for Efficient Piano Learning

This section describes the proposed method for adjustable dissimilarity measure. For this, we first describe a method for calculating average dissimilarity measure for joints in Sect. 4.1. In Sect. 4.2, we describe a procedure to initially adjust piano learning level, and in Sect. 4.3, we describe how to set piano learning level by automatically adjusting dissimilarity measure.

4.1 Calculating Average Dissimilarity Measure of Joints

In this section, we introduce a method called CalculateAv_DM, which calculates *Av_DM* of *JM*. Specifically, for comparing learner's and educator's data in piano education, Kinect sensors output 30 frames on average. CalculateAv_DM calculates the average value of learner's movements in 30 frames, and then, compares it with educator's data and detects dissimilarity measure.

The input of this function is *LJM* and the output is *Av_DM* of *ALL_J* during *Total_PT*. Pseudo code is as follows. First, we calculate *Av_DM* of *DM* during *Total_PT*. The proposed method contains two phases. There are following steps in first phase. Step 1.1 calculates *Av_LJM* in every second. Step 1.2 compares *LJM* with *EJM* in every second. Step 1.3 calculates *Av_DM* in every second. Step 1.4 calculates the *Av_DM* during *Total_PT*. Repeat step 1.1 – 1.4 on learner's *All_J*. Results of Step 1.1, 1.2, 1.3, 1.4 are stored in databases. In the second phase, we calculate *Av_DM* of *All_J* during *Total_PT*. Result of step 2 is stored in databases (Fig. 1).

Pseudo Code 1 : CalculateAv_DM method

Input. *LJM* : Learner's joint measure
Output. *Av_DM* of *All_J* during *Total_PT* : Average dissimilarity measure of all joints during total performance time
Pseudo Code :

1. Calculate the *Av_DM* during *Total_PT*

(Repeat step 1.1 – 1.4 on Learner's *All_J*)

   1.1. Calculate *Av_LJM* in every second
   1.2. Compare *Av_LJM* with *Av_EJM* in every second
   1.3. Calculate *Av_DM* in every second
   1.4. Calculate *Av_DM* during *Total_PT*

2. Calculate Learner's *Av_DM* of *All_J* during *Total_PT*

Fig. 1. Pseudo code for calculateAV_DM.

4.2 Adjusting Initial *PLL*

In this section, we describe AutoSetFirstPLL, which sets *PLL* automatically when the program is first accessed. It is important to note that piano learning level is decided by the average dissimilarity value. For example, the more learner makes mistakes, the lower his level goes down. Thus, in order to execute AutoSetFirstPLL, we need to know the result of CalculateAv_DM.

The input of this function is *JM* and the output is *LL*. Pseudo code proceeds as follows. Step 1 sets position of Kinect Sensor. Position of Kinect sensor consists of 5 types, 0 degree, 45 degree, 90 degree, 135 degree and 180 degree. In step 2, the system starts producing events as a learner starts playing the piano. Step 3 sets the first *PLL* automatically in two steps. Step 3.1 calls function 'Calculate the Av_DM(*LJM*)'. Step 3.2 set the *LL* of learning based on the result of step 3.1 (Fig. 2).

Pseudo Code 2 : AutoSetFirstPLL method

Input. *JM* : Joint Measure
Output. *LL* : Learner's Level
Pseudo Code :

1. Set position of Kinect Sensor (0°, 45°, 90°, 135°, 180°)
2. Learner start playing the piano
3. Set the *PLL* Automatically
   3.1. Call function 'CalculateAv_DM(*LJM*)'
   3.2. Set the *LL* of learning based on the result of step 3.1

Fig. 2. AutoSetFirstPLL method

4.3 Adjusting *PLL* by Automatic Configuring *DM*

In this section, we propose AdaptiveSetPLL method which sets the *PLL* by adjusting *DM* automatically. Specifically, using AdaptiveSetPLL, we overcome the limitations of the existing work by automatically adjusting a piano learning level without interrupting the piano education process and manually adjusting the dissimilarity measure.

We assume that the automatically set *LL* from a AutoSetFirstPLL method in Sect. 4.2 is Level 1, and the *EJM* is already stored in the database. The input of the AdaptiveSetPLL is *LL* and the output is a text which outputs a message, such as "Pass the stage". Here, passing a stage means passing from level 1 to level 10. Pseudo code proceeds as follows. In step 1, learner starts playing the piano. Step 2 sets *PLL* adaptively by adjusting *DM* automatically. Step 2 consists of following two steps. Step 2 repeats Step 2.1 and 2.2 until a learner passes the stage. Step 2.1 calls CalculatetheAV_DM(*LJM*) function. In step 2.2, if value of *DM* is more than 0.4 and less than 0.5 in 3 times, upgrade to Level 2 (Fig. 3).

Pseudo Code 3 : AdaptiveSetPLL method

Input. *LL* : Learner's Level
Output. Print text "Pass the stage"
Pseudo Code :

1. Learner start playing the piano
2. Set PLL Adaptively by Adjusting DM Automatically

(Repeat until master the stage)
   2.1. Call function 'CalculateAv_DM(*LJM*)'
   2.2. If DM is continuously more than 0.4 and less than 0.5 in 3 times, upgrade to Level 2

Fig. 3. AdaptiveSetPLL method

5 Conclusion and Future Works

In this paper, we have proposed an adjustable dissimilarity measure for efficient piano education. The existing work on piano education did not take into consideration a dissimilarity measure that can occur when comparing the students' and educator's data. Thus, we have proposed a procedure that can adjust dissimilarity measure and automatically set piano learning level for a learner. For this, we first describe preliminaries, including notations and definition on piano learning level. We further describe a method for calculating average dissimilarity measure for joints. Then, we describe a procedure to initially adjust piano learning level. Finally, we describe how to set piano learning level by automatically adjusting dissimilarity measure.

However, it is important to note that we report our initial observations without experiment results. In future work, we are planning to compare the proposed method with state-of-the-art methods in terms of accuracy and efficiency in piano education.

Acknowledgement. This work was supported by Institute for Information & communications Technology Promotion (IITP) grant funded by the Korea government (MSIP) (No. R7120-17-1007, SIAT CCTV Cloud Platform).

References

1. Chang, J., Chen, S.F., Huang, J.D.: A Kinect-based system for physical rehabilitation: a pilot study for young adults with motor disabilities. Int. J. Res. Dev. Disabil. **32**(6), 2566–2570 (2011)
2. El-laithy, R.A., Huang, J., Yeh, M.: Study on the use of Microsoft Kinect for robotics applications. In: Proceedings of the 2012 IEEE/ION Position, Location and Navigation Symposium (2012)
3. Park, S.H., Nasridinov, A., Park, Y.H.: A Kinect based piano education system for correction of pianist posture. Asia Life Sci. J. Suppl 12, 571–586 (2015)
4. Payeur, P., Nascimento, G., Beacon, J., Comeau, G., Cretu, A., D'Aoust, V., Charpentier, M.: Human gesture quantification: an evaluation tool for somatic training and piano performance. In: International Symposium on Haptic, Audio and Visual Environments and Games (2014)
5. Beacon, J.: Assessing 2D and 3D Motion Tracking Technologies for Measuring the Immediate Impact of Feldenkrais Training on the Playing Postures of Pianists. A Thesis Presented in Partial Fulfillment of the Requirements for the Degree Master of Arts in Piano Pedagogy (2015)
6. Hadjakos, A.: Pianist motion capture with the Kinect depth camera. In: Proceedings of the Sound and Music Computing Conference, pp. 303–310 (2012)
7. Lin, Y.H., Huang, S.Y., Hsiao, K.F., Kuo, K.P.: A Kinect-based system for golf beginners' training. Int. J. Inf. Technol. Convergence **253**, 121–129 (2013)
8. Marquardt, Z., Beira, J., Em. N., Paiva, I., Kox, S.: Super mirror: a kinect interface for ballet dancers. In: Proceedings of International Conference on Human Factors in Computing System, pp. 1619–1624 (2012)
9. Jegou, H., Hedi, H., Cordelia, S.: A contextual dissimilarity measure for accurate and efficient image search. In: Proceedings of International Conference on Computer Vision and Pattern Recognition (2007)

10. Birchfield, S., Tomasi, C.: A pixel dissimilarity measure that is insensitive to image sampling. IEEE Trans. Pattern Anal. Mach. Intell. **20**(4), 401–406 (1998)
11. Czerny, C.: 30 Etudes de Mecanisme, Op. 849. In: Ruthardt, A. (ed.) Peters, Leipzig (1892)
12. Czerny, C.: 100 Progressive Studies, Op. 139. In: Ruthardt, A. (ed.) Peters, Leipzig (1888)
13. Czerny, C.: 50 Exercises progressifs dans tous les tons majeurs et mineurs, Op. 840. Schott, Mainz (1855)
14. Faber, N., Faber, R.: My First Piano Adventure, Writing Book B, Steps on the Staff For the Young Beginner. Faber Piano Adventures (1996)
15. Faber, N., Faber, R.: Piano Adventures: Level 3A. Faber Piano Adventures (1996)

A Mapping Model to Match Context Sensing Data to Related Sentences

Lucie Surridge and Young-ho Park(✉)

Department of IT Engineering, Sookmyung Women's University, Seoul, Republic of Korea
{lucys,yhpark}@sookmyung.ac.kr

Abstract. Following current trends in language learning applications, a context-based language generating application was developed to aid learners in effective language acquisition. In an effort to not only match the user's situation with a relevant sentence, but also combine context information to create heterogeneous sentences, a new data model was devised. This paper describes the structure of this model based on a sensing data classifier as well as the corresponding language database. It also depicts a usage scenario with a procedural description of the underlying processes.

Keywords: Context awareness · Context-based learning · Experiential awareness · Mobile phone sensing · Ubiquitous learning

1 Introduction

As technology in the modern society develops, language learning is following suit. With the help of mobile technology, ubiquitous learning has become a new educational paradigm [1], allowing language learners to take a more personal approach towards learning. Moreover, situational approach to learning has shown that context is a crucial factor in language learning since it enhances learning motivation as well as efficiency [2]. According to [3], *context* is defined as information which can be used to characterize the situation of a person, place, or object relevant to the interaction between a user and an application, including the user and application themselves. Some of the types of context that are important for characterizing the situation of a particular entity include location, identity, time and activity. Vygotsky, who is considered to be the father of social constructivism, stressed that social context helps shape the construction of knowledge and is therefore an integral part of language learning [4].

Mobile technologies are now able to deliver learning materials which are time-, location-, and person-relevant [5]. With the use of sensors embedded in mobile devices, it is now easier to personalize language learning by transforming the learner's situational as well as social context into a context-aware learning environment. [6] defines *context-aware software* as one that adapts according to the current context. Such system is capable of examining the computing environment as well as reacting to its changes. In a language learning application, these responses may include displaying context-based language content to the user.

W. Lee et al. (eds.), *Proceedings of the 7th International Conference on Emerging Databases*, Lecture Notes in Electrical Engineering 461,
https://doi.org/10.1007/978-981-10-6520-0_13

Following this new trend in language learning, a new application called EXALL (Experiential Awareness Language Learning), which provides real life situational English, has been developed. With the EXALL application, learners can now experience language learning based on their immediate environment and the utilization of their personal information. This is achieved by sensing the learner's whereabouts, physical activity, current time and weather conditions, as well as data regarding the learner's plans, personal information, and English level of proficiency. Using big data analysis, all the gathered information is analyzed and processed in order to determine the learner's current situation. The application will then periodically offer the user a sentence related to the user's current experience in the form of a message. The user can not only read the sentence, but also listen to it using the built-in text-to-speech functionality. The design of the application resembles familiar chat services, such as KakaoTalk or Line.

This paper presents the data model applied in this project, which enables the generation of context-based sentences. This paper's contribution is threefold. First, it proposes a sensing data classifier, which classifies sensing data into heterogeneous groups based on their relatedness to the user's environment, situation, and personal information. Second, it defines a model that groups the classified data in order to create heterogeneous sentences based on multiple types of sensing data. Third, it provides a database model that includes sentence patterns with two types of variables. This novel technique allows the recombination of various types of sensed information and the mapping of full heterogeneous sentences that are based on the user's current situation. This data model can be easily used for languages other than English. The data model and the database structure would remain the same and only the actual language content in the database would have to be replaced. Therefore, this model is not only practical but also versatile.

2 Related Work

This paper draws upon several previous studies. [7] identified five sub-categories of context: environmental context, which captures the user's surroundings; personal context, which includes information about the user; social context, which describes the social aspects of the user; task context, which describes the user's activities and goals; and spatio-temporal context, which is concerned with time, location, and movement.

ZOE [8] is a continuous sensing wearable application. ZOE aims to continuously sense a comprehensive set of user behaviors and situational contexts. It does so by implementing a set of algorithms that analyze data from the accelerometer, gyroscope, WiFi, and microphone. These algorithms are capable of inferring three key areas of everyday life: Personal Sensing includes the actions, behaviors and the status of the user; Social Sensing monitors the user's social interactions; and Place Sensing tracks important locations for the user.

[9] proposed a system that produces context-aware English vocabulary. Language learner's context is defined here as current location, time, learning abilities and leisure degree. The system is composed of several agents. The learner locating agent first detects the user's location. Based on the user's location, the context analysis agent

analyzes the user's learning requirements and determines learning parameters that are consistent with the user's current context. The English learning materials searching agent then selects context-based learning materials from the database. Finally, the content delivery agent displays the contents to the user.

In [10], the language learning application named PALLAS was introduced. In this work, personalization is considered to be a part of contextualization, which is achieved using the learner's profile and environmental parameters. The profile includes personal information such as age, skill level, native language, etc. Environmental parameters comprise of location, time, day, and the user's mobile device. The application will notify the user of their target language-related events in their location. The users can also query the application's built-in dictionary to look up vocabulary of interest, which will display a related text accompanied by a glossary. The difficulty of the presented text is based on the user's profile, which is automatically built through the user's interactions with the application.

Our work builds upon these previous studies in terms of context data classification. Our model improves upon the previous models in two ways. First, our model allows combinations of context categories, thus creating heterogeneous context awareness that enables the system to produce content that is more specific to the user's current situation. Second, unlike the other aforementioned context-aware language learning applications, which only display context-based vocabulary, our model supports the generation of full sentences based on several types of context by incorporating variables in the sentences to fill in content with different types of context.

3 ESP Data Model

This section introduces the ESP model, explains ESP data classification with concrete examples, and describes how the model is implemented in the language database. Finally, it mentions what kind of applications this model is constrained to.

3.1 ESP Model Overview

The sensing data is classified into 3 categories: E (Environment), S (Situation), and P (Person). For each type of sensing data, a respective subcategory is created. The Environment category includes data based on the user's current location obtained from the Google Places API, current date and time obtained from the mobile, as well as weather and seasonal information, which is obtained through a Korean weather API. Subcategories for Location are selected based on the most common place types obtainable from Google Places, such as School, Airport, and Bus Station. Subcategories for Date include days of the week and some major holidays, such as Weekend, Christmas, and Valentine's Day. Subcategories for Time are predefined based on the time section of the day: Morning, Afternoon, Evening, and Night. Subcategories for Weather are either directly based on the information available from the API, such as Sunny, Rainy, and Cloudy, or they are based on predefined raw data ranges that determine whether the weather is Warm, Hot, Cold, Windy, or Humid.

The Situation category includes information about the user's scheduled plans obtained from the user's Google Calendar, as well as the user's physical activity, which is determined through an analysis of sensor data gained from the GPS, gyroscope and accelerometer embedded in the mobile phone. One of ten predefined subcategories is assigned to every Google Calendar entry that the user inputs through the application, such as Doctor's Appointment, Business Meeting, and Movie. As for the Activity data, there are five user activities that can be detected: Walking, Running, Still, in a Vehicle, or on a Bicycle. If the user's current movement is not identified as one of these five types of activity, the user's Activity is labeled as Unknown.

Finally, the Person category includes personal information about the user, such as the user's gender, age, line of work, and other personal information about the user gathered through the application's Settings panel. Based on the information provided by the user, the model can offer more personal language content that is relevant to the user's current situation. There are fifteen predefined types of jobs the user can select from, such as Educator, Student, and Medical Professional. The user's gender is stored in subcategories Male and Female and the user's birth date is stored in a subcategory named Birthday.

The abovementioned categories are grouped into combinations of two categories or a joint group of all three categories through the use of variables in the language database. The combination of Person and Environment categories forms an PE group that can cover topics that include personal as well as environmental information; the combination of Situation and Environment categories forms an SE group that can cover topics that include situational as well as environmental information; and the combination of Person and Situation categories forms and PS group that can cover topics that include personal as well as situational information. Examples of concrete data combinations for each group can be seen in Table 1. Furthermore, a joint group of Environment, Situation and Person is a combination of personal, situational, and environmental information, which covers all relevant topics. This combination of the basic categories creates a heterogeneous bundle that becomes the basis for sentence formation.

3.2 Language Database Structure

The language data is stored in a language database, or corpus. The ESP model is reflected in the corpus structure. The corpus has a hierarchical structure that enables the application to find the most appropriate sentence for the user's situation. At the top of the hierarchy, the corpus contains a sentence pattern section, a sensing data section, and a hyponym section, all of which are divided into E, S and P categories. These are further divided into sections based on the type of the sensing data, such as Location, Time, Weather, Calendar, Activity, etc., which were enumerated in Sect. 3.1. These sections are further divided into subcategories based on the actual data that can be sensed. Examples of these subcategories can also be found in Sect. 3.1 as well as in Table 1.

When the current sensing data input is classified according to the ESP model, it is mapped to the relevant corpus subcategory, from which it will randomly draw a sentence pattern. For instance, let us say that the user is currently on the move and it is

Table 1. Examples of data classification.

Category Set	
E	*school, hospital, café, windy, hot, Christmas, spring*
S	*business meeting, trip, exam, party, walking, running*
P	*male, student, journalist, athlete*

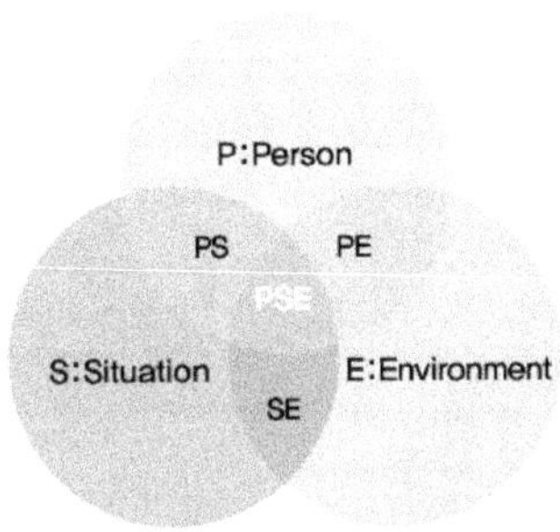

Grouped Categories	
SE	*running in a particular location*
PS	*a scheduled plan to meet a specific person*
PE	*the user's birthday*
PSE	*a scheduled meeting at a particular location with a particular person*

determined that the user is, in fact, walking. That corresponds to the Walking corpus subcategory, which belongs to the S category of the ESP model. From this subcategory, a random sentence pattern related to walking is drawn.

The sentence pattern can be combined with one hyponym and/or one sensing data value, thus creating SE, PS, PE, or ESP category combinations. Figure 1 shows a visual representation of this process. The joining of sentence patterns, hyponyms, and sensing data is achieved through the use of variables within the sentence patterns. Sentence patterns contain a # symbol, which acts as a placeholder for a hyponym, and/or an @ symbol, which acts as a placeholder for a sensing data value, such as temperature, date, time, name of a location, name of a person, etc. These variables allow the construction of sentences with heterogeneous information.

If the sentence pattern does not contain a variable symbol, the sentence will fall into the E, S or P category and will be displayed to the user as is. If the sentence pattern contains only a # symbol, the algorithm will locate any groups of hyponyms that are connected with the selected sentence pattern and insert a random hyponym from the selected groups into the sentence structure in place of the # symbol to form a complete sentence, which will then be displayed on the screen. Alternatively, if the sentence pattern contains only an @ sign, it will insert a relevant piece of sensing data in place of the @ sign and display the complete sentence. In both cases, the sentence will contain heterogeneous information and will fall in the PE, ES or PS combined categories. Finally, if the sentence pattern contains both a # symbol and an @ sign, it will include both a random hyponym from a hyponym group connected with the sentence pattern and a specific sensing data literal. The full heterogeneous sentence will fall into the joint ESP category and will therefore reflect all three types of situational context. Figure 2 shows a concrete example of how this model is implemented.

To continue the example above, the sentence pattern related to walking may contain, for instance, an @ symbol that points to sensing data regarding the user's current location, which is information that belongs to the E category. The combination

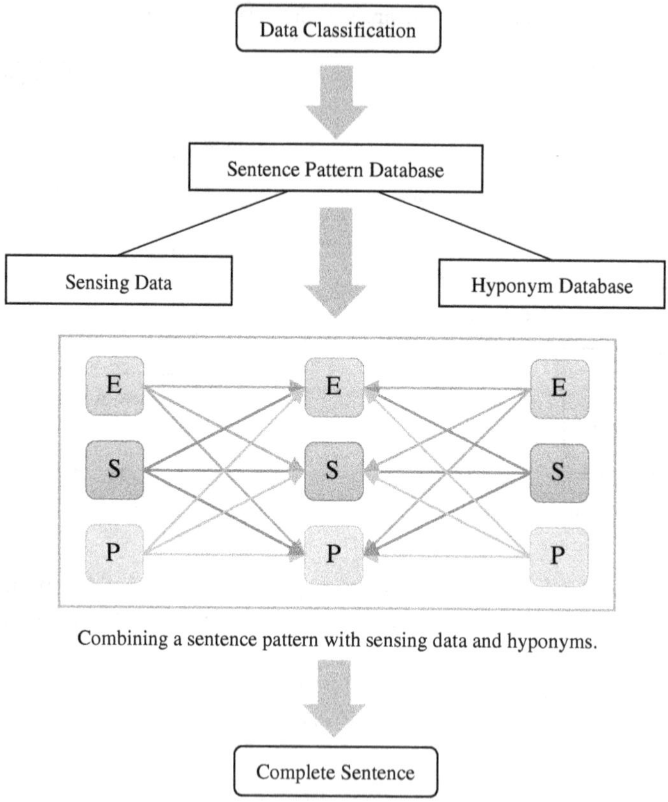

Fig. 1. ESP data model.

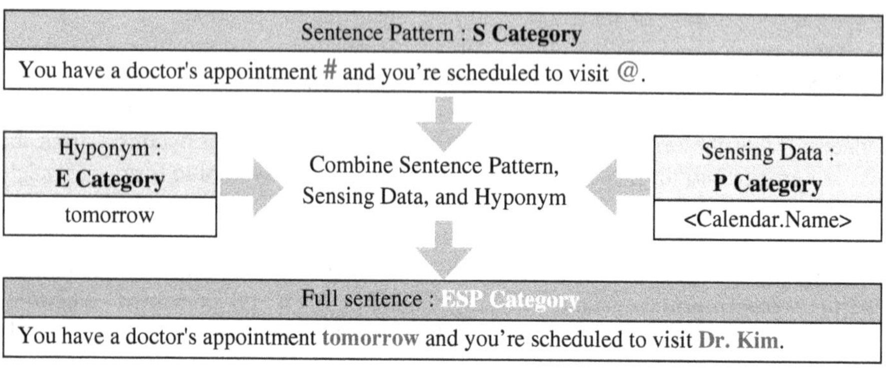

Fig. 2. An example of ESP model implementation.

of these two contexts would correspond to the SE group. To illustrate with an example sentence, the sentence pattern could be "Is @ a nice area to go for a walk?" When the user's location based on GPS information is inserted, the full sentence may be

"Is Hyochang Park a nice area to go for a walk?" This sentence would represent the SE group and therefore be relevant to 2 types of context.

3.3 Constraints of the ESP Model

The application of the ESP model is not specific to the EXALL software and can be implemented in other context-based language learning applications. This model is not, however, suitable for applications that require direct interaction with the user, such as chatbots and other dialog systems. The reason for this is because the corpus only classifies sentence patterns and hyponyms based on a particular situation, not based on a specific meaning. For instance, when the user enters a restaurant, the ESP model provides a method to offer the user a sentence that is specific to a restaurant setting. However, it does not distinguish between the different meanings of the sentences, nor can it choose a specific hyponym from the group of hyponyms that are connected with a sentence pattern. For this reason, this model's applications is constrained to software that only focuses on generating sentences that are specific to the user's current situation.

4 Usage Scenario for the ESP Model

The EXALL application continually senses contextual data, which is classified into the E, S or P category. The classified data are assigned priority values based on their variability, frequency, and relative priority weight. Then, in regular intervals, the highest priority data is mapped to a sentence related to the situation, which, following the ESP model, may contain heterogeneous information based on other known data sources. The following is an example usage scenario, which demonstrates how the ESP model is implemented in the EXALL application.

The user may start using the application on the way to school. While walking on the street, the application will sense that the user is on the move and, more specifically, that the user is currently walking, which is classified as situational information and assigned to the S category. The algorithm will then draw a sentence related to Walking from the corpus. If the sentence contains a # symbol, it will insert the relevant hyponym into the sentence. If the sentence contains an @ sign, which in this case may signify the current location or time, it will insert the relevant sensing data into the sentence in place of the @ sign. Both location and time fall under the E category, thus this process would create a heterogeneous sentence that includes both situational and environmental information that is matching the user's current situation.

On the way to school, the application may inform the user about today's weather forecast, which falls under the E category, since it is related to the state of the environment. It may display a homogenous weather-related sentence from the E category, or, based on the specific weather data received from the Korean weather API, it may classify this data further by determining how this weather condition would feel to the user. It will define the weather condition as hot, warm, cold, windy, or humid, and display a heterogeneous sentence that contains not only environmental, but also personal information.

While the user is at school, the application will be continually sensing data, such as the current time of day, calendar events, and the user's activity. If the user has a calendar event coming up that day or the following day, the application will construct a sentence describing the planned event. A calendar events falls under the S category, since it is related to the user's situation. The application will determine the type of calendar event based on 10 predefined calendar categories. If the user has a test scheduled, this event will be classified as Exam. The application may display a sentence informing the user about the upcoming test, or, if the sentence contains a variable, it will produce a sentence containing heterogeneous information, such as the name of the teacher, which would fall under the P category, or the location of the exam, which would fall under the E category, or the time of the exam, which would also fall under the E category. If the sentence contains information about, for instance, both time and the teacher's name, the generated sentence will represent the combined ESP category. An example of this usage scenario is depicted in Fig. 3.

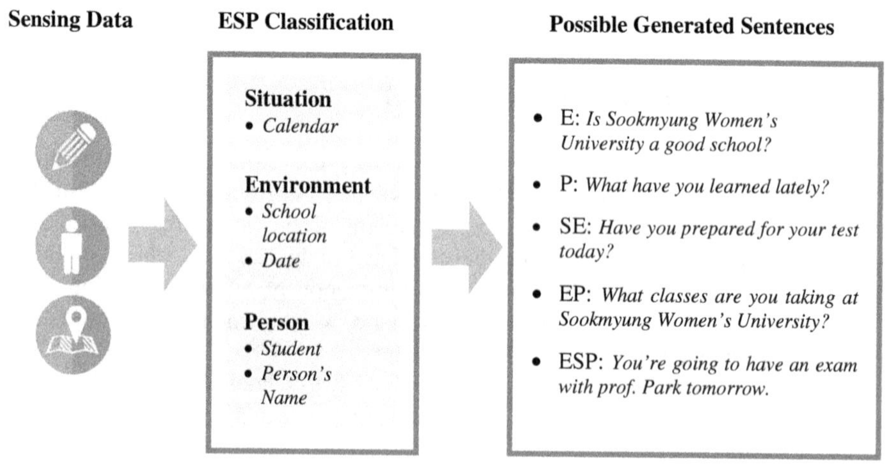

Fig. 3. An example ESP model usage scenario.

5 Conclusion

This paper presented a data model that maps heterogeneous mobile phone sensing data onto multiple context-based English sentences, which are displayed to the user in regular intervals, providing language content that matches the user's current situation. In this model, sensing data is classified into 3 categories: Environment, Situation, and Person. These categories are then grouped into combinations of two or three categories: Person-Environment, Situation-Environment, Person-Situation, and Environment-Situation-Person. This grouping is achieved through the use of variables in sentence patterns, which act as placeholders for either a hyponym that fits into the sentence pattern, or a sensing data value, such as temperature, date, time, name of a location, name of a person, etc. Both hyponyms and sensing data are also classified into E, S and P categories. The combination of the sentence patterns, sensing data and hyponyms from 3 different possible contexts allows the generation of complete English sentences based on multiple situational contexts.

The contribution of this work is threefold. It proposes a sensing data classifier for mapping sensing data to context-based sentences, defines a data model that combines the classified data into multi-context groups, and provides a database model complementing the data model, which maps the classified sensing data onto sentences that are based on several types of context, thus producing content more specific to the user's current situation.

Acknowledgment. This work was supported by Institute for Information & communications Technology Promotion (IITP) grant funded by the Korea government (MSIP) (No. R7120-17-1007, SIAT CCTV Cloud Platform).

References

1. Cope, B., Kalantzis M.: Ubiquitous learning: an agenda for educational transformation. In: Ubiquitous Learning, pp. 3–14 (2009)
2. Hornby, A.S.: The situational approach in language teaching (I). ELT J. **4**(4), 98–103 (1950)
3. Dey, A.K.: Providing Architectural Support for Building Context-Aware Applications. Diss, Georgia Institute of Technology (2000)
4. Vygotsky, L.S.: Mind in Society: The Development of Higher Psychological Processes. Harvard University Press, Cambridge (1980)
5. Kukulska-Hulme, A.: Language learning defined by time and place: a framework for next generation designs, pp. 1–13. Emerald Group Publishing Limited (2012)
6. Schilit, B.N., Adams, N., Want, R.: Context-aware computing applications. In: The First IEEE Workshop on Mobile Computing Systems and Applications, WMCSA (1994)
7. Kofod-Petersen, A., Mikalsen M.: Context: Representation and reasoning. Special issue of the Revue d'Intelligence Artificielle on Applying Context-Management (2005)
8. Lane, N.D., et al.: ZOE: a cloud-less dialog-enabled continuous sensing wearable exploiting heterogeneous computation. In: Proceedings of the 13th Annual International Conference on Mobile Systems, Applications, and Services. ACM (2015)
9. Chen, C.M., Li, Y.-L., Chen, M.-C.: Personalised context-aware ubiquitous learning system for supporting effective English vocabulary learning. Interact. Learn. Environ. **18**(4), 341–364 (2010)
10. Petersen, S.A., Markiewicz, J.-K., Bjørnebekk, S.S.: Personalized and contextualized language learning: choose when, where and what. Res. Pract. Technol. Enhanced Learn. **4**(01), 33–60 (2009)

Understanding User's Interests in NoSQL Databases in Stack Overflow

Minchul Lee, Sieun Jeon, and Min Song(✉)

Yonsei University, Library and Information Science, Seoul, South Korea
{bab2min, sieunjeon, min.song}@yonsei.ac.kr

Abstract. NoSQL (Not only SQL) has gained popularity with emerging demands of scalable database with big data. Despite the great interest of users toward NoSQL technology, an attempt to analyze how the actual users react to NoSQL has not been made yet. Thus, the present work utilizes question-answer data acquired from Stack Overflow, a question and answer site that works as a large knowledge repository to understand how people perceive NoSQL technology. To this end, LDA topic modeling techniques are used to find out the trend of NoSQL databases. In addition, we proposed topic discrimination value in attempt to find topics that distinguish each NoSQL databases.

Keywords: NoSQL · Database · Topic modeling · Question-answer data · LDA

1 Introduction

NoSQL (Not only SQL) refers to the next generation databases that are non-relational. Starting from the demands to have more flexible, scalable and less ACID-restricted database [1] rather than traditional Relational Databases, NoSQL databases fit for storing unstructured data and heavy read/write workloads [2].

Describing key features of NoSQL is generally done by reviewing the documents on NoSQL database, comparing them with Relational Database Management Systems, and checking if NoSQL databases share certain characteristics that RDBMS doesn't have. A number of works [3–7] tried to compare and evaluate NoSQL databases in this approach. On Table 1, popular NoSQL databases are listed with their data model categories suggested by NoSQL Archive [33] and descriptions gained from each website.

Stack Overflow is a website made for posting question and answer on wide range of topics in computer science field. Since 2008, Stack Overflow has gained its popularity as working as a platform for not only the community users but a lot of people from outside who still approach the website through search engine. According to the finding of Parnin and Treude [8], Stack Overflow appears on the first search results page for 84 percent of popular API search in Google search engine.

Various works have been done on Stack Overflow. Andrej et al. [9] used agglomerative clustering to cluster the dataset into N groups based on tag data in Stack Overflow. Barua et al. [10] analyzes the overall trend in Stack Overflow dataset using LDA topic modeling. Yang et al. [11] focuses on expert behavior on Stack Overflow.

W. Lee et al. (eds.), *Proceedings of the 7th International Conference on Emerging Databases*, Lecture Notes in Electrical Engineering 461,
https://doi.org/10.1007/978-981-10-6520-0_14

Table 1. Examples of NoSQL database

Data model	Name	Company/Institution	Description
Column-family model	Cassandra	Apache Software Foundation	A database with linear scalability and proven fault-tolerance [18]
	Hbase	Apache Software Foundation	A hadoop database designed for a distributed, scalable, big data store [19]
Document model	CouchDB	Apache Software Foundation	A database that designed to fit for the web [20]
	MongoDB	MongoDB, Inc.	A cross-platform document database [21]
	RavenDB	Hibernating Rhino, Inc.	A document database with scalability and extensibility [22]
Key-value model	Aerospike	Aerospike, Inc.	A enterprise-class, NoSQL database for real-time operational applications [23]
	DynamoDB	Amazon, Inc.	A fully managed NoSQL database that provides fast and predictable performance with scalability [24]
	LevelDB	Google (team)	A database that provides simple key/value data store [25]
	Redis	Redis labs	In-memory data structure store [26]
	Riak	Basho Inc.	A scalable, highly available database [27]
Graph model	ArangoDB	Company ArangoDB	A native multi-model database [28]
	Neo4j	Neo Technology, Inc.	A highly scalable, native graph database [29]
	OrientDB	OrientDB LTD	A multimodel, graph database [30]

LDA topic modeling method, generative probabilistic model for document collections, is proposed in 2003 by Blei [12]. It deduces topic distributions of each document and word distributions of each topic by Dirichlet distribution. According to topic distributions, it can be identified how each document contains topics semantically. Due to its latent characteristic, LDA has been widely used for text classification, document modeling, collaborative filtering, and so on. Recent applications of LDA [10, 13–15] are followed.

It is important to understand the interest, demand and problem of users who actually are interested in NoSQL database. In this paper, we aim to investigate the characteristics of NoSQL databases by analyzing question answer data written by NoSQL users-the programmers in Stack Overflow community. To analyze large amount of text data, Latent Dirichlet Allocation techniques are used to find out the major topics and trend in NoSQL related questions. Furthermore, we devised a Topic Discrimination Value that indicates the contribution of the topic to similarity between NoSQL databases.

The rest of the paper is organized as follows: the Methodology section describes how we collected and preprocessed data from Stack Overflow, why we used topic modeling and topic discrimination value to explain the difference between NoSQL databases. The Results section shows what users mainly talk about regarding NoSQL databases, and what differences exist in them. Finally, in the Conclusion section, we discuss the limitations of our methods and discuss how our research can be further developed.

2 Methodology

2.1 Data Collections and Preprocessing

There are total 8314 questions with "nosql" tag on Stack Overflow. Among these questions, some posts that are lack of certain information are excluded from the target. Total 7772 questions, 10747 answers, 20876 comments with "nosql" tag were collected from Stack Overflow in this stage. 1767 tags appeared on the whole dataset.

In preprocessing stage, we decided to use plain text part and discard the code area in the collected questions and answers since our motivation for this research is to find out the characteristics of NoSQL from question-answer data. All stopwords based on NLTK stopword list [31] were also removed to minimize the noise. Remaining words went through stemming stage by PorterStemmer given in NLTK library.

2.2 Latent Dirichlet Allocation (LDA) Topic Modeling

Topic modeling stage is done in several steps. First, questions and the following answers and comments are joined to form a single document. This is based on the assumption that the question and following answers and comments are under same topic. 7772 documents created in this stage are set as training dataset of LDA model. LDA model is implemented with $\alpha = 0.1$ and $\beta = 0.001$. We used python library, *gensim* [32] to train the model. We made 4 LDA models whose number of the topic is 10, 20, 30 and 40 respectively. On labeling stage, we choose LDA model with 40 topics. While it is well-accepted that there is no single value suitable for every situation [16], using 40 topic clusters is also proposed in [10] to build LDA topic modeling on Stack Overflow dataset.

Each topic cluster is labeled with appropriate topic name manually based on the word distribution in the cluster. In Table 2, an example of cluster #25, #32, and #35 and their top 15 words are given. Some clusters hard to assign to specific topic are left without cluster name and excluded in analysis stage.

By analyzing a set of documents using the LDA technique, we obtained the topic distribution for each document and the word distribution for the topic. The topic distribution vectors of the tag are calculated as following (Table 3):

$$T(t) = \frac{1}{|D_t|} \sum\nolimits_{d \in D_t} \theta_d \tag{1}$$

Table 2. Example of frequent word distribution of sample topic cluster

Topic	Words
(#25) Performance	performance(.021) query(.012) entity(.010) time(.009) store(.009) database(.009) record (.009) million(.009) much(.008) fast(.007) solution(.007) size(.007) write(.007) better (.007) read(.006)
(#32) Aggregation	aggregation(.039) value(.039) map(.033) reduce(.032) count(.031) function(.026) result (.023) field(.020) sort(.014) MapReduce(.013) number(.013) group(.011) output(.011) query(.010) sum(.010)
(#35) Transaction	transaction(.128) atomic(.045) counter(.030) operation(.026) account(.021) ACID(.021) increment(.019) update(.016) label(.014) commit(.013) support(.013) lock(.011) consist (.011) fail(.010) balance(.009)

Table 3. Parameters of topic distribution

Symbol	Meaning
K	Number of topics
M	Number of documents
θ_d	*K*-dimensional vector; distribution of topics in document *d* ($1 \leq d \leq M$)
D_t	A set of document ids that have tag *t*
$T(t)$	*K*-dimensional vector; distribution of topics in tag *t*

2.3 Topic Similarity and Discrimination Value of NoSQL DB Tags

We infer the characteristics of each NoSQL database through topic distribution of NoSQL database. To do this, tag that (i) is a name of the NoSQL database and (ii) appears more than 40 questions is selected as NoSQL DB tags. Those are 12 NoSQL DB tags: *aerospike, arangodb, cassandra, couchdb, dynamodb, hbase, mongodb, neo4j, orientdb, ravendb, redis* and *riak.* Documents that have same NoSQL DB tag are put together, forming average topic distribution of NoSQL DB.

Similarity between two NoSQL DB are calculated using cosine similarity. Discrimination value of topic *t* is a subtraction of similarity without *t* from total similarity. To devise a topic discrimination value (TDV), we modified the term discrimination, which is a measure of how useful keyword is as index, and TDV can be used to calculate which topic contributes to similarity. First, we use cosine similarity as a similarity measure. A cosine similarity between two tag set vectors, A and B is defined like:

$$\mathrm{Sim}(A,B) = \frac{A \cdot B}{|A||B|} = \frac{\sum_{k=1}^{K} A_k B_k}{\sqrt{\sum_{k=1}^{K} A_k^2}\sqrt{\sum_{k=1}^{K} B_k^2}} \tag{2}$$

TDV is defined as:

$$\mathrm{TDV}_k(A,B) = \mathrm{Sim}(A,B) - \mathrm{Sim}(A_k,B_k) \tag{3}$$

(*A* and *B* are average vectors of two tag, and *k* is a topic index ($1 \leq k \leq K$). A_k and B_k are defined as $K-1$ dimensional vector formed by removing *k*-th element from vector *A* and *B*.)

If *k*-th topic values of *A* and *B* are not significant, a cosine similarity between A_k and B_k would not be very different from one between *A* and *B*. But if $\mathrm{Sim}(A,B) > \mathrm{Sim}(A_k,B_k)$, then $\mathrm{TDV}_k(A,B) > 0$, and it means that *k*-th topic is contributing to the similarity between *A* and *B*. Thus, you can know they have a similar attitude towards *k*-th topic. Otherwise, if $\mathrm{Sim}(A,B) < \mathrm{Sim}(A_k,B_k)$, then $\mathrm{TDV}_k(A,B) < 0$, and it means that *k*-th topic reduces the similarity between *A* and *B*. So, it can be concluded that they may have an opposite attitude towards *k*-th topic.

3 Result

As suggested in 3.2 LDA modeling, 29 out of 40 topic clusters are labeled manually. The topic clusters can be divided into 3 categories: (i) topics that show the data model (type) of NoSQL (like the data models that appeared on Table 1), (ii) topics regarding the characteristics/evaluation criteria of NoSQL, and (iii) topics that are related to NoSQL tasks that users find difficulty in and applications that comes up with NoSQL techniques. Each category and its following topic clusters are listed on Table 4.

Table 4. Categorization of topic clusters

Category	Topic clusters
Data model	Column-family, Graph, Relational database, Key-value, Document
Characteristics	In-memory performance, Performance, Disk space, Transaction
Task & application	Update & change event, Index Search, DB Application, Time & date, Connection & run error, Query, Text & file format, Session & Expire, Array & Object, Tree & Hierarchy, Aggregation, View replication, In-memory, Board & Blog schema, Hadoop, Distributed systems, App, Java, Server & Multiprocessing

3.1 Topic Distribution

Since it has limits to find out clear border of NoSQL model on reviewing documents, we used the topic modeling of question-answer dataset which contains actual experience and perception of a lot of users. 5 clusters are selected as cluster that represent the theoretical model of the NoSQL database: Column-family, Graph, Relational database, Key-value and Document. These criteria are suggested in various researches [3–7, 17] On Table 5, we can find that the distribution of *neo4j, arangodb* and *orientdb* in Graph are relatively high compared to other NoSQL DB in other topic clusters.

Table 5. Top NoSQL DB with high distribution on each topic cluster representing data model

Topic Cluster	NoSQL DB				
Graph	Neo4j (0.199)	ArangoDB (0.176)	OrientDB (0.148)	HBase (0.011)	MongoDB (0.007)
Column-family	Cassandra (0.105)	HBase (0.061)	Riak (0.016)	OrientDB (0.012)	ArangoDB (0.009)
Document	MongoDB (0.135)	ArangoDB (0.113)	RavenDB (0.077)	CouchDB (0.068)	OrientDB (0.041)
Relational DB	CouchDB (0.129)	Riak (0.122)	Neo4j (0.111)	MongoDB (0.107)	Redis (0.095)
Key-value	Riak (0.095)	Redis (0.057)	Aerospike (0.05)	CouchDB (0.043)	RavenDB (0.019)

The distribution for the documents that are tagged with *Arango DB* and *OrientDB* is high both in Graph and Document Topic cluster. This supports the idea that they are considered as multi-model database. While *OrientDB* is said to be extremely versatile, including feature of all 4 data models-key/value, document, graph and object (Messina et al. 2016), the distribution results show that the users of *OrientDB* consider it mainly as Graph or Document model.

Relational DB cluster should be handled with care since documents in Relational DB cluster because the questioner tried to compare NoSQL DB with RDB.

Table 6 shows the distribution of NoSQL DB documents on topic clusters that are related to evaluation. The clusters are Performance, In-memory performance, Disk space and Transaction. On interpreting the result, one should note that the higher distribution does not always mean higher performance of NoSQL database in the topic cluster. Considering the motivation of questioner, high distribution in performance cluster can be considered in either as great interest or compliment on the performance or as great doubt on the performance.

On Table 6, *Aerospike* shows high distribution in several topic clusters. To find out if high distribution actually means higher performance, additional analysis based on each document in performance cluster is required. However, in this study, we didn't cover it.

Table 6. Top NoSQL DB with high distribution on each topic cluster representing characteristics

Topic Cluster	NoSQL DB				
Performance	Redis (0.055)	Aerospike (0.048)	Riak (0.036)	Neo4j (0.034)	Hbase (0.032)
In-Memory performance	Aerospike (0.074)	Redis (0.011)	DynamoDB (0.006)	Cassandra (0.004)	Riak (0.003)
Disk Space	ArangoDB (0.024)	Aerospike (0.021)	Cassandra (0.016)	OrientDB (0.011)	HBase (0.01)
Transaction	DynamoDB (0.015)	Aerospike (0.011)	Neo4j (0.009)	Riak (0.009)	OrientDB (0.008)

3.2 Topic Discrimination Value

Table 7 shows Topic discrimination values (TDVs) of 6 samples of NoSQL DB pairs, which include *Neo4j, ArangoDB, OrientDB, MongoDB, Aerospike, Redis* and *Cassandra*. We omitted columns with TDV values close to zero for readability. 3 graph model databases (*Neo4j*, *ArangoDB* and *OrientDB)* were selected to identify how TDV shows the difference between them. Since *ArangoDB* has high weight among both Graph and Document model, we chose the pair "*ArangoDB* and *MongoDB*" in order to see what is in accord and what is different. The pair "*Aerospike* and *Redis*" was selected because they are major tags in characteristic 'Performance' and 'In-Memory Performance'. The last pair, "*MongoDB* and *Cassandra*" was chosen since they are major tags that appears a lot(2529, 905 times each). Although not presented here, the following analysis can also be conducted for the other pairs to deduce commonalities and differences between the pairs.

First, topic ‘Query’ has generally high value for all pairs except *Aerospike* and *Redis*. Users in Stack Overflow doesn’t focus a lot on Query regarding *Aerospike* and *Redis*. Thus, when we discuss common features of a pair, we will not mention the topic ‘Query’. Next, you can see 3 pairs between *Neo4j*, *ArangoDB* and *OrientDB*. These pairs have obviously highest TDV value for topic ‘Graph Model’, because these DB are based on a graph model. *Neo4j* and *ArangoDB*’s lowest TDV value is −**.05** by topic ‘Document Model’, which means *Neo4j* differs from *ArangoDB* in terms of ‘Document Model’. In fact, *ArangoDB* is much more featured in the document model, with a ‘Document Model’ distribution value of *Neo4j* of about **.017** and *ArangoDB* of about **.113**. The same applies for *ArangoDB* and *OrientDB*.

Table 7. Topic discrimination value

	Update & Change	**In-Memory**	**Col-umn-family**	**Graph**	**Distribut-ed System**	**Time & Date**	**Connec-tion & Run Error**
Neo4j -ArangoDB	-.00	+.00	-.00	**+.10**	+.00	+.00	**+.01**
ArangoDB -OrientDB	-.00	+.00	+.00	**+.04**	-.00	-.00	**-.01**
Neo4j -OrientDB	**+.01**	+.00	-.00	**+.05**	+.00	-.00	**-.02**
ArangoDB - MongoDB	+.00	+.00	+.00	**-.16**	+.00	**+.01**	**+.01**
Aerospike -Redis	+.00	**-.09**	+.00	+.00	**-.01**	-.00	**-.02**
MongoDB - Cassandra	+.01	+.00	**-.05**	+.00	**-.03**	**+.01**	**+.01**

	Que-ry	**Array & Ob-ject**	**Ja-va**	**Relation-al DB**	**Server & Mul-tiprocessing**	**Key-val-ue**	**Docu-ment**
Neo4j - ArangoDB	**+.02**	**+.01**	-.00	**-.04**	+.00	+.00	**-.05**
ArangoDB - OrientDB	**+.02**	**+.01**	**-.01**	-.00	+.00	-.00	**-.03**
Neo4j - OrientDB	**+.02**	+.00	**-.01**	**-.03**	+.00	+.00	-.00
ArangoDB - MongoDB	**+.03**	**-.01**	+.00	**-.02**	+.00	+.00	**+.08**
Aerospike - Redis	+.00	+.01	+.00	-.01	**+.02**	**+.02**	+.00
MongoDB - Cassan-dra	**+.03**	**-.06**	+.00	**+.07**	+.01	+.00	**-.05**

Let's look at the pair *ArangoDB* and *MongoDB*. It shows a great similarity with **+.08** in the Document Model and a big difference with **−.16** in Graph Model. This allows us to see the mixed data model of *ArangoDB* well. Next, you can see Aerospike and *Redis* are somewhat similar in topic 'Key-value' and 'Server & Multiprocessing', but difference is shown in 'In-memory'. It shows the difference between *Redis*, mainly operating on the memory, and *Aerospike*, which is mainly operating on the disk.

Finally, with regard to *MongoDB* and *Cassandra*, they have a large common topic 'Relational DB', because both are widely used NoSQL DB and thus they are often compared with traditional RDB. They are different in topic 'Column-family', 'Document Model' and 'Array & Object'. 'Array & Object', which reflects one of *MongoDB*'s feature to accept data of array and object freely, is the largest different of TDVs with −.06 value. And 'Column-family' and 'Document model' have a value of −.05, which means a difference between them, because they belong to *Cassandra* and *MongoDB* respectively.

Using the TDVs above, we could find out what the two targets match or do not match, thus making it easy to look at the topic differences in the question and answer that are related with NoSQL in Stack Overflow.

4 Conclusion

In the present paper, we analyzed the NoSQL-related question and answer data gained from famous programming question and answer site, *Stack Overflow*. LDA topic modeling was used to find out topic distribution of each NoSQL databases. Furthermore, we proposed Topic Discrimination Value (TDV) to compare the distribution of topics between tag pair and confirm how similar or different NoSQL databases are accepted to actual users.

Applying LDA topic modeling to find out similarity and difference between NoSQL database by comparing topic distribution has advantage in figuring out the reason why some NoSQL databases are close to each other while others not. For example, by investigating topic clusters that are related to data model, we found out that NoSQL users in *Stack Overflow* generally accept each NoSQL database as data model provided by experts. This means that traditional NoSQL data model category matches actual user perception on data model.

However, due to the limitation of LDA, selecting the number of topics and labeling the topic were done manually. Another limitation of our research is that all documents are used equally to calculate the topic distribution, neglecting the significance of each questions or answers.

By using TDV, we find out that a great interest is on actual problems regarding NoSQL, like connection & run error and query problem. On further research, documents in this cluster can be focused on to figure out the problems each NoSQL users face.

Acknowledgement. This work was supported by the Ministry of Education of the Republic of Korea and the National Research Foundation of Korea (NRF-2015S1A3A2046711).

References

1. Kanwar, R., Trivedi, P., Singh, K.: NoSQL, a solution for distributed database management system. Int. J. Comput. Appl. **67**(2), 6–9 (2013)
2. Zhang, H., Wang, Y., Han, J.: Middleware design for integrating relational database and NOSQL based on data dictionary. In: 2011 International Conference on Transportation, Mechanical, and Electrical Engineering (TMEE), pp. 1469–1472. IEEE, December 2011
3. Hajoui, O., Dehbi, R., Talea, M., Batouta, Z.I.: An advanced comparative study of the most promising nosql and newsql databases with a multi-criteria analysis method. J. Theor. Appl. Inf. Technol. **81**(3), 579 (2015)
4. Tudorica, B.G., Bucur, C.: A comparison between several NoSQL databases with comments and notes. In: 2011 10th Roedunet International Conference (RoEduNet), pp. 1–5. IEEE, June 2011
5. Han, J., Haihong, E., Le, G., Du, J.: Survey on NoSQL database. In: 2011 6th International Conference on Pervasive Computing and Applications (ICPCA), pp. 363–366. IEEE, October 2011
6. Lourenço, J.R., Cabral, B., Carreiro, P., Vieira, M., Bernardino, J.: Choosing the right NoSQL database for the job: a quality attribute evaluation. J. Big Data **2**(1), 18 (2015)
7. Moniruzzaman, A.B.M., Hossain, S.A.: Nosql database: New era of databases for big data analytics-classification, characteristics and comparison (2013). arXiv preprint: arXiv:1307.0191
8. Parnin, C., Treude, C., Grammel, L., Storey, M.A.: Crowd documentation: exploring the coverage and the dynamics of API discussions on Stack Overflow. Georgia Institute of Technology, Technical report (2012)
9. Gajduk, A., Madjarov, G., Gjorgjevikj, D.: Intelligent tag grouping by using an agglomerative clustering algorithm (2013)
10. Barua, A., Thomas, S.W., Hassan, A.E.: What are developers talking about? An analysis of topics and trends in stack overflow. Empirical Softw. Eng. **19**(3), 619–654 (2014)
11. Yang, J., Tao, K., Bozzon, A., Houben, G.J.: Sparrows and owls: characterisation of expert behaviour in stackoverflow. In: International Conference on User Modeling, Adaptation, and Personalization, pp. 266–277. Springer International Publishing, July 2014
12. Blei, D.M., Ng, A.Y., Jordan, M.I.: Latent dirichlet allocation. J. Mach. Learn. Res. **3**, 993–1022 (2003)
13. Ramage, D., Heymann, P., Manning, C.D., Garcia-Molina, H.: Clustering the tagged web. In: Proceedings of the Second ACM International Conference on Web Search and Data Mining, pp. 54–63. ACM, February 2009
14. Celikyilmaz, A., Hakkani-Tur, D., Tur, G.: LDA based similarity modeling for question answering. In: Proceedings of the NAACL HLT 2010 Workshop on Semantic Search, pp. 1–9. Association for Computational Linguistics, June 2010
15. Zhao, W.X., Jiang, J., Weng, J., He, J., Lim, E.P., Yan, H., Li, X.: Comparing twitter and traditional media using topic models. In: European Conference on Information Retrieval, pp. 338–349. Springer, Heidelberg, April 2011
16. Wallach, H.M., Murray, I., Salakhutdinov, R., Mimno, D.: Evaluation methods for topic models. In: Proceedings of the 26th Annual International Conference on Machine Learning, pp. 1105–1112. ACM, June 2009
17. Messina, A., Storniolo, P., Urso, A.: Keep it simple, fast and scalable: a multi-model NoSQL DBMS as an (eb) XML-over-SOAP service. In: 2016 30th International Conference on Advanced Information Networking and Applications Workshops (WAINA), pp. 220–225. IEEE, March 2016

18. http://cassandra.apache.org/
19. https://hbase.apache.org/
20. http://couchdb.apache.org/
21. http://www.mongodb.org/
22. https://ravendb.net/
23. http://www.aerospike.com/
24. https://aws.amazon.com/dynamodb/?nc1=h_ls
25. http://leveldb.org/
26. https://redis.io/
27. http://basho.com/products/
28. https://www.arangodb.com/
29. http://neo4j.org/
30. http://orientdb.com/
31. http://www.nltk.org/
32. https://radimrehurek.com/gensim/
33. http://nosql-database.org/

MultiPath MultiGet: An Optimized Multiget Method Leveraging SSD Internal Parallelism

Kyungtae Song, Jaehyung Kim, Doogie Lee, and Sanghyun Park(✉)

Department of Computer Science, Yonsei University, Seoul, Korea
{skt8776, jaehyungkim, edoogie, sanghyun}@yonsei.ac.kr

Abstract. Accessing data with low latency is a major issue for big data platforms. Some platforms use a Multiget method when accessing data. The Multiget method can be used efficiently when a server should seek data with multiple keys or many single get requests are waiting in a command queue. The method is implemented to access multiple blocks of data using multiple keys. In this research, we developed MPMG, a novel multipath Multiget method for a solid state drive (SSD)-based key-value storage. In MPMG, we use the open source key-value storage RocksDB, which was developed by Facebook and is one of the most representative key-value NoSQL stores. Current RocksDB system cannot fully utilize the internal parallelism of SSD. The MPMG method was designed to leverage ssd internal parallelism and access data in SSD with parallel approach. The ability to support the parallel access inside SSD is efficient for performance. With multipath approach, the SSD can fully exploit its bandwidth, compared to the original approach. In an experiment, this method showed improvements in speed and throughput. The elapsed time for processing decreased up to 80.

Keywords: Flash-based SSDs · Big data · Key-value stores · LSM-tree · Database management · Parallel execution

1 Introduction

The speed at which data can be accessed is one of the most important features for a database in the big data era. Many highly data-intensive applications such as e-commerce and social networking services require almost immediate data access when users want to access the data. Because of its fast data access, key-value stores are widely used to process big data quickly in data-intensive industrial fields, including Dynamo [1] at Amazon, LevelDB [2] at Google, RocksDB [3] at Facebook, and Cassandra [4] or Hbase [5] at Apache. Key-value stores usually store keys and values with a hash data structure. Thus, they do not need a complicated interface compared with relational databases. However, even though the storage system is fast, the storage device becomes a bottleneck to system performance. Therefore, many key-value stores utilize a solid state drive (SSD) as a main storage device because of its fast data access speed. Unlike a hard disk drive (HDD), an SSD has a flash-based architecture that uses electronic signals instead of physical data access. Therefore, it offers speed and improves the read and write performance of a database. Other features of an SSD are

W. Lee et al. (eds.), *Proceedings of the 7th International Conference on Emerging Databases*, Lecture Notes in Electrical Engineering 461,
https://doi.org/10.1007/978-981-10-6520-0_15

wear out problem, fast random access, and internal parallelism. Recent studies have usually focused on the lifetime [6, 7] and fast random access [8] when developing key-value storages. Although an SSD has a fast random access speed, continuous random access cannot fully utilize an SSD's full performance. Therefore, there is a limitation when using the simple access pattern for random access. For these reasons, we attempted to fully utilize the capacity of an SSD by changing the simple data approach to parallel access. We focused on obtaining the full read performance of the NoSQL environment using our new method, multipath Multiget (MPMG). We conducted an experiment with our method using the open source platform RocksDB, which was developed by Facebook and is one of the most representative key-value NoSQL stores.

RocksDB uses log-structured merge (LSM) tree [9]-based sorted string table (SST) files for its data structure. Key-value stores that use an LSM tree like LevelDB guarantee a high write throughput but compromise the read latency because of the characteristics of the LSM tree. Many key-value stores use an LSM tree for their data structure because it is highly efficient for a write-intensive data workload. However, the read latency of the initial version of an LSM tree-based key-value store was too high because of its multi-layered tree structure. Therefore, many researchers tried to improve the read performance without impeding the write latency. bLSM [10] introduces bloom filters and merge schedulers, and LSMtrie [11] adopts a trie data structure for key-value storage. These studies attempted to use an LSM tree data structure to maintain a reasonable read latency but did not give much consideration to storage hardware architecture features.

In this study, we improved the Multiget method in RocksDB by changing from the multiple random access used by the original method to the bulk parallel access of the SSD. As a result, an experiment showed faster data access and greater bandwidth for the SSD. This paper consists of five sections: Sect. 2 explains the background of our study. Section 3 discusses the methods for implementing MPMG. Section 4 shows the experimental configurations and results. Section 5 suggests future work about another optimizing method for key-value store with SSD. Finally, Sect. 6 presents the conclusion of this paper.

2 Background

In this section, we describe the LSM tree, RocksDB, and internal parallelism of an SSD. The first section LSM tree is the base tree algorithm that motivated the development of many key-value stores like LevelDB, RocksDB, and Hbase. Then, we introduce the design of RocksDB, which is the SSD optimized key-value store of Facebook. RocksDB uses a parallel compaction algorithm for write amplification [12]. The last part explains the block-based and flash-based hardware architecture of an SSD.

2.1 Log-Structured Merge-Tree

The LSM tree was designed for write-intensive data workloads. It maintains a low write latency by sacrificing read amplification. An LSM tree is composed of one in-memory

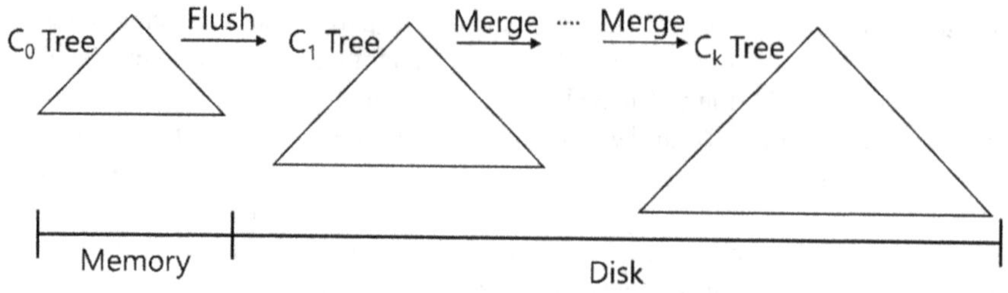

Fig. 1. Log-structured merge tree

data structure (e.g., RB tree) and many append-only data structures for disk storage. In an LSM tree, all of the data are first inserted into the in-memory data structure. As shown in Fig. 1, the LSM tree flushes the data to disk storage when the memory becomes full. To achieve a low write latency, the LSM tree does not instantly delete the data. Instead of deleting the data, the LSM tree inserts tombstone entries in the in-memory data structure to avoid accessing the deleted data. The on disk data are merged periodically to apply real deletion to the deleted data and reduce the read amplification. In this architecture, data manipulation workloads only occur in memory, which allows the LSM tree to quickly process the write operations. However, because of the many append-only data structures (trees), an LSM tree-based system encounters a high read amplification, which impacts the read performance. When reading data, the system should access all the trees on the disk for an existence check. Therefore, the LSM tree periodically merges the tree components when the size of each component becomes larger than a certain threshold. In this manner, the LSM tree controls the number of trees and performs real deletion of data. The merge operation is performed as follows:

1. Tree component C0 resides in memory and tree components C1 to Cn reside on the disk.
2. Merge operation occurs between Cn and Cn+1
3. The new tree component obtains data from Cn and Cn+1 using a merge sort method
4. When the merge operation is finished, Cn is joined to Cn+1 and becomes one unique tree

During the merge operation, the old data are deleted when the merge cursor meets a tombstone data marker.

2.2 RocksDB

This section explains RocksDB with its approach to store data and how RocksDB improved LevelDB. RocksDB is designed by Facebook and is one of the most popular open source key-value stores. RocksDB is based on LevelDB, which is developed by Google and is also a well-known open source LSM tree based embedded key-value store. LevelDB uses an LSM tree for high write throughput and aims to optimize the performance of the key-value store based on memory and SSD characteristics. RocksDB applies bloom filters to improve read performance and utilizes parallel programming for compaction, which corresponds to a merge operation on the original LSM tree. At first data are stored in an in-memory table named 'memtable' as shown in Fig. 2. If the size of a memtable reaches its threshold, the memtable moves its data to a different structure called 'immutable memtable' which is not affected by data insertion. Immutable memtable does not apply those commands, since this structure should be prepared for flushing pipeline whereas memtable applies update or deletion immediately. The flush pipeline transfers contents of immutable memtable to the disk based table structures. A LevelDB-based system uses SST files, which are maintained in the multi-level structure of the LSM tree, for its file system. Each level corresponds to a tree component of the LSM tree and the data on each level are stored in SST files in a sorted order. SST files are sequentially appended after each file, and this makes efficient write operations on flash devices. An SST file has metadata that includes the minimum and maximum values of the data sets. Therefore, when the system accesses the data in a file, the metadata works as an index and the system does not have to access redundant files in the disk (or memory). As the file size grows, RocksDB uses a compaction method (similar to the merge method of the LSM tree) for deleting and merging data. During the compaction, target-level SST files are merged and saved as a single large SST file at the next level. In this manner, compaction decreases the number of files and data size. The LSM-tree based feature allows RocksDB to access data faster with flash devices because this structure saves data with sequential order. In RocksDB, the compaction range is divided by the number of threads and each thread processes the compaction method with assigned key-value pairs. Also, RocksDB uses block-based approach, since SSD controls data with page blocks. This multi-thread compaction and optimization of the memory and SSD allowed RocksDB to outperform then existing frameworks when it was first designed. These features motivated our study. Although RocksDB is designed for flash storage, but its Multiget method does not fully utilize

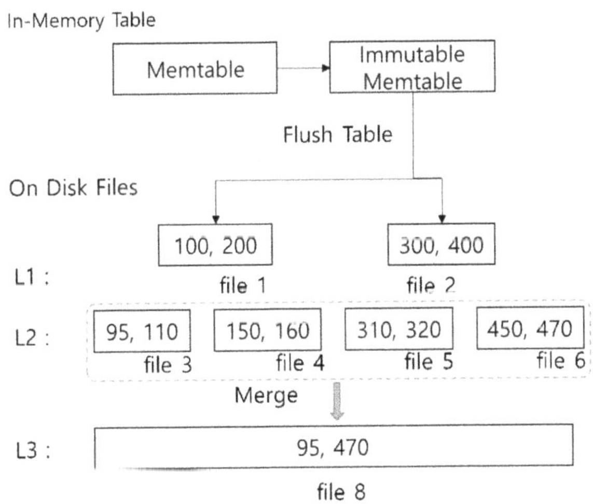

Fig. 2. Basic RocksDB architecture

the characteristics of SSDs, which allowed us to optimize the key-value store with a flash device with a different perspective.

2.3 Architecture of SSD

SSD architecture is described in Fig. 3. An SSD consists of a DRAM, a controller, multiple channels for data transfer, and multiple sets of flash chips. This architecture allows an SSD to access multiple blocks of data in parallel. The controller sends multiple I/O requests to multiple flash chips and the flash chips deliver data to the controller using multiple channels. This unique parallel I/O operation of an SSD is called an internal parallelism.

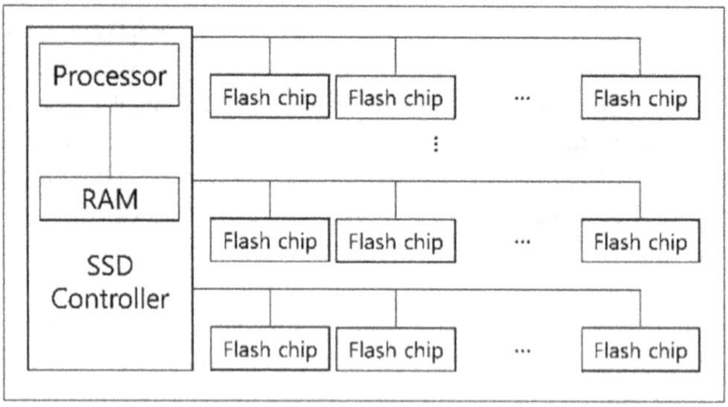

Fig. 3. SSD architecture

In a recent study, Lee et al. [13] suggested a new external merge sort method using the internal parallelism. They used psync [14] I/O, which delivers multiple I/O requests with a single I/O request function and obtained the result immediately. Through the psync I/O interface, the researchers could improve the merge sort method for an SSD. Psync I/O was first introduced through the research of Roh et al. [14].

The SSD saves data as page-structured blocks inside each flash chip. Usually, each page size is 4 KB or 8 KB, and multiple pages are stored in a single block. In this block-based architecture, the SSD reads and writes data page-by-page. In page based approach, it is more efficient to approach data with sequential access. There have been many researches about changing a random access method to a sequential access method for this reason. [15–17]. Moreover, S.W. Lee suggested in- page logging for flash memory database servers [18] to overcome limitations of previous databases. Recently, many non-relational databases log based structures like LSM tree are widely used for managing data to improve performances. In the same manner, RocksDB uses log structured merge tree to write data sequentially. Also, RocksDB uses block based access to optimize the read pattern of SSDs. Even though RocksDB efficiently uses block-based architecture, there was an improvement point with regards to reading operations.

3 Implementation

In this section, we propose a MultiPath MultiGet (MPMG) method for SSDs, which leverages internal parallelism to boost SSD read performance. Unlike the original method, the new method does not use the simple for-loop of a get function, but accumulates the materials for the read operation and sorts them according to the offsets of files. Using this approach, we could conduct parallel accesses and exploit the bandwidth of SSDs.

3.1 Original Multiget Method

The Multiget method needs a list of keys to retrieve and it returns a list of found values according to this key list. The original Multiget method used in RocksDB is quite simple. The method is shown in Figs. 4 and 5. The internal algorithm of the Multiget method calls a simple get function repeatedly until it approaches the size of the list. When the key exists in memory, the function returns data from the memory table, and if the key does not exist, the function should seek this key in the disk table. This method is inefficient, especially when block size of RocksDB is smaller than page size of SSD. The get method is implemented using a binary search and filtered approach as shown in Figs. 5 and 6. When the data are filtered with bloom filters, the system starts searching inside the SST file using a binary search. RocksDB can use a binary search because an SST file is sorted in key order. The binary search returns the iterator of the file block index, which leads us to the block iterators. We can fetch the real data block into memory using block iterators. However, this method sends only one I/O request at a

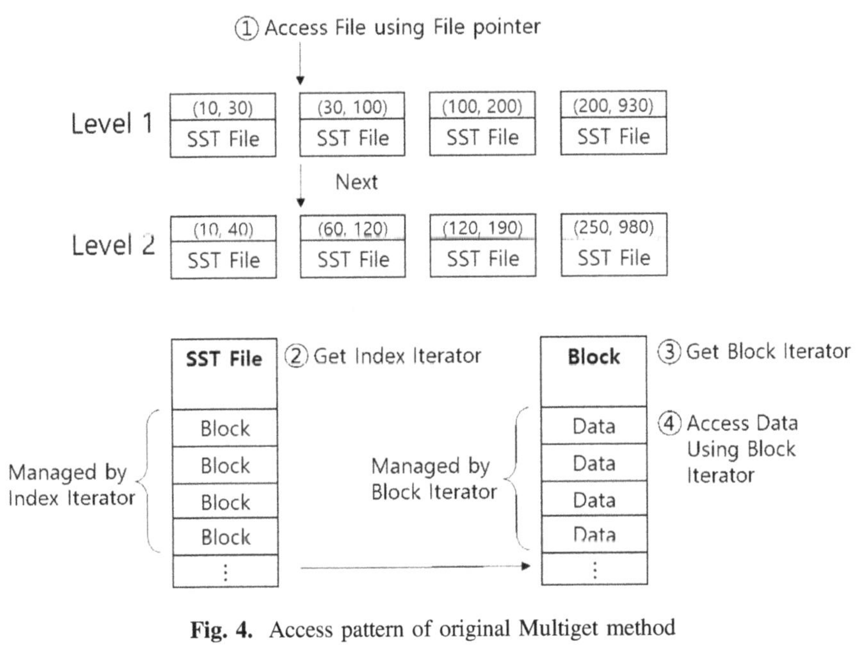

Fig. 4. Access pattern of original Multiget method

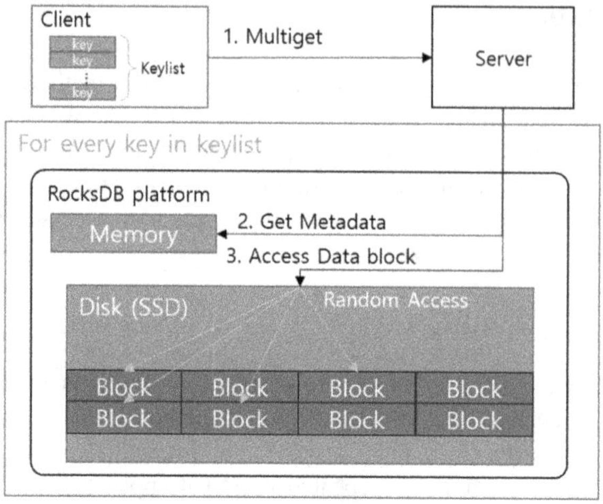

Fig. 5. Access pattern of original Multiget method

time and requires numerous random accesses of the SSD, preventing it from exploiting the full performance of the SSD.

3.2 MultiPath MultiGet Method

We designed a new Multiget method, multipath Multiget(MPMG) that uses internal parallelism of SSD. Figure 6, Algorithms 1 and 2 describes this new method. The Algorithm 2 explains MPMG and it is divided into three parts. The first part collects metadata for disk access in memory (Line 5 to Line 15). Then the second part sorts metadata (Line 16) and the third part sends I/O request to disk using the metadata (Line 17 to Line 25). First, the new method gathers all the information for accessing the data. This information is composed of file descriptor, offset, and size of the data. SST files on the same level in RocksDB have metadata that includes minimum and maximum values in sorted order. Therefore, it is easier to find the appropriate file to access the data if the key list is maintained in sorted order. When a file is determined to be unsuitable, it will never be chosen during the file selection step in line 7 of Algorithm 2. Suitable files can be accessed with file pickers. File picker chooses a file and makes an index iterator (Line 10). Material list starts accumulating the materials after the index iterator has been created (Lines 12 to 14). When the accumulating process is done, mlist begins sorting with each fd and offset component (Line 16). The sorting used in this method is stl::sort, which is quick sort. As a result of these processes, the datablock in SSD can be read in sequential access. Psync I/O begins with makeBlockIterator() function. The process of psync I/O is described in Algorithm 1. First of all, variables for the psync I/O interface(buffer, io context, pointers) are prepared. Then, makeBlockIterator() function uses materials from mlist and prepare async I/O operations. When async I/O operations are ready, psync interface sends async I/O operations to the SSD and starts reading. Each read operation of async I/Os works same as the original method. The read operation returns data and these data work as block meta

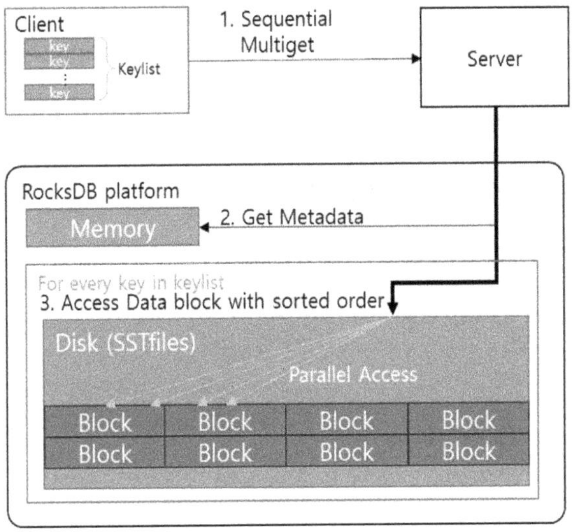

Fig. 6. Access pattern of multipath Multiget method

data and they are used to make a blockiterator in makeBlocklerator method. Blockiterator has the information about the offset of key-value data, compression type, and properties. It performs binary search when finding appropriate data with keys. If the key is matched with value, valuelist stores the value. The client receives valuelist from MPMG method after the function is done. MPMG method separated the data obtaining process into two parts, one is obtaining metadata and the other is data requests. As a result, when the data request phase starts, the method sends I/O request phase starts, the method sends I/O requests continuously without asking metadata, and this leads to performance improvement.

Algorithm 1 . makeBlockIterator Method

```
Prepare psync I/O variables
for (i:0 to sizeof(mlist))
  for (j:0 to ITER)
    Prepare async I/O with mlist.contents[j+i*ITER];
  end for
  Submit async I/Os
  Get the result of async I/O
end for
Free the allocated variables
```

Algorithm 2 . Sequential Multiget Method

Input
keylists : list of keys that needs Multiget batch operation
Output
valuelist : the result of Multiget method is value according to key

```
Get metadata for locking
Material mlist;
/* material consists of list of file descriptor(fd), block size(size), and block offset(offset)*/
if(keylist contents exists in memory)
   Get(keys,value)
else
   Sort keylist with key order
   Get the appropriate file list by using keylist and metadata
   Make (key, file) set with keylist and the matched filelist
   for(i : 0 to sizeof(filelist))
      FilePicker fp = filelist[i];
      TableReader t = GetTableReader(fp.file_descriptor);
/* TableReader contains materials for read access */
      t.appendMaterials(keylist[i], mlist);
      IndexIter iiter = makeIndexIterator(t,key);
      iiter.appendMaterials(keylist[i], mlist);
   end for loop
   Sort mlist with fd and offsets
   for (i:0 to sizeof(mlist))
      BlockIter biter = makeBlockIterator(mlist[i]);
/* makeBlockIterator function performs psync operation with fd, size, offset from mlist */
      while(biter.valid())
         biter.seek(keylist[i]]);
         if(biter.value == available)
            valuelist[i] = biter.value;
            break;
      end while loop
   end for loop
   Release metadata for locking
   Return with stats of the operation
   End
```

4 Experimental Results

4.1 Experiment Settings

Experiments were held to compare the original method to Sequential Multiget approach in terms of execution time of RocksDB benchmark. We used a single machine having Intel Core i7-6700 K CPU @ 4.00 GHz with 8 cores to test the RocksDB source code and used the benchmark method in RocksDB version 4.4.1 environment. To figure out the difference of actual disk performance, we disabled the OS buffer in the experiments. The specific hardware and software settings are as follows (Tables 1 and 2).

Table 1. Hardware settings

Hardware	Environment
CPU	Intel Core i7-6700 K CPU @ 4.00 GHz
Memory	64 GB
SSD	Samsung 850-pro 512 GB

Table 2. Software settings

Software	Environment
RocksDB version	4.4.1
OS	Linux OS, kernel version 3.1 Centos 7.3
File system	Extended file system 4 (ext4)

4.2 Multiget with *db_Bench*

RocksDB provides its own experimental benchmark tool, called db_bench. The db_bench has various methods that can check the performance of RocksDB. We used fillseq, fillrandom, and multiget benchmarks in the experiments. The fillseq method fills sequential data, while the fillrand fills data with random order. To fill data, keys are created using 16 byte random integers and values are 512 bytes each using the key. We experimented with 8 KB of a block size because SSD fetches data with a flash page size of 8 KB. The total number of key is 1000000. The number of batch operation for the internal parallelism was changed from 1 to 10000. This shows effect of the internal parallelism. Figures 7 and 8 shows the experimental results. In both experiments, the gain from the internal parallelism increased as the number of batch size gets bigger. The result was the best when batch size was 10000. Figure 7 illustrates the results when multiget ratio was varied. Multiget ratio is calculated as follows:

$$number\ of\ the\ multiget\ keys/number\ of\ total\ keys$$

Multiget ratio was changed from 1% to 100%. The value size was fixed to 8 KB, which is the same as the page size of SSD. The result was almost same though the multiget ratio was changed. In the result, traditional multiget performed better when the

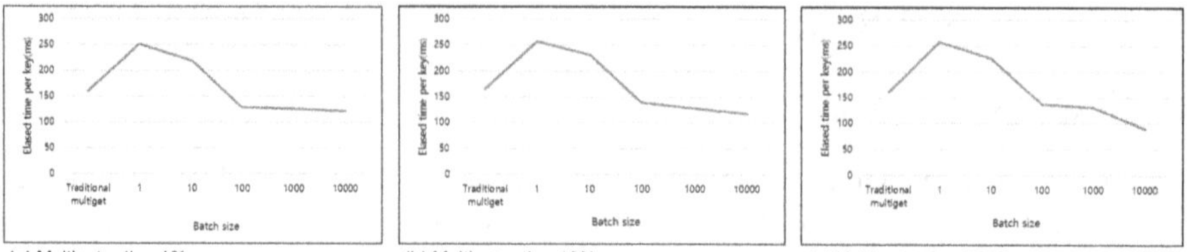

(a) Multiget ratio : 1% (b) Multiget ratio : 10% (c) Multiget ratio : 100%

Fig. 7. Elapsed time of MPMG on 8 KB value size

batch size of psync I/O was small. Psync I/O needs preparation phase and this becomes overhead when the batch size is small. As the batch size gets bigger, MPMG performed better than the original method. The best gain is shown on Fig. 7(c). In the result, MPMG performed 45% better than the original method when batch size is 10000.

In Fig. 8, the value size of RocksDB was varied from 256 to 1024. The result tendency was similar to the result of Fig. 7. The result of the original method was 160 ms in each experiment, which was the same result of Fig. 7. As batch size gets bigger, MPMG performed better and the original method was better when batch size was small. Value size affects the database size and data fragmentation inside SSD. Due to the page based structure of SSD, the data fragmentation becomes severe when the value size is smaller. This affects the performance of MPMG. As the value size gets bigger, the gain from the internal parallelism decreased. The best result was when the value size is 256 byte and MPMG performed 80% better than the original method. As a result, MPMG is useful when storing small data such as metadata.

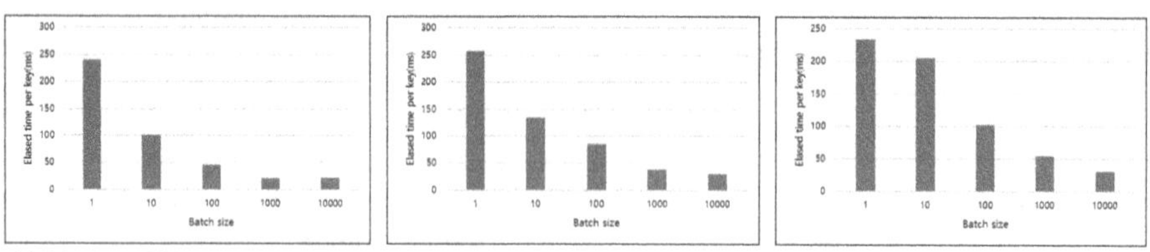

(a) Value size : 256 byte (b) Value size : 512 byte (c) Value size : 1024 byte

Fig. 8. Elapsed time of MPMG on 100% multiget ratio

5 Future Work

We are expecting to utilize the internal parallelism with compaction operation in RocksDB. Compaction operation needs continuous I/O for the disk contents. Though compaction progresses in background, it still affects the performance of database. When the compaction occurs, the system need to read all the files of the selected level, and write back to the disk after merging the files. In this mechanism, compaction makes high read amplification and write intensive work, which leads I/O penalty. Therefore, we can say that it is an appropriate point for ssd optimization. Using internal parallelism, the read and write operation does not need a lot of threads to control the compaction method. With psync I/O, the CPU does not care about multiple I/O operations and regards the I/O as a single I/O operation. In this way, we are looking forward to develop compaction algorithm with internal parallelism of SSD.

6 Conclusion

On Big-data platforms, I/O Performance is highly important since it produces a bottleneck when both CPU and RAM are relatively faster. Therefore, inducing the disk I/O to its full performance impacts the efficiency of the entire system. In this research, we proposed our new method, MPMG, that enhances read performance of the database. The MPMG method works efficiently when a server should process a lot of multiple get operations or when numerous read jobs are waiting in a job queue. The major difference of the proposed algorithm compared to its original is the access pattern of a file — MPMG method sends multiple batch asynchronous I/Os to the disk to take advantage of the internal parallelism of flash devices. By changing simple random access pattern to the parallel access pattern, we could deliver multiple I/O request faster. This method effectively reduced the query execution time of RocksDB benchmark up to 58%. The performance was best when the entry size of multiget was the greatest. The sort cost takes about 0.1 ms per key, which doesn't affect a lot to the elapsed time. MPMG approach can also be used for scan operation and compaction operation, since they also require access to multiple data. We anticipate breaking the limitation of the hardware bottleneck and optimizing for SSDs more with the MPMG method.

Acknowledgement. This research was supported by the MSIP, Korea, under the "SW Starlab" (IITP-2017-0-00477) supervised by the IITP (Institute for Information & communications Technology Promotion).

References

1. DeCandia, G., et al.: Dynamo: Amazon's highly available key-value store. ACM SIGOPS Oper. Syst. Rev. **41**(6), 205–220 (2007)
2. LevelDB. https://github.com/google/leveldb
3. RocksDB. https://github.com/facebook/rocksdb
4. Cassandra. https://cassandra.apache.org
5. Hbase. https://hbase.apache.org
6. Lee, C., Sim, D., Hwang, J., Cho, S.: F2FS: a new file system for flash storage. In: 13th USENIX Conference on File and Storage Technologies (FAST 15), pp. 273–286 (2015)
7. Min, C., Kim, K., Cho, H., Lee, S.W., Eom, Y.I.: SFS: random write considered harmful in solid state drives. In: 10th USENIX Conference on File and Storage Technologies (FAST 2012), p. 12 (2012)
8. Lim, H., Fan, B., Andersen, D.G., Kaminsky, M.: SILT: a memory-efficient, high-performance key-value store. In: Proceedings of the Twenty-Third ACM Symposium on Operating Systems Principles (SOSP 2011), pp. 1–13 (2011)
9. O'Neil, P., Cheng, E., Gawlick, D., O'Neil, E.: The log-structured merge-tree (LSM-tree). Acta Informatica **33**(4), 351–385 (1996)

10. Sears, R., Ramakrishnan, R.: bLSM: a general purpose log structured merge tree. In: Proceedings of the 2012 ACM SIGMOD International Conference on Management of Data, pp. 217–228 (2012)
11. Wu, X., Xu, Y., Shao, Z., Jiang, S.: LSM-trie: an LSM-tree-based ultra-large key-value store for small data items. In: 2015 USENIX Annual Technical Conference (USENIX ATC 2015), pp. 71–82 (2015)
12. Moshayedi, M., Wilkison, P.: Enterprise SSDs. Queue **6**(4), 32–39 (2008)
13. Lee, J., Roh, H., Park, S.: External mergesort for flash-based solid state drives. IEEE Trans. Comput. **65**(5), 1518–1527 (2016)
14. Roh, H., Park, S., Kim, S., Shin, M., Lee, S.W.: B+-tree index optimization by exploiting internal parallelism of flash-based solid state drives. Proc. VLDB Endow. **5**(4), 286–297 (2011)
15. Li, Y., He, B., Yang, R.J., Luo, Q., Yi, K.: Tree indexing on solid state drives. Proc. VLDB Endow. **3**(1–2), 1195–1206 (2010)
16. Tsirogiannis, D., Harizopoulos, S., Shah, M.A., Wiener, J.L., Graefe, G.: Query processing techniques for solid state drives. In: Proceedings of the 2009 ACM SIGMOD International Conference on Management of Data, pp. 59–72 (2009)
17. Manning, C.: YAFFS: Yet another flash file system (2004)
18. Lee, S.W., Moon, B.: Design of flash-based DBMS: an in-page logging approach. In: Proceedings of the 2007 ACM SIGMOD International Conference on Management of Data, pp. 55–66 (2007)

An Intuitive and Efficient Web Console for AsterixDB

SoYeop Yoo, JeIn Song, and OkRan Jeong(✉)

Gachon University, Seongnam 13120, Republic of Korea
bbusso90@gmail.com, wpdls601@gc.gachon.ac.kr,
orjeong@gachon.ac.kr

Abstract. Enabling users to create and consume data anywhere and anytime, the development of Internet technology changed a large part of data ecosystem. Along with the changes in environment including data form and storage method, NoSQL-based database system began to receive more attention than the existing RDMS (Relational Database Management System). AsterixDB, an open source big data management system specialized in big data analysis and processing, has many characteristics to process unstructured data efficiently. However, the existing AsterixDB Web Console has provided users with very inefficient environment in terms of UI/UX. In our proposed solution, we propose and implement new web console that provides efficient and intuitive interface on which the characteristics are well applied to increase the utilization of AsterixDB.

Keywords: AsterixDB · Web console · User interface · UI/UX

1 Introduction

The rapid growth in Internet technology brought many changes in database ecosystem including data form and storage method. While traditional data, which mainly had text-based structured form, had relatively simple form, recent data has rather free form and the amount of collected data becomes copious [1, 2].

Besides user generated contents such as social network, blog and news, there is a lot of sensing data around. The Internet, which is available anywhere and anytime, enables pumping huge amounts of digital information out in real time opening the big data era. In big data era, the data is created and used in unprecedented scale. To manage unstructured data efficiently, NoSQL-based database system is made and used more than traditional RDMS (Relational Database Management System) [1–4].

Many NoSQL-based database systems, which aim to manage big data, reflect their own features for efficient data processing. Intuitive console is necessary to make better use of database system and provide more user-friendly service environment. As the existing RDMS mainly uses structured data, it shows data and structure based on simple table such as 'phpMyAdmin [5]' or 'MySQL Work-bench [6]' and enables management and processing of data. As NoSQL database has to express unstructured data, however, it expresses data in JSON (Javascript Object Notation) form rather than tables. In this case,

W. Lee et al. (eds.), *Proceedings of the 7th International Conference on Emerging Databases*, Lecture Notes in Electrical Engineering 461,
https://doi.org/10.1007/978-981-10-6520-0_16

it is easy to process various types of data but difficult to identify numerous big data at a glance.

Among many other NoSQL-based database systems to process big data, AsterixDB, which is a kind of Apache Project, is a BDMS (Big Data Management System) provided with open source and has appropriate characteristics to store and analyze unstructured big data, especially social data [7]. In addition, though it has Web Console for user's convenience, it lists all the information simply in JSON form upon running a query making intuitive identification harder. To make better use of AsterixDB, which is an open source, and guarantee many users' convenience, we need new web console that can show main characteristics of AsterixDB efficiently and provide intuitive interface. We propose and implement a web console that can provide AsterixDB users with intuitive experience.

2 AsterixDB

AsterixDB is an extensible open source BDMS. Based on NoSQL Style data model that extends object-oriented database and JSON, it features easy processing of data of semi-structured form. It has wide range of queries and provides analysis of collected data. For easy processing of big data, it can construct cluster easily for distributed processing and is extensible into more than 1000 cores and 500 discs. To process realtime data efficiently, it supports B+ tree, R tree, inverted keyword and other indexing methods [7].

ADM (AsterixDB Data Model) can be explained roughly with dataverse, datatypes, and datasets. Dataverse is an abbreviation of data universe and the most top-level concept of AsterixDB, and it is similar with database of the existing RDBMS. To store data, you have to create a dataverse and then use datatypes and datasets. Datatype takes the role of specifying what kind of data to be stored in advance. In addition, Dataset has similar concept with data table of the existing RDBMS [7–9].

As AsterixDB targets semi-structured data, ADM provides Datatypes of open type and closed type. If you declare it as open type upon declaring data type at first, it permits additional contents when necessary; however, if declared as closed type, it will strictly restrict contents other than declared. If not otherwise declared, it is declared as open type to respond to various data types. Moreover, AsterixDB supports Nested type for flexible data storage and processing [7–9].

In this way, AsterixDB is designed to be flexible and extensible for big data especially for unstructured data just like social data. Just like many other database systems, we would like to give convenience to system users through web console. However, we cannot feel many advantages of AsterixDB with the existing web console and it does not even provide intuitive interface. As seen in the Fig. 1, you have to enter all the queries to see information included in the Dataset, and it is hard to see intuitive result as provided in JSON form. Moreover, it provides inefficient interface as main characteristics of AsterixDB, such as Nested type or open/closed type are not applied at all.

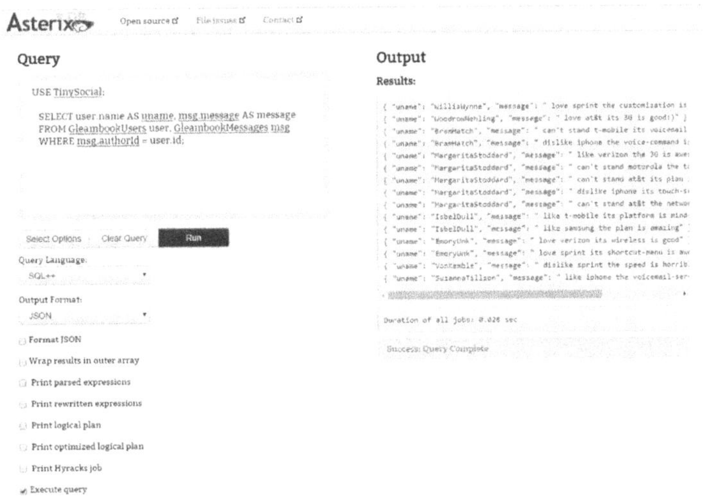

Fig. 1. The existing web consoler for AsterixDB

3 AsterixDB Web Console

The biggest problem of the existing web console is that it did not consider user's convenience in implementation. It does not show core functions of AsterixDB and is hard to check the result as it outputs query result in unintuitive method. By supplementing these problems, we propose and implement new AsterixDB web console that provides intuitive interface.

3.1 System Structure

In our work, we propose new console that improves the problems of the existing AsterixDB web console, applies the characteristics of AsterixDB, and provides users with convenient interface. The system of proposed web console is structured as in the Fig. 2. Figure 2(a) is the system structure diagram of the web console used by the existing AsterixDB, and Fig. 2(b) is the structure diagram of proposed console. The existing system is of simple structure in which the user has to enter query to get the result. Proposed new web console consists of Components part, which is in charge of main functions, and Services part, which is in charge of communication with AsterixDB system and sharing between components.

Components are mainly composed of three core parts: Browse function to show records existing in Dataset, Datatype showing function, and Query function with which a user can build query directly. Besides core components, NavBar, Sidebar, and TabMenu components take the roles of connecting to core components and expressing the information of current state. Movement between components is made through Routing. All the components can share variables through Global Service. Also, Query Service is used to communicate with AsterixDB so as to access and use necessary data.

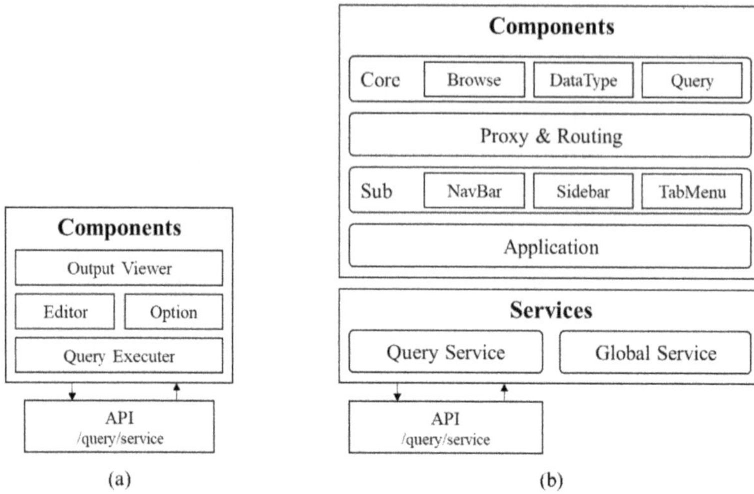

Fig. 2. Comparison of system architecture between the existing web console and the proposed web console; (a) the existing web console, (b) the proposed web console

3.2 Main Functions

The existing web console of AsterixDB did not have user-oriented design. As it targets unstructured data, it shows records in JSON form just like other NoSQL-based database system. As users want to check records at a glance, however, it is better to utilize table form rather than JSON form for that purpose. Showing records in table form just like the existing RDBMS would provide the users with result more intuitively; however, there are following issues to apply the characteristics of AsterixDB.

- Nested Type: AsterixDB stores and manages data in JSON form. On this occasion, it supports Nested type data for flexible storage and management of data. However, if these nested type data were shown in JSON form just like the existing web console, it would be hard to intuitively identify in which field the nested type exists. We may show Nested type data by adding a table inside the other but we designed other method as it uses space inefficiently.
- Open/Closed Type: open/closed type is decided upon declaring the type, and closed type will permit strictly declared contents only for data input while open type permits free addition by the user when necessary. Upon record lookup, it is an important issue how to show field name (column name) of stored data.
- Special Type: AsterixDB has special data types just like Unordered List or Ordered List besides previously mentioned Nested, open/closed type. It should be showed that a user can recognize data type easily upon record lookup.
- Query: the existing web console is composed of textarea basically provided by html, and so it could make good influence in the UI/UX aspect. We improved it to enable utilization that is more efficient by adding Syntax Highlighting function upon building query rather than simply providing space to enter query.

- Pagination: in case of the existing web console that calls and shows whole query results, there has been an issue of long latency time to execute query result due to insufficient memory on the browser as well as difficulty in expression when there is a large number of record applicable to the result value. It was necessary to improve such an issue by reducing the amount to be loaded at once, and we could do it by implementing pagination for query result value. We mainly used PrimeNG [10] Library for UI of Angular2 [11] which was used as framework upon implementation but only the callback, which can send request to database whenever pressing the page upon implementing pagination, had been implemented then. After all, there was a weakness of the longer loading time as the dataset size gets bigger upon using PrimeNG's pagination that results in a few seconds of loading time for a page in the worst case. We designed and implemented separate Table Component that can import data of a few pages in one chunk and improve reactivity between pages.

4 Implementation

The proposed system has been implemented by using web front-end framework Angular2 and Angular2's UI component library Primeng in ubuntu 14.04 environment in which Apache AsterixDB(v0.9.1) is installed. Angular2 is a framework that supports component-based system design with good maintainability in development and that's why we used it upon implementing our system. As for the server of frontend application developed with Angular2, we used Node.js.

4.1 Detailed Web Console

The web console implemented based on main issues to supplement the problems of the existing web console is as in the Fig. 3. New users can see all the Dataverse in the left with navigation bar, and select specific Dataverse/Dataset to see records and Data type of relevant Dataverse/Dataset, and transmit query to see the result value.

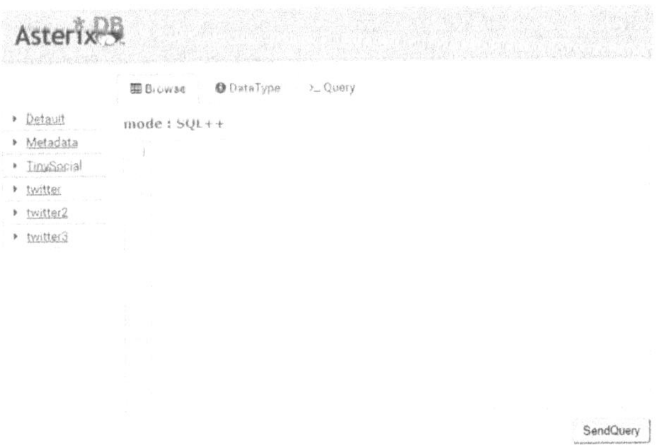

Fig. 3. The proposed web console for AsterixDB

```
Global fetchPageNum(10), columns, records, allData, expandedArea list

getChunk( fetchPageNum){
 result = `USE ${dataverseName}
          SELECT VALUE ds FROM ${datasetName} ds
          LIMIT ${limit * fetchPageNum} OFFSET ${offset}`
 columns = bulid_maximum_columns(result/2)
 allData = result.records
 getPageData(1)
}

getPageData(pageNum){
 pageNum = pageNum – ((chunkNum – 1) * fetchPageNum))
 limit += offset
 records = allData.slice(offset, limit)
}

getNextChunk(){
 getChunk(limit * (chunkNum++) * fetchPageNum)
}

expandCell(row_index, col_index){
 expandedArea[row_index] = records[row_index][col_index]
}
```

Fig. 4. Pseudo code for browsing

Browse. Browse function shows records of selected dataset by using table. Figure 4 is a pseudo code to show record. When the component is initialized, getChunk function is called to import the first chunk of currently selected data set; to respond to open/closed type data, it reads half of imported chunk to build a column with maximum number. After this, it calls getPageData to fill the page with data. fetchPageNum is the number of pages relevant to one chunk, and if all the pages are browsed, it gets the next chuck with getNextChunk. expandCell is browse function of nested data, and it shows detailed look of nested data applicable to clicked cell by expanding it right below clicked row. We implemented separate expandAll function so that all relevant nested data can be expanded just by clicking the top Column name (Fig. 5).

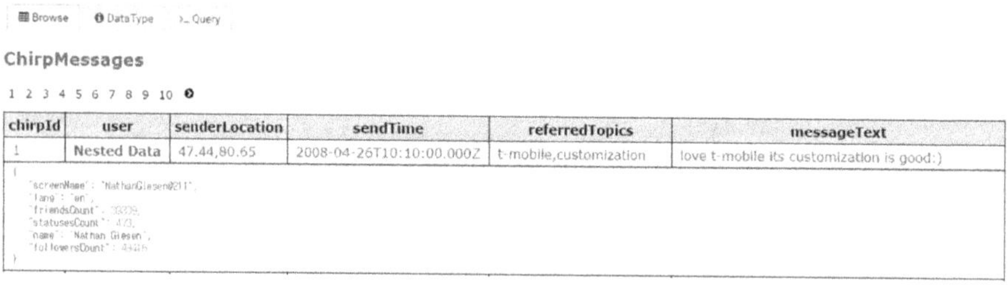

Fig. 5. Implementation result for browsing

DataType. Datatype function shows detailed data type of currently selected dataset. If there is any nested data type, it is showed with separate table. It reads record of Metadata.Dataset dataset to find data type of currently selected dataset. Since then, it looks for detailed implementation of data type found before in the Metadata.Datatype dataset. Here, it performs processing of special case and nested data type. If the user clicks nested type field, it is showed in a table below the table that expresses current data type (Fig. 6).

Browse | DataType | >_ Query

type_tweet (isOpen:true)

FieldName	FieldType	IsNullable
create_at	datetime	false
id	string	false
uid	uuid	false
text_msg	string	false
in_reply_to_status	int64	false
in_reply_to_user	int64	false
favorite_count	int64	false
geo_location	point	true
retweet_count	int64	false
lang	string	false
is_retweet	boolean	false
hashtags	{{ string }}	true
user_mentions	{{ int64 }}	true
user	type_twitter_user	false
place	type_twitter_place	false

Fig. 6. Implementation result for data type

Query. Query function is for a user to build query to get result from Database. We expanded codemirror library and added AsterixDB's SQL++ and AQL keywords in a list implementing syntax highlighting. The search result used the browse component implemented before (Fig. 7).

Fig. 7. Implementation result for query

5 Experiments

We found a problem of the existing AsterixDB web console that its UI/UX is not user-friendly and implemented a web console that can provide more intuitive interface by considering various supplementary measures. We also performed experiments to verify that the proposed measure actually provides efficient result.

The experiment targets are datasets of the same Datatype but with different sizes of 1 MB, 10 MB, 100 MB, …, 10,000 MB data. We performed comparative measurements on the difference in reaction speeds per click between PrimeNG's Pagination and newly designed Component according to AsterixDB's characteristics in Browse component, which is one of main functions of proposed web console. The comparison result of average latency of 'getPageData' between the existing PrimeNG's Table component and proposed table component is in the Fig. 8. Figure 8 is a graph expression of average latency per page click by the user measured in millisecond unit.

In case of record browse, the bigger the total data size contained in dataset, the slower the response speed (latency) to query. By bringing chunk as much as 10 pages to be included in a table that shows record at a time, we could reduce average latency per page click maximum 10 times than the table component of PrimeNG. And we did the same experiments on the different query types, but the result was almost same with each other.

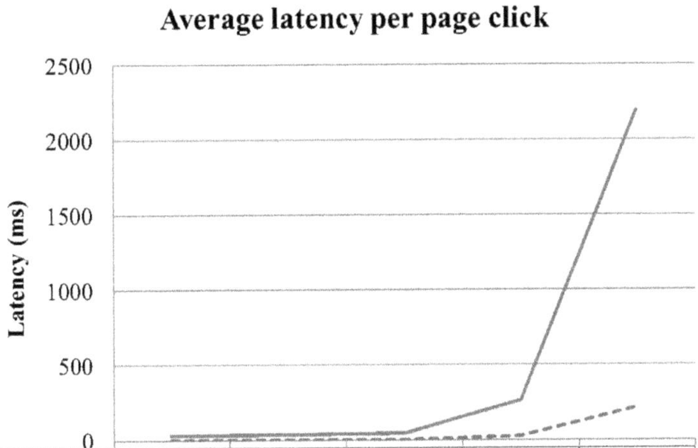

Dataset Size (MB)	1	10	100	1,000	10,000
PrimeNG component	31ms	38ms	45ms	267ms	2,189ms
Proposed component	5ms	4ms	4.5ms	26ms	216ms

Fig. 8. Comparison of average latency per page click between PrimeNG and the proposed method

6 Conclusion

In this work, we have proposed new web console that provides efficient and intuitive interface for AsterixDB, and open source BDMS. AsterixDB reflects unstructured data characteristics of big data well and therefore, it is characterized to manage various kinds of data flexibly, but the existing web console is unable to manifest such a feature. We show intuitive expression methods for Nested type, Open/Closed type, and many special data types of AsterixDB system. Especially, we made pagination in Chunk unit to speed up the response speed upon processing huge amount of big data, and could improve average latency reduced to 1/10 compared to the existing PrimeNG Library increasing user's convenience.

Acknowledgments. This research was supported by Basic Science Research Program through the NRF(National Research Foundation of Korea), and the MIFP(Ministry of Science, ICT & Future Planning), Korea, under the National Program for Excellence in SW(2015-0-00932) supervised by the IITP(Institute for Information & communications Technology Promotion (Nos. NRF- 2015R1C1A2A01051729, 2015-0-00932).

We would like to thank the Big Data Research Group of Prof. Michael J. Carey and Prof. Chen Li in University of California, Irvine (UCI) for supporting us to use AsterixDB they had developed.

References

1. Lawton, G.: Knowledge management: ready for prime time? IEEE Comput. **34**(2), 12–14 (2001)
2. McCune, J.C.: Data, data, everywhere. Manage. Rev. **87**(10), 10 (1998)
3. Gessert, F., Wingerath, W., Friedrich, S., Ritter, N.: NoSQL database systems: a survey and decision guidance. Comput. Sci.-Res. Dev., 1–13 (2016)
4. Deka, G.C.: A survey of cloud database systems. IT Prof. **16**(2), 50–57 (2014)
5. phpMyAdmin (2017), https://www.phpmyadmin.net/
6. MySQL Workbench (2017), https://www.mysql.com/products/workbench/
7. Alsubaiee, S., Altowim, Y., Altwaijry, H., Behm, A., Borkar, V., Bu, Y., Gabrielova, E.: AsterixDB: a scalable, open source BDMS. Proc. VLDB Endowment **7**(14), 1905–1916 (2014)
8. Grover, R., Carey, M.J.: Data ingestion in AsterixDB. In: EDBT, pp. 605–616 (2015)
9. Agrawal, D., Chawla, S., Elmagarmid, A.K., Kaoudi, Z., Ouzzani, M., Papotti, P., Zaki, M. J.: Road to freedom in big data analytics. In: EDBT, pp. 479–484 (2016)
10. PrimeNG (2017), http://primefaces.org/primeng/
11. Angular 2 (2017), http://www.angular2.com/

Who Is Answering to Whom? Finding "Reply-To" Relations in Group Chats with Long Short-Term Memory Networks

Gaoyang Guo[1], Chaokun Wang[1(✉)], Jun Chen[1], and Pengcheng Ge[2]

[1] School of Software, Tsinghua University, Beijing 100084, China
{ggy16,chenjun14}@mails.tsinghua.edu.cn, chaokun@tsinghua.edu.cn
[2] Lenovo Information Technology Ltd, Beijing, China
gepc@lenovo.com

Abstract. Social networks enjoy great popularity among Internet users while generating large volumes of online short-text conversations every day. It leads to a huge number of free-style asynchronous conversations where multiple users are involved and multiple topics are discussed at the same time in the same place, e.g., an instant group chat in WeChat. Here emerges an interesting problem: As a result of a large number of users and topics, the conversation structure may get into a mess, which interferes with our access to the messages we are interested in. For example, when we open the chat records, we do not want to read all the historical messages. We just want to get the messages that are the most relevant with the messages we care about. Therefore, it is an essential task to understand the logical correlations among messages, which benefits the text mining, the natural language processing and the web intelligence techniques.

In this paper, we present the concept of "reply-to" relations to capture most kinds of logical correlations between messages, such as Q&A or complement. Also, we propose a model called LSTM-RT to predict the "reply-to" relations between messages, which is based on the high-quality vector representations of words and LSTM networks. In addition, we give two versions of LSTM-RT based on word level and sentence level, respectively. Experiments conducted on two real-world group chat datasets demonstrate the effectiveness of our proposed models.

Keywords: Group chat · "reply-to" Relations · LSTM network

1 Introduction

In recent years, social networks have been seen with rapid growth of group chats. The emergence of new social media, e.g., microblog and instant group chats, allows people to share ideas and feelings freely in real time. Hence, social instant messaging apps like WeChat (the largest standalone messaging communication service developed by Tencent in China) have catalyzed the formation of social

W. Lee et al. (eds.), *Proceedings of the 7th International Conference on Emerging Databases*, Lecture Notes in Electrical Engineering 461,
https://doi.org/10.1007/978-981-10-6520-0_17

group chats, bringing people a stronger sense of community [13] and connection compared with the traditional text messaging [1]. As a result, a large quantity of group chats are generated every single minute in online social networks. Typically, there are large quantities of group chats in WeChat, which produces in average 2.3 million new group chats every day. Also, the life cycle of these group chats has a regular pattern based on statistics in [9]. As more and more group chats grow, it is greatly interesting to have a preliminary insight into their structure.

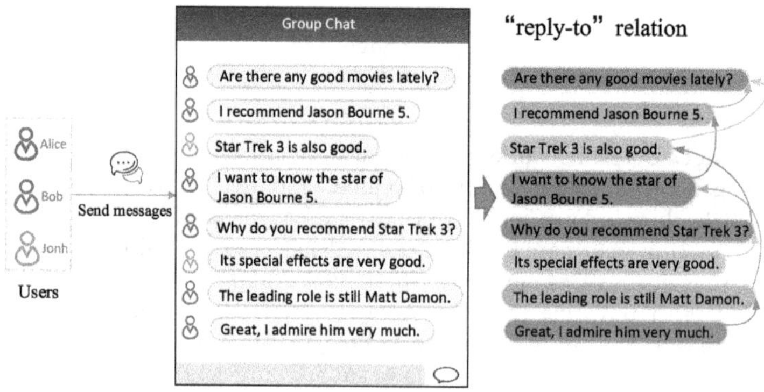

Fig. 1. "reply-to" relations between messages. Alice, Bob, and John are talking about the latest movies in the current group chat. We use arrows to denote "reply-to" relations between messages sent by them.

Generally speaking, there are "reply-to" relations between messages in group chats. That is, someone sends a message in a group chat, then a later message in this group chat replies to the previous one that has just been sent. For example, in a group chat of the instant messaging app WeChat, someone raises a question at some point, and another user sends an answer to the question at the next moment. Then, we can say that the answer has a "reply-to" relation with the question. We recognize the "reply-to" relations between messages in group chats so that users can understand the logic and meaning of group chats better.

Figure 1 denotes an example of "reply-to" relations in a group chat. Alice, Bob, and John constitute a set of users participating in the current group chat. They are free to send messages. Suppose they are talking about the recent movies. Alice presents a question for the latest films. Bob and John recommend her Jason Bourne 5 and Star Trek 3, respectively. Then, Alice continues asking some relevant questions about these two movies. Bob and John go on to answer the questions. Under this scenario, there exist "reply-to" relations among these messages. For example, the second message from Bob has a "reply-to" relation with the first from Alice. The arrows in Fig. 1 denote "reply-to" relations.

The structure of the group chat in Fig. 1 is relatively simple. However, the structure of most group chats is complicated, especially when there are large numbers of users participating in and large numbers of topics discussed [11]. Unlike the conversation between two people in which each message replies to the last message in most cases, a message may reply to any of previous messages in a group chat. We say such conversations are asynchronous.

In this paper, we attempt to predict "reply-to" relations between messages in group chats. The inherent value of this work is the better understanding of the logic of group chats. We bring forward a neural network model based on neural word representations. We summarize our main contributions as follows:

1. We present a novel problem of predicting "reply-to" relations between messages in a group chat involving multiple topics. To the best of our knowledge, this is the first time that the problem is studied.
2. We propose a model called LSTM-RT based on the high-quality vector representations of words and LSTM networks to solve the problem of predicting "reply-to" relations in group chats. We give two versions of LSTM-RT: WL-LSTM-RT and SL-LSTM-RT. The former is based on the word level, while the latter is based on the sentence level.
3. We acquired a small group chat corpora from WeChat and a large comments corpora from Douban Group which can be regarded as a group chat. We use these two group chats corpora to evaluate our LSTM-RT model. The experimental results show that our LSTM-RT model outperforms the baselines in the prediction accuracy.

The rest of this paper is organized as follows. Section 2 introduces the related work. Section 3 gives an overview of the problem we study in this paper. In Sect. 4, we propose LSTM-RT model which has two versions to predict "reply-to" relations. We evaluate the effectiveness and analyze the parameters sensitivity of our model on two group chat corpuses in Sect. 5. We conclude the paper in Sect. 6.

2 Related Work

There have been many studies on RNN or LSTM networks for NLP tasks such as language modeling [6] and machine translations [4]. The research in [8] presents a siamese adaptation of the Long Short-Term Memory (LSTM) network for labeled data comprised of pairs of variable-length sequences. The model consists of two networks LSTMa and LSTMb where each of them processes one sentence in a given pair. The model takes the word2vec embeddings from [7] as input. Kiros et al. [5] propose the skip-thoughts model, which feeds each sentence into an RNN encoder-decoder which attempts to reconstruct the immediately preceding and following sentences. To adapt their model to the sentence similarity task, besides feeding a sentence into the RNN encoder to obtain a sentence vector, they train a separate classifier to predict the similarity between a pair of sentences. Our model is a little similar to the above two models, but we train a softmax_linear classifier to predict "reply-to" labels between messages in group chats. In addition, we can take our model as an encoder fed with sentences.

Meanwhile, there has been some research work studying the message clustering problem in Internet Relay Chat (IRC) conversations [3]. The problem has been stated as *chat disentanglement* in the literature. The research in [2,3] uses a maximum-entropy classifier to decide whether two messages are of the same topic based on timestamps, user-mention relation, cue words and other content features. The work in [12] enriches the TF-IDF feature of a single message by combining it with the TF-IDF features of related messages having similar timestamps or sharing the same usernames in their contents. Then a single-pass clustering algorithm is used to group messages into topics based on the augmented context.

Unlike the problem of chat disentanglement, the ultimate goal in this paper is to learn the logical correlations between messages by predicting the "reply-to" relations between messages. The most related work to ours is the thread prediction problem [14] which predicts how each message in a newsgroup style conversation is related to each other. Therefore, we use its method as our baseline in subsequent experiments. However, our work differs from it in several aspects: (1) We are dealing with messages in group chats where messages are much shorter and more informal than the newsgroup conversations in that work; (2) The work in [14] only redefines the TF-IDF features based on bag-of-words (BOW) which ignores semantic relation between words, while our model is an order-sensitive function of the sequence of words because of inherent characteristics of RNN. Thus, we are tackling a much more challenging problem here and some advanced techniques are used in the proposed method.

3 Problem Overview

We first give two basic definitions about our problem.

Definition 1 (Group Chat Corpus). *A group chat corpus is a list of messages* $\mathscr{M} = [m_1, m_2, \ldots, m_{|\mathscr{M}|}]$ *which are sorted by sending time in a group chat. The message length, i.e. the number of words, of each* $m \in \mathscr{M}$ *is short (e.g. less than 15 words each). The words and phrases in* $\mathscr{M}$ *are usually used in an informal way (e.g. many symbols, abbreviations and Internet words).*

Definition 2 (Group Chat Users). *Group chat users are a set of users* $\mathscr{U} = \{u_1, u_2, \ldots, u_{|\mathscr{U}|}\}$ *who participate in a group chat. Users are free to send messages. For* $\forall m \in \mathscr{M}$ *and* $\forall u \in \mathscr{U}$*, we say* $m \lessdot u$ *if* m *is sent by* u.

$\forall m_i, m_j \in \mathscr{M}, m_i \neq m_j$, we say $m_j \prec m_i$ if m_i has a "reply-to" relation with m_j. When m_i starts a new topic in a group chat, it has no relation with any of preceding messages. In this special case, we specifically define that m_i has a "reply-to" relation with itself to facilitate research on our problem. Thus, we say $m_i \prec m_i$ iff m_i starts a new topic.

In this paper, the problem we are going to solve is to predict the "reply-to" relations between messages in a group chat corpus $\mathscr{M}$ involving multiple topics. More specifically, suppose we acquire an $\mathscr{M}$ which is a time-ordered list of messages sent by U over a past period of time. When noticing an interesting message $m_i \in \mathscr{M}$, we want to find out which message $m_j \in \mathscr{M}$ that $m_j \prec m_i$ where $m_i = m_j$ if m_i starts a new topic.

4 Proposed Method

In this section, we propose our method to solve the problem of predicting "reply-to" relations between messages.

Based on statistics, we observe that the current message replies to another message which is always within the range of last $t+1$ preceding messages including the current message, which can greatly simplify our research. When we predict the "reply-to" relation for a message m in a group chat corpus $\mathscr{M}$, we can narrow the search range from all preceding messages to the last $t+1$ preceding messages including itself. The size of t may be related to the number of users participating in the group chat or the category of topic discussed. In this paper, t is a constant for a certain group chat corpus. Based on this observation, we present a heuristic method to solve the predicting "reply-to" problem.

Given $\forall m_i \in \mathscr{M}$, we intercept m_i and the last t preceding messages from m_i. They constitute $\mathscr{T} = [m_{i-t}, m_{i-t+1}, ..., m_i]$, a time-ordered message list with the size of $t+1$. Then, we just need to predict the "reply-to" label $0 \leq y_i < t+1$ of m_i so that $m_{i-y_i} \prec m_i$. The specific predicting process is based on LSTM-RT model as described below.

4.1 LSTM-RT Model: Word Level

In this paper, we propose the LSTM-RT (Long Short-Term Memory Reply-To) model to predict "reply-to" relations between messages. We have two versions of LSTM-RT model. The first version is outlined in Fig. 2. There is one LSTM network which processes a time-ordered message list $\mathscr{T} = [m_{i-t}, m_{i-t+1}, \ldots, m_i]$ with the size of $t+1$. We first put $t+1$ messages to word segmentation. After word segmentation, $\forall m_j \in \mathscr{T}$, m_j consists of a sequence of words $\mathscr{W} = [w_1^j, m_2^j, ..., m_{l_j}^j]$,

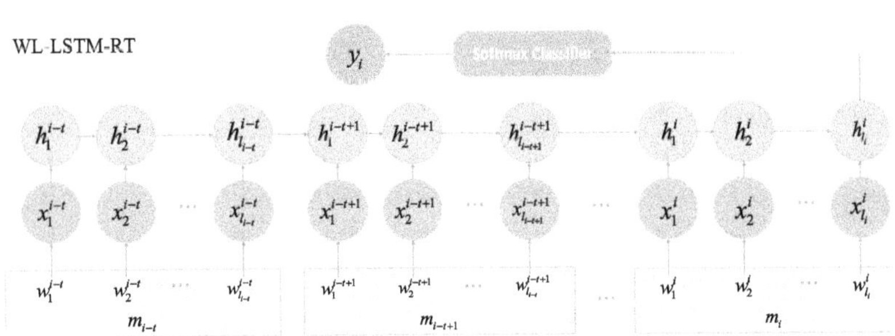

Fig. 2. The first version of LSTM-RT. This technique is based on word level. We name it WL-LSTM-RT. This model directly feeds all word vectors of $t+1$ messages into LSTM to predict the "reply-to" label.

where l_j denotes the length of m_j. Then, we map each word of $\mathscr{W}_j$ into a d_{in}-dimensional vector ($d_{in} = 200$ in this work). After that, $\forall m_j \in \mathscr{T}$, m_j consists of a sequence of word vectors $\mathscr{V}_j = [v_1^j, v_2^j, ..., v_{l_j}^j]$. Finally, we feed all these sequences of word vectors representing $t+1$ messages into LSTM. Considering that the model directly feeds all word vectors as input into LSTM, we say this version of LSTM-RT model is based on word level. We call this version the Word-Level LSTM-RT (WL-LSTM-RT).

Please note that the time steps of this LSTM is not fixed-length as a result of variable length of messages. Thus, the WL-LSTM-RT learns a mapping from the space of variable length sequences of d_{in}-dimensional vectors into $\mathscr{R}^{d_{out}}$ ($d_{out} = 50$ in this work). More concretely, segmented $t+1$ messages (represented as sequences of word vectors) are passed into LSTM. Then, we train a softmax classifier with $t+1$ classes, which maps d_{out}-dimensional vectors of LSTM final state into $\mathscr{R}$. Finally, the output y_i of the softmax classifier represents the "reply-to" label of m_i.

Our loss function is the average negative log probability of the target "reply-to" labels as follows:

$$\text{loss} = -\frac{1}{N}\sum_{i=1}^{N} \ln p_{y_i} \tag{1}$$

where N is the batch size of the training set, and y_i is the target "reply-to" label for m_i. The objective of the whole training process is to minimize the loss.

4.2 LSTM-RT Model: Sentence Level

The above model WL-LSTM-RT, which is based on word level, feeds all sequences of word vectors representing $t+1$ messages into only one LSTM network directly. Another idea is to feed each message of $t+1$ messages into the same LSTM network separately so that we obtain a sentence vector of each message first. After that, we get a list of sentence vectors of $t+1$ messages, and then feed them into a second LSTM network to predict "reply-to" relations. We say this technique is based on sentence level compared to the above model.

The second version of LSTM-RT model is outlined in Fig. 3. There are two LSTM networks totally, i.e. LSTM_1 and LSTM_2. In general, LSTM_1 takes word vectors as input, while LSTM_2 takes sentence vectors as input.

Similarly, there is a time-ordered message list $\mathscr{T} = [m_{i-t}, m_{i-t+1}, \ldots, m_i]$ with the size of $t+1$. $\forall m_j \in \mathscr{T}$, m_j consists of a sequence of words $\mathscr{W}_j = [w_1^j, m_2^j, \ldots, m_{l_j}^j]$, which can be mapped into a sequence of word vectors $\mathscr{V}_j = [v_1^j, v_2^j, ..., v_{l_j}^j]$.

We can understand LSTM_1 as an encoder. We feed $\mathscr{V}_j$ into LSTM_1 for each $m_j \in \mathscr{T}$, whose output is denoted by s_j. Then, LSTM_1 encodes each $m_j \in \mathscr{T}$ into s_j, which is a sentence vector in fact. We get a time-ordered sentence vector list $\mathscr{S} = [s_{i-t}, s_{i-t+1}, \ldots, s_i]$ with the size of $t+1$.

We feed this sentence vector list $\mathscr{S}$ into LSTM_2. Then, we train a softmax classifier with $t+1$ classes, which maps d_{out}-dimensional ($d_{out} = 50$ in this

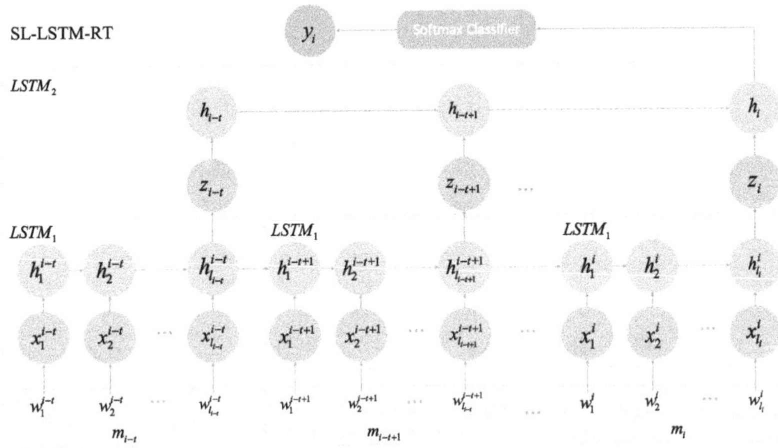

Fig. 3. The second version of LSTM-RT. This technique is based on sentence level. We name it SL-LSTM-RT. This model feeds word vectors of each message into LSTM_1 to get a sentence vector first, and then it feeds sentence vectors of $t + 1$ messages into LSTM_2 to predict the "reply-to" label.

work) vectors of LSTM_2 final state into $\mathscr{R}$. Finally, the output y_i of the softmax classifier represents the "reply-to" label of m_i. Considering that the model feeds the word vector of each message into LSTM_1 to get a sentence vector first, we call this version Sentence-Level LSTM-RT (SL-LSTM-RT).

The loss function of SL-LSTM-RT is the same with WL-LSTM-RT as shown in Eq. 1.

5 Experiments

5.1 Datasets

Due to privacy concerns, there are no public group chat corpuses from WeChat to be accessible. Then, we collect a group chat corpus in WeChat from ourselves as a dataset with the other users' consent. There are about 1500 continuous messages in this group chat corpus, where the average length of these messages is about six. Although the size of this corpus is limited, we believe that it is typical for this study since it comes from the most realistic chat situation and could capture the most authentic "reply-to" relations between messages. We mark the "reply-to" labels manually as the ground-truth.

Meanwhile, we investigate Douban Group, a popular Chinese Web forum. In Douban Groups, users can publish topics for discussion. When someone replies to a comment c, the content of c is automatically quoted by the new comment. This is how we obtain the ground-truth of "reply-to" relations in our experiments. That is also the main reason why we choose it as a dataset. We crawl 51,734 messages in chronological order, which we can regard as a big group chat corpus.

5.2 Results and Discussion

In the experiments, we have three baselines. The first is a text similarity measuring method based on the TF-IDF approach [10]. We denote it by "TS". The second method in [14] which redefines the TF-IDF feature to compute text similarity is denoted by "RTS". In the experiments, we choose the top 80% of the messages as the training set and leave the rest as the test set for each group chat corpus. We use the following metric of "reply-to" relation prediction accuracy as the major measurement, i.e., $\frac{\#\text{correctly predicted “reply-to” relations}}{\#\text{total “reply-to” relations}}$.

Predict "Reply-To" Relations. Figure 4(a) and (b) show the results of accuracy performance on datasets from WeChat and Douban Group, respectively. In corpus from WeChat, we set t to 7 since the corpus is small, while in corpus from Douban Group, we set t to 11. Therefore the former is an eight classification problem, and the latter is a twelve classification problem. According to the results, we can see: (1) Both WL-LSTM-RT and SL-LSTM-RT have nearly the best accuracy performance compared with "TS" and "RTS". They both achieve nearly 56.7% accuracy on the dataset from WeChat, which has around 41 % points better than the baselines. They both achieve nearly 49.3% accuracy on the dataset from Douban Group, which has around 35 % points better than the baselines. (2) In fact, the accuracy of SL-LSTM-RT is a little better than WL-LSTM-RT. SL-LSTM-RT has 0.15% points and 0.35% points better than WL-LSTM-RT on the dataset from WeChat and Douban Group, respectively. Though the difference of accuracy between two models are small, we say that SL-LSTM-RT is a little better than WL-LSTM-RT to a certain extent. We can partly explain it from the aspect of the model. WL-LSTM-RT directly takes all word vectors of $t + 1$ messages as a sequence of inputs. It ignores the separation between messages, which means it does not know which word that each message ends with. This drawback may influence the accuracy performance of WL-LSTM-RT. While SL-LSTM-RT feeds word vectors of each message into LSTM_1 separately and feeds sentence vectors of $t + 1$ messages into LSTM_2. It considers the separation between messages because it represents each message

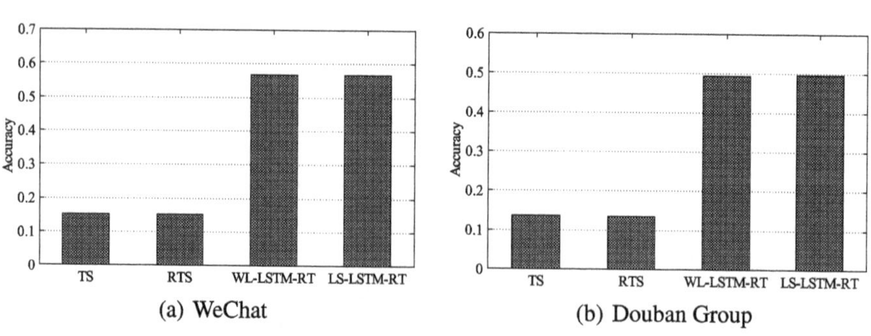

Fig. 4. Comparisons of the accuracy performance.

with a sentence vector first and each input of time step in $LSTM_2$ denotes a message. Based on this analysis, SL-LSTM-RT is more reasonable than WL-LSTM-RT. (3) “TS” and “RTS” have low accuracy performance on both two datasets, which are just a little better than random prediction. We give two reasons for this result. First, text similarity measuring methods are not suitable for the predicting “reply-to” problem we define in this paper, since “TS” and “RTS” are both used to measure text similarity. This point is easy to understand, since the “reply-to” relations in this paper are a little complicated. It cannot be decided simply just by text similarity. It should be determined by meaning of content and logic between messages, since it captures the latent relevance between messages. Second, both “TS” and “RTS” are based on bag-of-words model. They will suffer sparse text feature problem since messages in our datasets are short. They are also insufficient to capture the semantics of messages due to the model is order-insensitive.

Sensitivity of Parameters. We analyze the sensitivity of three parameters in WL-LSTM-RT including the size of each mini-batch B, the size of hidden layer H and the size of training epochs E. We also analyze the sensitivity of two parameters in SL-LSTM-RT including the size of hidden layer H and the size of training epochs E. The experiments on the dataset from Douban Group show that the accuracy performance varies little with the change of parameters, which is similar with the result in the last section on the dataset from Douban Group. So we omit its results in this part. The following experiments are all based on the dataset from WeChat. If not explicitly specified, the default settings of parameters in WL-LSTM-RT are $B = 10$, $H = 50$, and $E = 1$. The default settings of parameters in SL-LSTM-RT are $B = 1$, $H = 50$, and $E = 1$.

Figure 5(a) shows the accuracy performance with different B values in WL-LSTM-RT. The value of B determines the number of training examples used in one step of update. From the results, we can see that the performance rises first and then falls. The optimal setting of B value should be around 30. This is because when B is too small, the training examples used in each step of update cannot represent overall training examples. Therefore, the gradient direction generated by B training examples is not accurate, which leads to relatively low accuracy. When B is too big, the model will fit the training examples excessively, which leads to overfitting. This also results in the decline of accuracy on the test set.

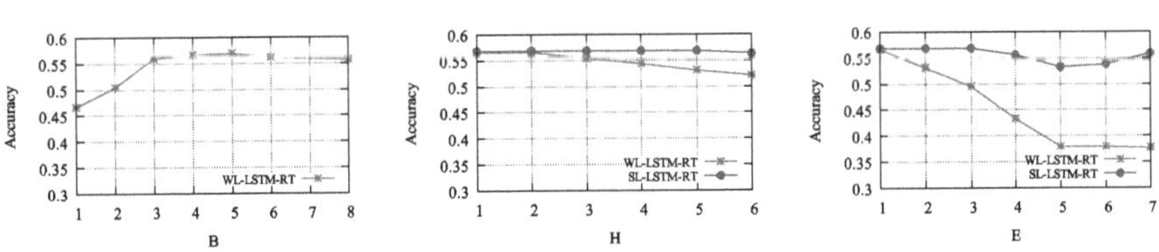

(a) (b) (c)

Fig. 5. Accuracy performance under different settings of parameters.

Figure 5(b) shows the accuracy performance with different H values in WL-LSTM-RT and SL-LSTM-RT, respectively. The value of H determines the number of units in the hidden layer of LSTM. We can see that the accuracy of WL-LSTM-RT decreases with H increasing. The decrease in accuracy may be caused by overfitting. The larger the value of H is, the more complex LSTM is. In other words, LSTM fits the training set more sufficiently. It may lead to overfitting, which results in the decline of accuracy of WL-LSTM-RT. From the results, we can see that the optimal setting of H value in WL-LSTM-RT should be less than 50. We can also see that the accuracy of SL-LSTM-RT almost has no change. It shows that SL-LSTM-RT has a stable generalization ability in different H values.

Figure 5(c) illustrates the performance of WL-LSTM-RT and SL-LSTM-RT by changing E values, i.e., the number of loops on the training set. The figure shows that the accuracy performance of WL-LSTM-RT declines as the value of E increases. This is also due to the overfitting. The larger the value of E is, the more sufficiently LSTM fits the training set. According to the results, we can see that $E = 1$ is a good choice for WL-LSTM-RT. We can also see that the accuracy of WL-LSTM-RT still keeps nearly unchanged. It shows that SL-LSTM-RT is robust under different E values.

Overall, SL-LSTM-RT has more stable ability of generalization than WL-LSTM-RT.

6 Conclusion

In this paper, we present a novel problem of predicting "reply-to" relations between messages in a group chat involving multiple topics. We define three types of "reply-to" relations systematically. We present a heuristic method to solve the predicting "reply-to" problem based on an important observation. Then a model called LSTM-RT based on the high-quality vector representations of words and LSTM networks are proposed. We give two versions of LSTM-RT. One version called WL-LSTM-RT is based on word level. The other version called SL-LSTM-RT is based on sentence level. We evaluate these two versions of LSTM-RT model through experiments on two group chat corpuses from WeChat and Douban Group, respectively. The evaluation shows the effectiveness of our model compared with two baselines. We also analyze sensitivity of parameters of WL-LSTM-RT and SL-LSTM-RT, respectively. The experimental results show that SL-LSTM-RT has more stable ability of generalization than WL-LSTM-RT under different values of parameters. In the future, we will consider recovering the structure of group chats in the form of trees or graphs based on predicted "reply-to" relations between messages to understand group chats further.

Acknowledgments. This work was supported in part by the National Natural Science Foundation of China (No. 61373023) and the China National Arts Fund (No. 20164129).

References

1. Church, K., de Oliveira, R.: What's up with Whatsapp?: comparing mobile instant messaging behaviors with traditional SMS. In: MHCI, pp. 352–361. ACM (2013)
2. Elsner, M., Charniak, E.: You talking to me? a corpus and algorithm for conversation disentanglement. In: ACL, pp. 834–842 (2008)
3. Elsner, M., Charniak, E.: Disentangling chat. Comput. Linguist. **36**(3), 389–409 (2010)
4. Kalchbrenner, N., Blunsom, P.: Recurrent continuous translation models. In: EMNLP, vol. 3, p. 413 (2013)
5. Kiros, R., Zhu, Y., Salakhutdinov, R.R., Zemel, R., Urtasun, R., Torralba, A., Fidler, S.: Skip-thought vectors. In: NIPS, pp. 3294–3302 (2015)
6. Mikolov, T.: Statistical language models based on neural networks. Presentation at Google, Mountain View, 2 April 2012
7. Mikolov, T., Dean, J.: Distributed representations of words and phrases and their compositionality. In: NIPS (2013)
8. Mueller, J., Thyagarajan, A.: Siamese recurrent architectures for learning sentence similarity. In: Thirtieth AAAI Conference on Artificial Intelligence (2016)
9. Qiu, J., Li, Y., Tang, J., Lu, Z., Ye, H., Chen, B., Yang, Q., Hopcroft, J.E.: The lifecycle and cascade of Wechat social messaging groups. In: WWW, pp. 311–320 (2016)
10. Salton, G., Buckley, C.: Term-weighting approaches in automatic text retrieval. Inf. Process. Manage. **24**(5), 513–523 (1988)
11. Wang, C., Xin, X., Shang, J.: When to make a topic popular again? a temporal model for topic re-hotting prediction in online social networks. IEEE Trans. Sig. Inf. Process. Netw. (2017). doi:10.1109/TSIPN.2017.2670498
12. Wang, L., Oard, D.W.: Context-based message expansion for disentanglement of interleaved text conversations. In: Proceedings of the 2009 Annual Conference of the North American Chapter of the Association for Computational Linguistics, pp. 200–208 (2009)
13. Wang, M., Wang, C., Yu, J.X., Zhang, J.: Community detection in social networks: an in-depth benchmarking study with a procedure-oriented framework. Proc. VLDB Endowment **8**(10), 998–1009 (2015)
14. Wang, Y.C., Joshi, M., Cohen, W.W., Rosé, C.P.: Recovering implicit thread structure in newsgroup style conversations. In: Proceedings of the 2nd International Conference on Weblogs and Social Media (2008)

Search & Update Optimization of a B$^+$ Tree in a Hardware Aided Semantic Web Database System

Dennis Heinrich[1](✉), Stefan Werner[1], Christopher Blochwitz[1], Thilo Pionteck[2], and Sven Groppe[1]

[1] Universität zu Lübeck, 23562 Lübeck, Germany
{heinrich,werner,groppe}@ifis.uni-luebeck.de, blochwitz@iti.uni-luebeck.de
[2] Otto von Guericke University, 39106 Magdeburg, Germany
thilo.pionteck@ovgu.de

Abstract. This paper presents a hybrid architecture for accelerating search and update operations on Semantic Web indices. This database system uses a B$^+$-tree index structure distributed in a Field Programmable Gate Array (FPGA) and a CPU-based host system. The index is divided into two parts. The host system stores the values and the keys of the lower levels of the B$^+$-tree while a certain amount of the frequently accessed levels including the tree root is stored in the FPGAs internal and attached memory. Inside the FPGA we accelerate search operations by exploiting the parallel nature of the FPGA. By this, update operations can benefit from the speed up of their necessary searches. Furthermore, we estimate the performance based on the given experiments in a worst case scenario.

Keywords: Hardware acceleration · Semantic web · Index structures

1 Motivation

Today computers aid people to collect and find the information they desire. This is a big challenge because the amount of data collected grows steadily. This growth is not only caused by humans, sensors embedded in everyday things collect data, like in the Internet of Things. However, the most important factor for most humans is the World Wide Web. Only after a few decades the knowledge inside the Web seems to grow limitless. This growth is caused by the development of the Web 2.0 where everyone in the Internet can participate in the Web. This leads to the need of bigger and faster databases to handle the continuous flow of information. Still the aid of computers to find the important data for their users is limited because most data is lacking some sort of context to help the computer understand the connections between different information. At this point the idea of the Semantic Web arouse. Enhancing data with context sensitive annotations

W. Lee et al. (eds.), *Proceedings of the 7th International Conference on Emerging Databases*, Lecture Notes in Electrical Engineering 461,
https://doi.org/10.1007/978-981-10-6520-0_18

made it possible to process data in a way that machines could evaluate if certain information belong together or contradict each other. With more complex data structures the need for memory and processing power grows further. A hybrid index structure for a Semantic Web database running on a CPU-based host system and a Field Programmable Gate Array (FPGA) providing a much higher parallelism than multi-core CPUs is our contribution.

2 Related Work

FPGAs are well-known for their good performance in parallelizable tasks in such a way that they are used in many scientific areas to exploit this behavior. Although there is a wide range of research that considers FPGAs we will focus on researches that try to accelerate databases with the help of FPGAs. First attempts to accelerate databases with the help of specialized hardware were made in the late 70's of the past century. De Witt presented the DIRECT [1] design which is a multiple-instruction multiple-data stream architecture using microprocessors and charge-coupled device (CCD) memories to access relational data in parallel and provide concurrent updates by locks. In these times hardware limits were reached quite fast and even nowadays it is easy to exceed the limits of FPGAs with resource excessive designs or inadequate tasks. Still there is much more research around databases focusing on sorting [2,3], merging [4] and searching [5] the data in a database with the help of an FPGA.

Casper and Olukotun focus in their paper [4] on three primitive database operations: selection, merge join, and sorting. They propose new designs for these operations and combine them in an FPGA cluster to perform a join on two tables by first sorting each table on a separate FPGA and then merging and selecting the results on another FPGA.

Mueller and Teubner cover a wide field of topics on FPGAs in database context. First, they proposed some database designs with FPGA support [6]. Furthermore, they focused on sorting [2], especially on sorting networks [3] which is an efficient way to sort data parallel on an FPGA. With Glacier [5] they proposed a library and compiler that transforms continuous queries into logic circuits.

Werner et al. present a hardware accelerated query engine [7]. At system runtime, the proposed approach dynamically generates an optimized hardware

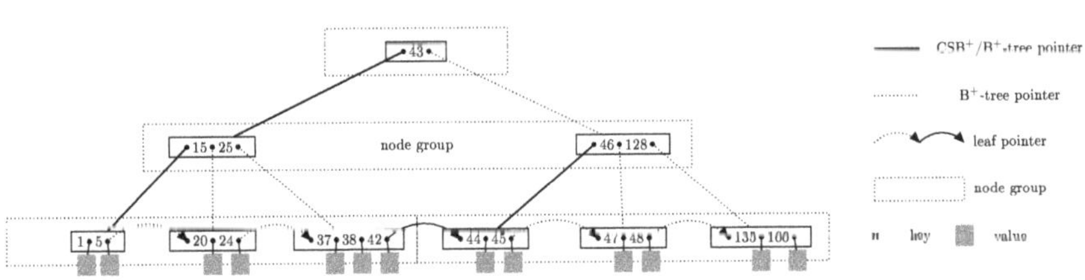

Fig. 1. Example of a B+-tree/CSB+-tree with order 2.

accelerator in terms of an FPGA configuration for each individual query and transparently retrieves the query result to be displayed to the user.

Blochwitz et al. optimized the data structure of a radix tree for a FPGA and exploited its characteristics to accelerate the dictionary generation for Semantic Web databases [8].

We introduce the FPGA-acceleration of B$^+$-trees in [9]. While the main focus in the referenced paper is the parallel search operation in a B$^+$-tree, we focus in this contribution on update operations and how the ratio between searches and updates influences our system.

3 Basics

This section introduces some of the basics needed to understand our hybrid index structure approach. Starting with the general structure of a B$^+$-tree and a CSB$^+$-tree. Thereafter the Semantic Web database system is introduced.

3.1 B$^+$-tree

B$^+$-trees are widely used balanced search trees in database indices and are derived from B-trees [10]. In contrast to its ancestors it saves values only in leaves which are nodes that do not have children. All other nodes, called interior nodes, only contain keys. In Fig. 1 a B$^+$-tree is shown on the left. The numbers inside the nodes are keys while the values are represented by gray boxes beneath the leaves. The lack of values inside the interior nodes makes it possible to hold more keys per node and therefore less nodes are needed to be loaded while searching. The order of a tree k is the minimal number of keys a node requires. The maximum number of keys in a node is $2 \cdot k$ to prevent nodes from growth without limits. At maximum, the node can be split into two halves meeting the minimum number of nodes. Each key in an inner node has a pointer to a child with keys less or equal to its own. This means a node can have $k + 1$ to $2 \cdot k + 1$ pointers because there is an additional pointer to a child that contains all keys that are bigger than the keys in its parent node. The only exception is the root which has no minimum. The leaves in a B$^+$-tree have their own order k'. This makes it possible that interior nodes can have a different number of keys than the leaves. This can be necessary to fit the nodes better into the block size of a hard drive. The search inside the nodes is mostly performed as a binary search on uncompressed keys or a linear search when the keys are compressed by differential encoding and have to be decoded. We have shown in [9] that FPGAs allow to search in parallel in nodes by comparing all keys at once in the node with the search key.

3.2 CSB$^+$-tree

CSB$^+$-trees [11] are modified B$^+$-trees with only one pointer per node to the children. The basic idea behind this is that a B$^+$-tree has per node one more pointer than keys (see Fig. 1, left side). If the child nodes are scattered in different address spaces it makes sense to address each node independent but also limits the capacity of the parent node by holding many pointers. CSB$^+$-trees (see Fig. 1 on the right) avoid this problem by putting child nodes into *node groups* and storing nodes in a node group contiguously. While searching inside the tree the eliminated pointers can easily be restored by adding an offset to the one leftover pointer. We use a CSB$^+$-tree as an index structure inside our FPGA to use less memory for pointers and more space for keys.

3.3 Semantic Web

The Semantic Web is an extension of the World Wide Web. The basic idea described by Tim Berners-Lee [12] is to enable machines to process a web of data in a more context aware way to interpret not only the data but also the connection to other data and resulting implications. To achieve this, the World Wide Web Consortium (W3C) promotes the standards for data formats and exchange protocols in the Web.

Data Representation. The standard way to describe a web resource in the Semantic Web is the use of the Resource Description Framework (RDF) [13]. A statement to a resource in the Web is always a RDF triple consisting of a subject (s), predicate (p) and object (o) resembling a simple sentence that describes the subject. The formal definition of a triple is $(s, p, o) \in (I \cup B) \times I \times (I \cup B \cup L)$ where I stands for Internationalized Resource Identifiers (IRIs), B are blank nodes and L are literals. In our example we use two literals to describe that an IRI has a title and a publication date.

Queries. To retrieve and manipulate data from RDF triples the SPARQL Protocol and RDF Query Language (SPARQL) [14] is used. The basic concept of SPAQL queries is the triple pattern, which consist of constants and variables. Triples matching the constants of the pattern at the right position bind the given variables with the values of its components. The results of triple patterns can be further combined with operators of the relational algebra like joins, union etc.

3.4 LUPOSDATE

The LUPOSDATE project [15,16] is a Semantic Web database system developed by the Universität zu Lübeck. LUPOSDATE uses a full B$^+$-tree structure for indexing. The data inside LUPOSDATE is stored as RDF triples. The triples are stored in all six different combinations (*spo*, *sop*, *pos*, *pso*, *osp*, *ops*) to maximize the number of possible merge joins [17,18]. To avoid large strings while searching, the SPARQL engine of LUPOSDATE uses a dictionary like Hexastore [17] or RDF-3X [18]. In this dictionary the RDF-terms are mapped to integer ids (using 96 bits per triple, 32 bit each component) and vice versa. By using ids the intermediate results and the result consume less memory. On the other hand the mapping from ids to strings is time consuming for large query results. Further details about LUPOSDATE are presented in [15].

4 Architectural Overview

In the following a short introduction of the entire system with its concept and ideas is given. Thereafter the worst case scenario of the update operations are described, focusing on the possible acceleration.

4.1 Basic Concept

The basic concept of this hybrid design is that the software side uses partial solutions for search operations precomputed by the hardware. On a simplified view the basic concept is a B^+-tree in LUPOSDATE whose upper levels are transferred to the FPGA. With these levels, the FPGA can perform a search in the upper levels of the tree. The result can be delivered to LUPOSDATE as an entry point inside the tree rather than starting from the root. Both components of the system benefit from each other. The FPGA can perform parallel searches inside the upper levels. Still the hardware resources on the FPGA are limited and storing all values on the FPGA is not necessary if the host system already stores them. As a consequence, only the parallel search task in the upper levels of the tree is assigned to the FPGA. The remaining operations (insert, delete, etc.) are handled by the LUPOSDATE system. In [9] we describe the data structure of the upper levels of the FPGA where pointers are eliminated for communication and recreated at the FPGA to a CSB^+-tree.

4.2 Acceleration of Update Operations

Besides accelerating search operations, our system is optimized to enhance insert and delete performance as well. If a value is inserted/deleted in a B^+-tree the first step is always to find the correct position inside a leaf. This means, there is a search operation before any update inside the tree. The structure of the tree will be changed only if a node has more than $2k$ or k' keys and needs to be split in case of insertion. In case of the performed operation removes a key the structure is only changed if there are less keys left in the node than the order *k or* k' and two nodes can be merged. Since we only transfer the root and the most upper inner nodes from the tree to the FPGA, updates inside leaves and the lower inner nodes on the host system do not invalidate the data located on the FPGA. With this in mind, it is possible to create a scheduler that decides if a transfer of the tree to the FPGA has a certain chance of accelerating the operations of the tree. At first we should look at the number of update operations that are possible. The simplified Eq. 1 shows the important factors for this:

$$\begin{aligned} possibleNumberOfUpdates = (UpdatesInLeaves \\ \cdot UpdatesInInnerNodesSoftware \cdot EntryPointsFPGAtoHost) \end{aligned} \quad (1)$$

We always start update operations in the leaves ($UpdatesInLeaves$). After filling a leaf to its maximum, the leaf is split and a new key is inserted into the inner nodes of the software part ($UpdatesInInnerNodesSoftware$) in the case of an insert operation. For a delete operation it is the opposite, since if a leaf has less than the minimum number of keys it is merged with another leaf and a key in an inner node is removed. The last factor are the number of entry points from the FPGA to the host system. This can be seen as multiple B^+-subtrees that are located on the host and be addressed by the FPGA. With Eq. 1 in mind we can look at the worst case:

The number of times an update operation can be performed before the data in the FPGA needs to be updated is only one ($possibleNumberOfUpdates = 1$). For the insertion case a key is entered into a maximum filled leaf that causes splittings in the inner nodes which are also filled up to the maximum and lead to the point that the data inside the FPGA expires.

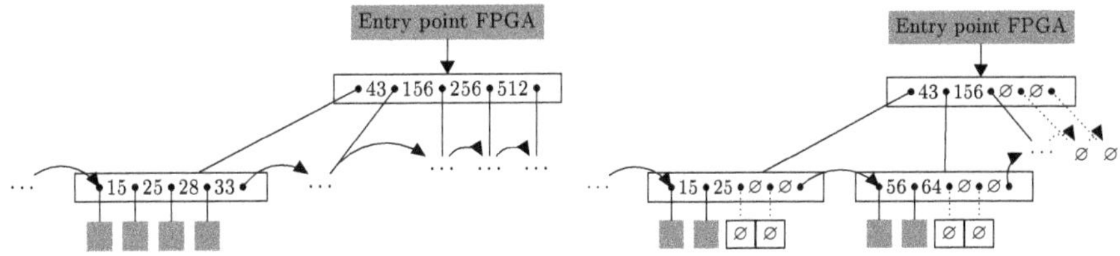

Fig. 2. Examples of B^+-trees with the orders $k' = k = 2$. The keys are in the nodes (numbers) and the values are only in the leaves (gray boxes). This represents a worst case scenario for insert operations (left) and for delete operations (right)

Figure 2 represents this scenario on the left. The values $k = 2$ and $k' = 2$ are the orders of the leaves and the inner nodes in the host system. If another key is entered into the leaf it needs to be split. This leads to a leaf with 3 keys and a new leaf with 2 keys. After this split the inner node needs to be updated with a new key resembling the highest key in the node with 3 keys. The inner node has more keys than the order so another split happens where a new inner node is created. However, the FPGA can only address the old node while the new node can not be reached by any entry point from the FPGA. In this case the structure on the FPGA needs to be rebuilt to address the new inner node.

When the number of update operations ($possibleNumberOfUpdates$) is determined the scheduler still needs an equation to calculate the potential benefit which leads to Eq. 2:

$$t_{gain} = possibleNumberOfUpdates \cdot t_{Op} \cdot r_{Op} - t_{setup}(x) \tag{2}$$

The important part is the t_{gain} which is the time we gain if we are using the FPGA. If it is negative, this means the setup of the system ($t_{setup}(x)$) takes more time than the acceleration of the operations can compensate before the data in the FPGA expires and needs to be updated. If it is positive, it is expected that the use of the FPGA boosts the database system. Since the setup of the tree depends on the number of triples that are transferred to the FPGA $t_{setup}(x)$ is a function to determine the time for a given number of triples x. Furthermore, there are three factors that influence the time we gain using the FPGA. The first is the *possibleNumberOfUpdate* we can perform. For each special case we determined the number we can enter here. The time per operation we gain is t_{Op} which needs to be measured first before the scheduler can use heuristics for this factor. The ratio between different operations (search, update) is also important. A pure search gives the hybrid system the opportunity to save time while the structure of the B$^+$-tree will not change. And while an update most likely will add or remove a key in the tree and therefore has the possibility to change the tree structure, there is also a search for the correct position of the key. This leads to Eq. 3 and the following limitations of the ratio r_{Op}.

$$r_{Op} = \frac{number\ of\ searches}{number\ of\ updates} \tag{3}$$

First the value cannot be below 1 because each update operation implies another search operation to find the correct position. If it is exactly 1 ($r_{Op} = 1$) this means there are only update operations i.e., after each search there is an insertion in a leaf. Between 1 and 2 ($1 < r_{Op} < 2$) there are at least some searches while the most parts are updates. As an example $r_{Op} = 1.5$ means that per two updates there is one search. For a value of two the search and update operations are even, meaning after every second search there is an update in a leaf. Every value of $r_{Op} > 2$ means that there are more search operation than updates. For our worst case scenario it is clear that $r_{Op} = 1$ applies to represent an "insert only" scenario.

5 Evaluation

The evaluation is divided into two parts. First, we recapitulate parts of our experimental setup from [9] in which we focused on search operations. Thereafter, when we focus on the insert operations, which are important for our scheduler and therefore are examined further.

5.1 Experimental Setup and Searches

Our experimental setup consists of a Dell Precision T3610 workstation [19] with 40 GB of RAM and an Intel Xeon E5-1600 v2 processor with 3.0 GHz. The FPGA inside the workstation is a Xilinx Virtex-6 XC6VHX380T [20] using only its BRAM to store the triples. The communication between workstation and FPGA takes place via PCIe 2.0 connection with 8 lanes.

There are three different groups of trees we measured. The L_{B+} group which is an single B$^+$-tree with an order of 500 for all nodes in software. The L_{MM} group which resembles our hybrid approach in software to make a more direct comparison. And the L_{FPGA} group which is our hybrid system. The last two

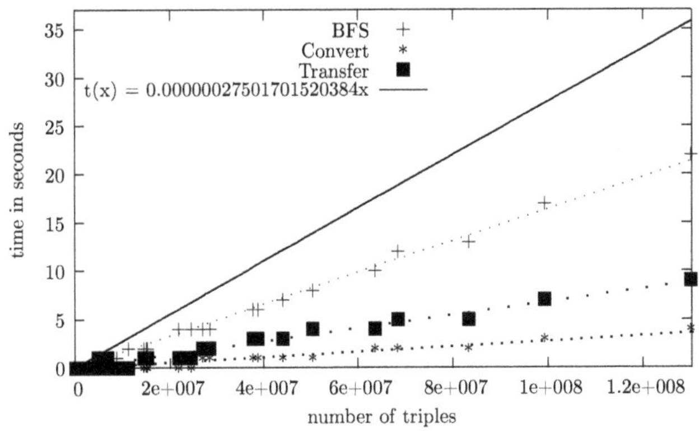

Fig. 3. Computation time for setting up the hybrid system compared to the number of triples inside the tree

Table 1. The needed minimum ratio of search and insert operations for worst case

Order	Ratio
3	30,112.9
4	34,210.9
5	73,021.3
6	125,228.9
7	205,270.3
8	280,292.6
9	486,721.4

groups have variable orders from 3 to 9 for their inner nodes. For further notes check [9].

In our experiments we reached a maximum *speed up* of 2.3 against the typical B$^+$tree representation of our Semantic Web system LUPOSDATE (L_{B^+}) and were still faster compared to the L_{MM} group.

Further we describe in [9] how we manage to create trees with differently filled nodes. The same trees are used in this evaluation and for the description of this processes we refer to this contribution due to space limitations.

5.2 Impact of Updates Inside the B$^+$-tree

After measuring the times in this evaluation scenario we can get back to the Eq. 2 in Sect. 4.2 to look if it is even possible for our scheduler to make a decision to transfer data to the FPGA or not.

First we will look at the common parts of Eq. 2 before we come to the special cases. The t_{Op} is the difference between the execution times of the L_{B^+} and the L_{FPGA} group. Since we looked at different filling states of the nodes we will take the worst times for the worst case scenario. The setup time of the system t_{setup} is the accumulated time of three processing steps shown in Fig. 3.

First a modified Breadth-first search (*BFS*) is performed inside the upper parts of the tree located on the host system. This will also determine the entry point from the FPGA to the host system. After gathering all information the tree is converted (*Convert*) to reduce the data that needs to be transferred to the FPGA. This takes about a quarter of the time the BFS took. The last step is the *transfer* of the data via PCIe to the FPGA which takes the double of the time of the conversion and half the time of the BFS. Adding all three times we get the total amount of time to setup the FPGA with the tree data which we equate with $t_{setup}(x)$ of the Eq. 2. As it can be seen all times grow linear to the amount of triples that are inserted into the tree. Using linear regression for our experiments $t_{setup}(x)$ can be approximated as:

$$t_{setup}(x) = 2.7501701520384 \cdot 10^{-7} \cdot x \text{ seconds} \tag{4}$$

With Eq. 4 we can calculate the time $t_{setup}(x)$ for a given number of triples x. Further we know the orders of the tree. Since we are looking at the lower levels of the tree in the host system the typical values for LUPOSDATE are $k' = 500$ and $k = 500$. The levels inside the FPGA are four in our experiments, hence $L_F = 4$ applies and the total number of levels is five $L_{tree} = 5$. There are two variables left. The ratio between updates and searches r_{Op} and the time we gained for using our hybrid system (t_{gain}). For our scheduler it is important to know this ratio in order to decide, whether or not it should migrate the B$^+$-tree to the FPGA. To get the minimum worst case ratios for the different numbers of triples used, the value will be set to zero $t_{gain} = 0$ so the hybrid system will be at least as fast as the software solution.

After switching the variables from the Eq. 2 we get Eq. 5.

$$\frac{t_{setup}(x)}{possibleNumberOfUpdates \cdot t_{Op}} = r_{Op} \tag{5}$$

The $possibleNumberOfUpdates$ variable will be replaced by the corresponding number of the case.

In the worst case we can only perform one update which leads to Eq. 6.

$$\frac{t_{setup}(x)}{t_{Op}} = r_{Op} \tag{6}$$

Therefore the worst case is independent from the orders k_{FPGA}, k' and k and only determines how many searches need to be performed before a single update can take place.

In Table 1 the minimum ratio r_{Op} is shown with the restriction to be at least as fast as the software system. In the worst case the ratio is increasing by every order. This is plausible since the order k_{FPGA} has no impact on Eq. 6 and $possibleNumberOfUpdates$ is constant. With a linear growing setup time $t_{setup}(x)$ the ratio can only grow.

Table 1 shows the ratios for the worst case in which need much more search operations than inserts into the tree. Still, there are multiple million values stored in the trees so even in the worst case the needed search ratio is only a fracture of the possible searchable values. Moving away from the worst case, the average case quickly shows much smaller ratios, for example with 500 possible updates in an order 9 tree the ratio shrinks down to 70. Details for this case had to be left out due to page limitations.

6 Conclusion

In this paper we have shown a hybrid index for big data systems using an FPGA and a traditional computer, first introduced in [9]. Since the setup of our system takes some time, we need to compensate this loss with our acceleration. We introduced our scheduler which decides whether it is beneficial to transfer the tree structure to the FPGA or not. For this we evaluated the worst case scenario and developed equations for our scheduler. We came to the assumption that our proposed scheduler needs to avoid the worst case scenarios of the updates for a better performance. However, it is most likely possible to accelerate the database system even in worst case scenarios.

Acknowledgment. This work is funded by the German Research Foundation (DFG) project GR 3435\9-1.

References

1. DeWitt, D.J.: Direct - a multiprocessor organization for supporting relational data base management systems. In: Proceedings of the 5th Annual Symposium on Computer Architecture, ISCA 1978, USA, pp. 182–189. ACM (1978)
2. Mueller, R., Teubner, J., Alonso, G.: Data processing on FPGAs. Proc. VLDB Endow. **2**(1), 910–921 (2009)
3. Mueller, R., Teubner, J., Alonso, G.: Sorting networks on FPGAs. VLDB J. **21**(1), 1–23 (2012)
4. Casper, J., Olukotun, K.: Hardware acceleration of database operations. In: Proceedings of the 2014 ACM/SIGDA International Symposium on Field-Programmable Gate Arrays, FPGA 2014, USA, pp. 151–160. ACM (2014)
5. Mueller, R., Teubner, J., Alonso, G.: Streams on wires: a query compiler for FPGAs. Proc. VLDB Endow. **2**(1), 229–240 (2009)
6. Mueller, R., Teubner, J.: FPGAs: a new point in the database design space. In: Proceedings of the 13th International Conference on Extending Database Technology, EDBT 2010, pp. 721–723. ACM, New York (2010)
7. Werner, S., Heinrich, D., Groppe, S., Blochwitz, C., Pionteck, T.: Runtime adaptive hybrid query engine based on FPGAs. OJDB **3**(1), 21–41 (2016)
8. Blochwitz, C., Joseph, J.M., Pionteck, T., Backasch, R., Werner, S., Heinrich, D., Groppe, S.: An optimized Radix-Tree for hardware-accelerated index generation for Semantic web databases. In: ReConFig, Cancun, Mexico, 7–9 December 2015
9. Heinrich, D., Werner, S., Stelzner, M., Blochwitz, C., Pionteck, T., Groppe, S.: Hybrid FPGA approach for a b+ tree in a semantic web database system. In: ReCoSoC 2015, pp. 1–8, June 2015
10. Bayer, R., McCreight, E.: Organization and maintenance of large ordered indices. In: SIGFIDET 1970, USA, pp. 107–141. ACM (1970)
11. Rao, J., Ross, K.A.: Making b+- trees cache conscious in main memory. SIGMOD Rec. **29**(2), 475–486 (2000)
12. Berners-Lee, T., Hendler, J., Lassila, O.: The semantic web. In: Scientific American, pp. 29–37, May 2001
13. World Wide Web Consortium (W3C). Rdf 1.1 concepts and abstract syntax (2014). https://www.w3.org/TR/2014/REC-rdf11-concepts-20140225/
14. World Wide Web Consortium (W3C). SPARQL 1.1 Overview (2013). http://www.w3.org/TR/sparql11-overview/
15. Groppe, S.: Data Management and Query Processing in Semantic Web Databases. Springer, Heidelberg (2011)
16. Groppe, S.: LUPOSDATE (2013). https://github.com/luposdate

17. Weiss, C., Karras, P., Bernstein, A.: Hexastore: sextuple indexing for semantic web data management. Proc. VLDB Endow. **1**(1), 1008–1019 (2008)
18. Neumann, T., Weikum, G.: Rdf-3x: a risc-style engine for RDF. Proc. VLDB Endow. **1**(1), 647–659 (2008)
19. Dell. product website (2015). http://www.dell.com/de/unternehmen/p/precision-t3610-workstation/pd
20. Xilinx. data sheet virtex family (2012). http://www.xilinx.com/support/documentation/data_sheets/ds150.pdf

Multiple Domain-Based Spatial Keyword Query Processing Method Using Collaboration of Multiple IR-Trees

Junhong Ahn, Bumjoon Jo, and Sungwon Jung(✉)

Department of Computer Science and Engineering, Sogang University, 35 Baekbeom-Ro, Mapo-Gu, Seoul 04107, Korea
{junhongahn, bumjoonjo, jungsung}@sogang.ac.kr

Abstract. In this paper, we propose multiple domain-based spatial keyword query, called the MDR (Multiple-Domain based Range) query, and query processing algorithm. The MDR query consists of two sub-queries. The first sub-query, called the primary range query, identifies a group of geo-textual objects that are within a query range. The second sub-query, called the refining range query, identifies the geo-textual objects whose neighboring geo-textual objects contain the desired keywords from the results of the primary range query. The problem is that the keywords domains of each sub-queries are different. Existing methods for spatial keyword queries cannot handle the MDR query efficiently because they are designed to handle the keywords of a single domain. In order to accommodate the keywords of various kinds of domains concurrently, we divide the geo-textual database by the domain of the keyword and construct IR-trees for each domain. Based on this approach, our query processing algorithm traverses multiple IR-trees simultaneously, in order to reduce the number of the refining query. Our performance studies show that our searching strategy significantly reduces the query response time.

Keywords: Multiple-domain based range (MDR) query · Spatial keyword search · IR-tree collaboration algorithm

1 Introduction

As the popularity of micro blog data has grown in recent years, various services on geo-textual data have received significant attention. Geo-textual data refers to data that contains both geographical information and textual contents. For example, tweets, reviews, news comments, and other social media content are tagged with their locality and keywords of context. In recent years, several methods have been developed to retrieve geo-textual data by considering both spatial and textual relevance simultaneously [1–5]. However, most of these methods assume that the keywords of geo-textual data belong to a single domain. In the geo-textual world, the keywords of different domains can also be used for a single query.

In this paper, we propose a multi domain-based spatial keyword query, called the Multiple-Domain based Range (MDR) query. The MDR query consists of two sub-queries. The first sub-query is called the *primary range query*. It identifies a group

W. Lee et al. (eds.), *Proceedings of the 7th International Conference on Emerging Databases*, Lecture Notes in Electrical Engineering 461,
https://doi.org/10.1007/978-981-10-6520-0_19

of geo-textual objects that are within query range and contain keywords of the query. The second sub-query is called the *refining range query*, and it refines the results of the primary range query, and then identifies the geo-textual objects whose neighboring geo-textual objects contain desired keywords. For example, "*Find the hotels that provide 'free Wi-Fi' and 'twin bed room' within* 1 km *from the bus station. There should be a Chinese restaurant selling 'Beijing roast duck' nearby the hotel.*" In this example, the keywords '*Wi-Fi*' and '*twin bed room*' for the primary range query belong to the '*hotel*' domain, and the keyword '*Beijing roast duck*' for the refining range query belongs the 'restaurant' domain.

The problem is that the existing methods for spatial keyword queries cannot concurrently handle the keywords of various kinds of domains. Suppose that we construct a single IR-tree on a geo-textual database that consists of multi-domain keywords. Then the geo-textual objects having the keywords of the query are grouped together with unrelated objects. In terms of objects that contain the keywords to be retrieved, such a spatial partitioning method does not accurately reflect the distribution of these types of objects. Moreover, when the number of keywords increases, the index structure becomes more complex and the size of the inverted files that describe the textual information of the objects also increases.

Therefore, we divide the geo-textual database according to different keyword domains and construct IR-trees [2] within each domain, in order to efficiently retrieve the geo-textual object for that particular domain. Constructing IR-trees for each domain significantly reduces the complexity of the index and the size of textual information. Based on this approach, we propose a collaboration algorithm for traversing multiple IR-trees efficiently. Our proposed algorithm reduces the number of candidate set of primary range query through its cooperation with the index for other domains.

This paper is organized as follow. In Sect. 2, we provide a formal definition for the MDR query. In Sect. 3, we propose the collaboration algorithm of multiple IR-trees for efficient processing of the MDR query. In Sect. 4, we evaluate the performance of our method through experiments. Finally, Sect. 5 concludes this paper.

2 Problem and Background

2.1 Problem Statement

Geo-textual data can be viewed as a spatial object that contains a set of keywords. Let O be a database consisting of a set of geo-textual objects. Each geo-textual object $o \in O$ is defined as a pair $(o.\rho, o.\psi)$, where $o \cdot \rho$ is a 2-dimensional geographical point location and $o \cdot \psi$ is a set of keywords. The MDR query retrieves geo-textual objects that satisfy three conditions- the objects are located within query range, contain a set of query keyword and they must pass an environmental evaluation. Two sub-queries, called *primary range query* and *refining range query*, are used to examine these conditions. The primary range query takes two arguments, a set of keywords and query range in order to retrieve the geo-textual objects that are located within query range and contain query keywords. The refining range query examines the environmental evaluation of objects. Similar to primary range query, it takes a keywords-range pair as an argument, and investigates the results of the primary range query in order to identify the geo-textual objects whose nearby geo-textual objects contain query keywords. We formally define the MDR query and its sub-queries as follow.

Definition 1 Multi-domain based range query (MDR query): An MDR query $q = \langle p, r, k, \{(k'_1, r'_1), (k'_2, r'_2), \ldots, (k'_m, r'_m)\}\rangle$ takes four arguments, where query point p, query range r, as well as set k of keywords for primary range query and a set of refinement conditions (k'_i, r'_i) for refining range queries. The domain of keywords used in primary range query is called **DPR** (domain of the primary range query). Each set k'_i of keywords and range r'_i pair for refining range queries belongs to different keyword domain, called **DRR** (domain of the refining range query). The result of a MDR query, $q(O)$, is then the intersections of the result of primary range query and the results of all the refinement range queries corresponding to the given refinement conditions.

Definition 2 Primary Range Query (PRQ): The primary range query for MDR query is similar to Boolean range query. PRQ $q = \langle p, r, k\rangle$ takes three arguments, where p is a query point, $q.k$ is a set of keyword and $q.r$ is a range of query. The result of PRQ, $q(O)$, is a set of objects such that $\forall o \in q(O)(dist(q.p, o.\rho) < r \wedge q.k \subseteq o.\psi)$.

Definition 3 Refining Range Query (RRQ): A refining range query $q = \langle k', r'\rangle$ accept a set k' of keywords and range r' pair as arguments. The result of a RRQ, $q(O)$, is a subset of O containing objects such that $\forall o \in q(O)((\exists o' \in O \backslash q(O))(dist(o'.\rho, o.\rho) < r) \wedge q.k \subseteq o'.\psi))$.

A running example of an MDR query is shown in Fig. 1. There are three keywords, k_1, k_2, and k_3, each from three keyword domains respectively. Suppose that the MDRQ $q = \langle q, r, k_1, (r_1, k_2)\}\rangle$ is issued. In this example, the number of refining queries is one (i.e., $m = 1$). For the primary range query, objects $\boldsymbol{d}_1$ and $\boldsymbol{d}_9$ are returned because they are located within range r and contain the keyword k_1. Then, the refining range query is processed using this result set. Object $\boldsymbol{d}_1$ satisfies the condition because object $\boldsymbol{d}_7$ of keyword k_2 exists within the range r_1. However, the object $\boldsymbol{d}_9$ does not satisfy the condition of the refining query because there is no object within the query range r_1. Therefore, object $\boldsymbol{d}_1$ is returned as the result of the MDR query.

2.2 Background of the IR-Tree

The IR-tree [2] used in our algorithm provides an index that supports geographic document search and ranking. This method constructs an R-tree using spatial information of the geo-textual data, and associates each node with an inverted file and a document summary that describe the textual information for the objects in their sub-tree. The inverted file indicates which keyword is associated with which child node. It is represented as a list of keywords and lists of objects that contain each keyword. The document summary provides a summary of textual information of the documents in sub-tree of node in order to calculate TF-IDF values of the words.

$$\left\langle A_i, |D_i|, \bigcup\nolimits_{w \in W_i} \{df_{w,D_i}, TF_{w,D_i}\}\right\rangle$$

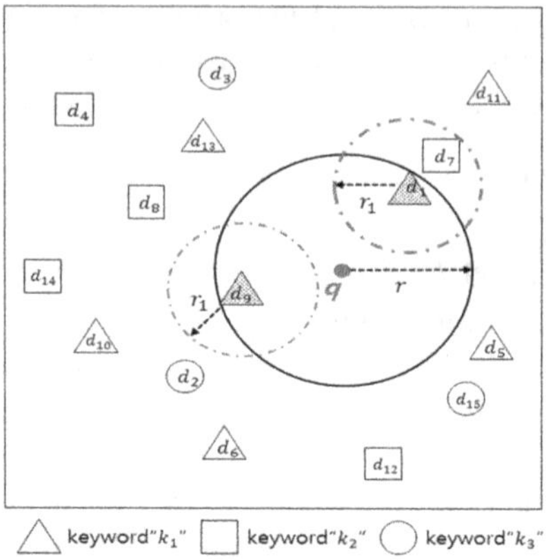

Fig. 1. A running example of an MDR query

$\boldsymbol{A_i}$ is the range of the MBR (minimum bounding rectangle) and $|D_i|$ is the number of the document in set $\boldsymbol{D_i}$. df_{w,D_i} represents the number of objects containing the keyword $\boldsymbol{w}$ in the set $\boldsymbol{D_i}$, and TF_{w,D_i} is the frequency of the word $\boldsymbol{w}$. This information can be used to define the TF-IDF score, which is a technique for processing keywords. The IR tree uses the keyword score and the document summary of the inverted file to derive the results required by the query. The limitation of the IR-tree is that the index becomes further complicated when the number of keyword increases. As the number of keywords increases, the size of the inverted files should also increase, and the textual relevance of the data, which are grouped together, should decrease.

3 Multiple Domain Based Range Query Processing Techniques

3.1 Approximate Refinement for Multiple Domain Based IR-Trees

As implied by the definition of the MDR query, the MDR query consists of a primary range query, and refining range query for refining of the candidate set that derives from the primary range query. Since each query searches only the geo-textual objects contained in a specific domain, maintaining a single index for all the domains of various keywords not only increases the complexity of the index, but also causes the searching for irrelevance objects. Therefore, we divide the spatial-textual database according to the domain of the keywords and construct multiple IR-trees using the geo-textual objects belonging to each domain. This method reduces the overall execution time of the MDR query because it reduces the size of the database required to process the primary and refining range queries.

The number of refining range queries in the MDR query is equal to the size of the result set in the primary range query. If we filter out the objects that will be removed by the refining range query during the primary range query processing, we retrieve objects more efficiently because the number of refining range queries is reduced. In other words, the node of the IR-tree for the DPR does not necessary to be traversed if there are no nearby objects that are used to refine. Based on this approach, we process the *approximate refinement* at each phase of the primary range query processing. Figure 2 is an example of an expanded range for the approximate refinement. Based on the size and position of the current node $\in$ to be traversed in DPR, the expanded range is derived from expanding the refinement range R_i to the MBR. If there is a node of IR-tree in the DRR (domain of refining range query) within the expanded range, the children of current traversed node in the DPR are continuously searched because there is the possibility that an object of the refinement domain exists nearby it. If there is no object within the expanded range in the DRR, the current traversed node is removed from the searching plan without waiting for the results of the primary range query. Figure 3 shows an example of an approximate refinement for DRR.

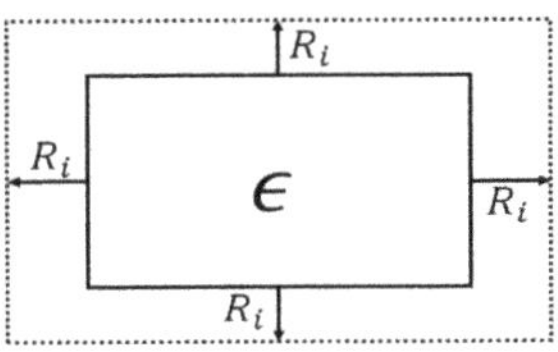

Fig. 2. Expanded range of the node in DPR

3.2 Multiple IR-Tree Collaboration Algorithm for the MDR

An effective strategy for searching the IR-trees for primary and refining range queries is through the cooperation of multiple IR-trees using an approximate refinement. Algorithm 1 shows the collaboration technique-based MDR query processing algorithm.

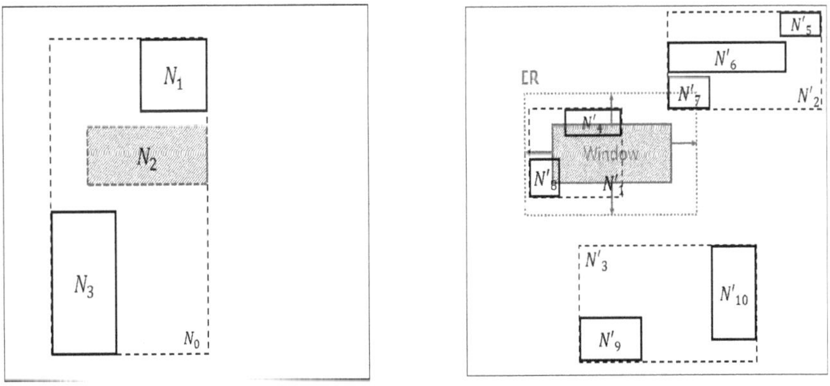

(**a**) MBRs of IR-tree in the DPR

(**b**) Extended Range ER based on N_2 for approximate refinement

Fig. 3. Example of an approximate refinement

To describe brief, the candidate set S (lines 1–2) containing candidate nodes is derived from the primary range query. If the candidate node is not a leaf node, the approximate refinement is processed by expanding the range of the MBR in lines 8–9. In this process, if the window query does not satisfy the approximate refinement, the node is removed from S. Conversely, if it satisfies the approximate refinement, the algorithm inserts its children into the candidate set S (lines 10–11). The Boolean function *ApproximateRefining*() represented in Algorithm 2 traverse the IR-trees for DRR, and returns true if there exist a node or an object covered by expended window range of node.

If the candidate node is a leaf node, the algorithm inserts the object into the result set V if it satisfies the conditions of the refining queries (lines 15–16). The function *RefiningRangeQuery*() shown in Algorithm 3 performs similar with *ApproximateRefining*(). This function returns a subset of set C that includes the objects which passed the refining range query. As a result, the result set of MDR query contains elements within query range r, contains the keyword of k, and satisfies the conditions of the refining range queries.

Algorithm 1: Generate a set of Multiple Domain based Range Query result
Input : Query point $\boldsymbol{q}$, primary query range $\boldsymbol{r}$, primary query keywords $\boldsymbol{k}$, keywords $\boldsymbol{k_i}$ of ith domain, query range $\boldsymbol{r_i}$ for i^{th} domain
Output : A set of geo-textual objects

---

1. $\boldsymbol{V} = \emptyset, \boldsymbol{S} = \emptyset$ // $\boldsymbol{V}$ is a set of results, $\boldsymbol{S}$ is a set of candidate nodes
2. $\boldsymbol{S} = \boldsymbol{S} \cup$ {root node of IR – tree for DPR}
3. while $\boldsymbol{S}$ is not empty do
4.   for each node $\boldsymbol{n}$ in $\boldsymbol{S}$
5. $\boldsymbol{S}$=$\boldsymbol{S}$-{$\boldsymbol{n}$}
6.   if $\boldsymbol{n}$ is not a leaf node then
7.   for each child node $\boldsymbol{c}$ of $\boldsymbol{n}$
8.   if(*dist*(*c*, *q*)< *r*& *ContainsKeywords*(*c*, *k*)) then
9.   for each domain $\boldsymbol{i}$
10.   window range $\boldsymbol{W} = expand(\boldsymbol{c.MBR}, \boldsymbol{r_i})$
11.   if(*ApproximateRefining*(*W*, *k_i*)) then
12. $\boldsymbol{S} = \boldsymbol{S} \cup \{\boldsymbol{c}\}$
13.   else if $\boldsymbol{n}$ is a leaf node then
14.   set of objects $\boldsymbol{C}$= *WithinRange*(objects in $\boldsymbol{n}$, $\boldsymbol{q}$, $\boldsymbol{r}$, $\boldsymbol{k}$)
15.   for each domain $\boldsymbol{i}$
16. $\boldsymbol{C}$=*RefiningRangeQuery*(*C*, r_i, k_i)
17. $\boldsymbol{V} = \boldsymbol{V} \cup \boldsymbol{C}$
18. return $\boldsymbol{V}$

Algorithm 2: Approximate Refining
Input : Query window ***W***, keywords ***k*** of ith domain
Output : Boolean value for approximate refining

```
S = ∅   // S is a set of candidate nodes
S = S ∪ {root node of IR − tree for DRR}
while S is not empty do
    for each node n in S
        S=S-{n}
        if n is not a leaf node then
            for each child node c of n
                if W covers c.MBR & ContainsKeywords(c, k)then
                    return true
                else if W overlap with c.MBR then
                    S = S ∪ {c}
        else if n is a leaf node then
            for each object o of n
                if W covers o & ContainsKeywords(o, k)then
                    return true
return false
```

Algorithm 3: RefiningRangeQuery
Input : Set of object ***B***, query range ***r***, keywords ***k*** of ith domain
Output : A subset of object ***B***

```
R = ∅  //R is a set of result
for each object b in B
    S = {root node of IR − tree for DRR}  // S is a set of candidate nodes
    while S is not empty do
        for each node n in S
            S=S-{n}
            if n is not a leaf node then
                for cach child c of n do
                    if dist( b, c) <= r & ContainsKeywords(c, k)then
                        S = S ∪ {c}
            else if n is a leaf node then
                for each object o of n
                    if dist( b, o) <= r & ContainsKeywords(o, k)then
                        R = R ∪ {b}
return R
```

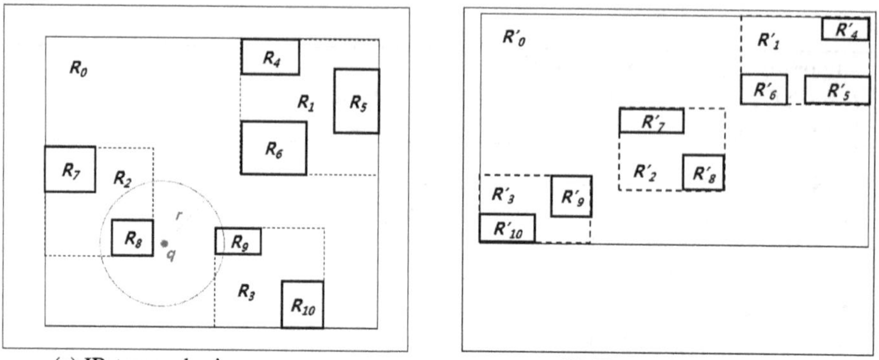

(a) IR-tree and primary range query in the DPR

(b) IR-tree in the DRR

Fig. 4. Example of an MDR query

Figure 4 shows an example of an MDR query. The algorithm starts by inserting the root node of IR-tree constructed in the DPR. There are two children, R_2 and R_3, which overlap to the query range. Therefore, the algorithm expands the range of the MBR for R_2 and R_3, and processes the approximate refinement. Because the expanded range of R_2 overlaps that of R'_3 in the DRR, the children of R_2 are inserted into searching queue. However, node R_3 is removed from the searching queue because there are no nodes overlapping with R_3 in the DRR. In the next iteration, the children of R_2, R_7 and R_8, are examined using primary query range and keywords. R_7 is located outside of the range. Therefore, only node R_8 is evaluated through the refining range query. As a result, the result of refining range query on R_8 is inserted into the result set, and the algorithm is terminated because there are no more nodes to traverse.

4 Performance Evaluation

In this chapter, we evaluate our algorithm through experiments. We compare the response time of two methods. The first method, tagged *IRTreeSeq* in the figures, refers to the processing primary range query and the refining range query sequentially. The second method, tagged *CMIRTRee* in the figures, refers to our collaboration algorithm for multiple IR-trees. Our experiments were performed on a physical machine consisting of a 3.60 GHz quad-core, 16 GB memory, and a 1 TB HDD that operates on a 64bit Linux operating system. The database used in experiment was generated synthetically. It consisted of 10 domains, and contained 10,000 geo-textual objects per domain.

4.1 Effect of the Number of the DRR for MDR Query

In order to evaluate the effect of the number of the DRR for the MDR query, we increased the number of domains from 2 to 10. Figure 5 shows the experimental results of *IRTreeSeq* and *CMIRTRee* in terms of the average search time for 100 randomly

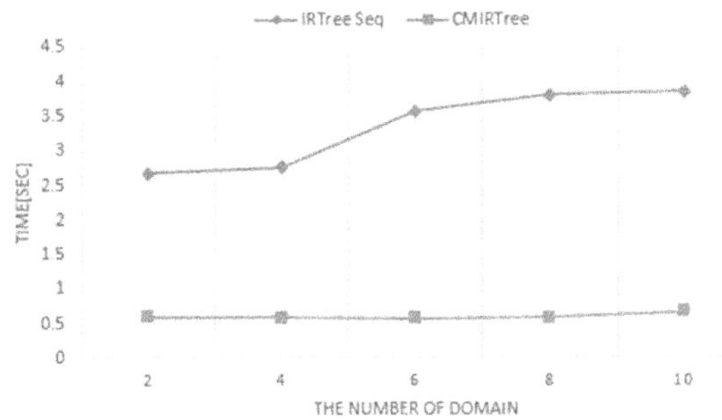

Fig. 5. Effect of the number of the DRR on the query execution time

generated query. As shown in the figure, the collaboration algorithm not only demonstrated better performance, but was also affected by the number of domains of the query. On the other hand, the performance of the sequential querying method degraded as the number of domains increased.

4.2 Effect of the Number of Data

In order to evaluate the effect of the size of the data, we increased the number of data from 10,000 to 40,000 per domain. The number of domains used in the query was fixed at four. Figure 6 shows the experimental results of the effect of the number of data on query execution time. Similar to the above experiment, the collaboration algorithm showed better performance than the sequential querying algorithm in all cases. Regarding the collaborative algorithm, the performance degradation was slightly higher because the number of nodes required to be examined with the expanded window increased as the data size increased. This means that the split threshold of the MBR should be defined appropriately according to the data size.

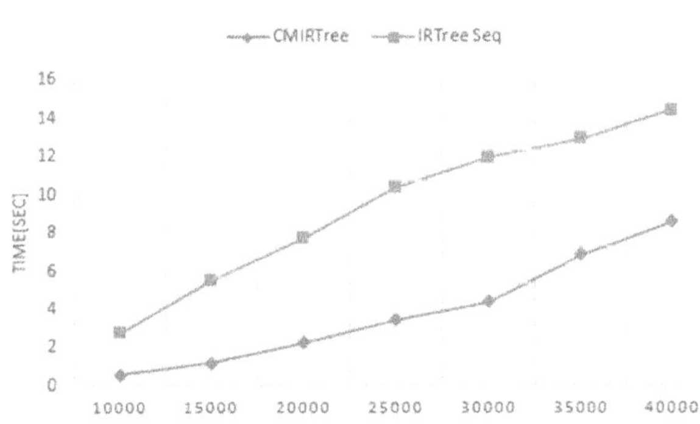

Fig. 6. Effect of the number of data on query execution time

5 Conclusion

In this paper, we propose the MDR query. The MDR query is a more selective query than traditional spatial keyword queries used for geo-textual data. We also propose a query processing algorithm that uses the collaboration of multiple IR trees. The proposed scheme simultaneously traverses multiple IR-trees constructed in each domain. Our collaboration algorithm allows additional pruning during the processing of the primary spatial keyword query. Our proposed algorithms significantly reduce the query execution times, by reducing the number of refining queries. Experimental results demonstrate that our proposed algorithms outperform the sequentially processing algorithm conventionally used.

References

1. Zhou, Y., Xie, X., Wang, C., Gong, Y., Ma, W.: Hybrid index structures for location-based web search. In: Proceedings of the 14th ACM International Conference on Information and Knowledge Management, pp. 155–162 (2005)
2. Li, Z., Lee, K.C., Zheng, B., Lee, W., Lee, D., Wang, X.: IR-tree: an efficient index for geographic document search. IEEE Trans. Knowl. Data Eng. **23**(4), 585–599 (2011)
3. Vaid, S., Jones, C.B., Joho, H., Sanderson, M.: Spatio-textual indexing for geographical search on the web. In: Advances in Spatial and Temporal Databases. Lecture Notes in Computer Science, pp. 218–235 (2005)
4. Salton, G., McGill, M.J.: Introduction to Modern Information Retrieval. McGraw-Hill Book Company, New York (1983)
5. Zhang, D., Chee, Y.M., Mondal, A., Tung, A.K., Kitsuregawa, M.: Keyword search in spatial databases: towards searching by document. In: 2009 IEEE 25th International Conference on Data Engineering, pp. 688–699 (2009)

Exploring a Supervised Learning Based Social Media Business Sentiment Index

Hyeonseo Lee, Harim Seo, Nakyeong Lee, and Min Song(✉)

Library and Information Science, Yonsei University, Seoul, South Korea
{hs_lee, seohr415, nakyolee, min.song}@yonsei.ac.kr

Abstract. We investigate the relationship between public sentiment on specific issues including Sewol ferry disaster and the outbreak of MERS and economic performance in Korea. For this work, we implement two tasks: (1) after training sentiment classifiers with several sources of social media datasets, we consider three kinds of feature sets such as feature vector, sequence vector and combination of dictionary-based feature and sequence vectors. Then, the performance of six classifiers including MaxEnt-L1, C4.5 Decision Tree, SVM-kernel, Ada-boost, Naïve Bayes and Max Ent is assessed. Hence, MaxEnt-L1 shows the highest performance amongst other classifiers. (2) we collect datasets pertinent to a number of critical events that public explicitly express their opinions on such as Sewol ferry disaster, the outbreak of MERS, etc. Then, we calculate sentiment value of the collected data with trained classifiers and compare it with economic indices.

Keywords: Supervised learning · Social media · Sentiment analysis · Maximum entropy L1

1 Introduction

With much attention towards the prediction of economic performance, the studies about economic indicators are important to be taken into consideration. There are many factors that affect economic performance and financial market. Especially social media by communication technologies based on the web, smartphones etc. might be one of the important factors that affect the stock market and economic activities because of rapid growth of social media usage. According to the reports by Pew Research Center [1], nearly two-thirds of American adults (65%) use social networking sites as of 2015. Hence, the use of social network becomes a critical tool for business to interact and share information [2]. [3–5] also find that social media and economic market have a significant relationship. In this paper, we investigate whether there is a linkage between public opinion on certain topics on social media and economic performance including financial market using newly developed economic sentiment indicator (MESI) in Korea. As the first step, we train sentiment classifiers with several sources of social media datasets including movie review, twitter, and news articles. We then consider three different feature sets including feature vector and sequence vector with positive and negative word dictionary, emoticons, lexical properties of sequence of words.

W. Lee et al. (eds.), *Proceedings of the 7th International Conference on Emerging Databases*, Lecture Notes in Electrical Engineering 461,
https://doi.org/10.1007/978-981-10-6520-0_20

We evaluate the performance of 6 classifiers such as MaxEnt-L1, C4.5 Decision Tree, SVM-kernel, Ada-boost, Naïve Bayes and MaxEnt.

Our results indicate that MaxEnt-L1's performance is consistently more than other classifiers. The second step is to collect datasets pertinent to a number of critical events that public explicitly express their opinions on such as Sewol ferry disaster, the outbreak of MERS, etc. The third step is to predict sentiment of the collected datasets with the trained classifiers, and compare it with economic index. The rest of the paper is organized as follows. Methodology section provides how we collect data and experiment method such as choosing feature sets and machine learning algorithms. Result section illustrates performance on sentiment classification, possibility of sentiment prediction for MESI, the Pearson correlation coefficient, analysis of sentiment graph by terms, and overall characteristics in this study.

2 Methodology

2.1 Collected Data

To create a MESI that can be used to identify public economy from social media data, we sought to index consumer responses to the welfare economy based on simple frequency of economic keywords. We collected 52 words of Twitter, 49 words of blog, and 48 words of news for each medium. In detail, in this study, we consider 73,229 news articles, 860,445 Naver blogs and 9,749,893 tweets from tweeter covering from January 1st 2014 to October 31 2015. The reason why we consider the periods between 2014 and 2015, the Sewol ferry disaster happened back in 2014 and the MERS virus was running rampant during 2015. When collecting data, terms of economic situation and event related words were collected as query terms as in the following Table 1. Terms related to Korean social and economic situation are selected by an economic expert.

Table 1. Queries used for data collection

Economic terms
Household loans, recession, prices, livelihoods, universal welfare, liabilities, depression, poverty gap, debt, economy for the ordinary people, consumption contraction, small business owners, Delinquent borrower, unemployment, owner-operator, hardships of life, deterioration, jeonse crisis, traditional markets, fiscal deficits, policy distrust, entrepreneurship, youth employment, business sentiment, chicken, coffee shops, strike, closure rate, boom
Event related words
four-river refurbishment project, Ministry of Strategy and Finance, drought, infection, restructuring, national disaster, bribery, deprivation, corruption, pessimistic suicide, MERS, Sampo generation[a], the Sewol ferry disaster, year-end tax adjustment, divorce, typhoon, flood, Hell-Joseon, Chilpo-generation

[a] Generation who gives up courtship, marriage and children.

2.2 Method

Through sentiment analysis, we made an index into consumers' sentiment to the public welfare, and analyzed how the events such as the Sewol ferry disaster, MERS, and H1N1 influenced the media index. In other words, it is possible to present semantic analysis results based on the text of social data through sentiment analysis. To be specific, it was performed sentiment analysis of economic keywords and made it to the index by adding a frequency related to simple economic results. For economic keywords, we conducted a correlation analysis between economic index and economic index based on simple frequency and sentiment analysis of economic keywords. Then, if the coefficient of correlation with economic variables is more than 6, the keyword is substantially related to the economy. Also, we analyzed the correlation between the sentiment response index of the keywords and the frequency and sentiment analysis results of the keywords related to the events such as Sewol ferry disaster, MERS, and H1N1. As a result, we tried to analyze how the event affected the Media Economic Sentiment Index.

2.2.1 Selection of Feature Set

The following three different feature sets were fed into classifiers to predict sentiments.

(1) Feature vector: there are six different features built by the following sources: Korean positive word dictionary (11461 words), Korean negative word dictionary (13767 words), curse word dictionary (3863 words), positive emoticon dictionary (49), negative emoticon dictionary (52), Korean SentiWordNet (105178 words). Each input text is transformed and represented by these six features. Dictionaries are made by human manually.

(2) Sequence vector: Each input text is split into a set of sentences, and each sentence is transformed into feature sequence. During the transforming process of feature sequence, lexical properties of tokens are identified and used as part of feature sequence.

(3) Combination of dictionary-based features and sequence vector: We combine dictionary-based feature vector and sequence vector into augmentable feature vectors.

2.2.2 Machine Learning Algorithms

In this paper, we concentrate on selecting a right classifier based on a variety of feature set generation methods. Therefore, we apply six kinds of machine learning-based classification algorithms used for evaluation; MaxEnt L1, Decision-tree, SVM-kernel, Ada-boost, Naïve Bayes, and MaxEnt. MaxEnt, that is called Max Entropy, is a probabilistic classifier that is one kind of exponential model by figuring out the probability distribution of maximum entropy [6]. MaxEnt is based on the principle of maximum entropy and can be applied to language detection, topic classification, and sentiment analysis. Focusing on the fact that we contribute to the performance of MaxEnt, we use MaxEnt L1. According to [7], MaxEnt model is a one-to-one relationship between subsets of variables that emerge model's parameterized factors and subsets of variables to use in constraints. While, MaxEnt L1 that adapts generalized expectation criteria for semi-supervised learning has a flexibility to break out the one-to-one relationship due to the fact that the generalized expectation criteria are defined from model that contains generalized expectation terms. In addition, generalized expectation criteria have lots of advantages such as easiness of use and simplicity [8]. The generalized expectation criteria don't need to have an additional process like making inverted index for pre-clustering unlabeled data. In this regard, we add the MaxEnt L1 for evaluate measures. Also, we use C4. 5 decision tree classifier that is used for approximating discrete valued functions by decision tree and is the most popular in inductive inference algorithms [9]. As another classifier we use, ada-boost has many benefits in the way that it is not only fast but also simple to programs [10]. Besides, ada-boost doesn't require prior knowledge about the base learner, so it can be combined with any other method for finding base classifiers. We also use Naïve Bayes that is a probabilistic classifier based on Bayes theorem [11]. Using training data, Naïve Bayes predicts that the category of documents by cue words that occur in classified target document. Lastly, we utilize SVM [12] which is a method that can find hyperplane divided by maximal margin in positive subset and negative subset. For SVM, we set the complexity parameter (-C) to 0.1 that SVM uses to build the hyperplane between any two target classes. Also, we set the gamma parameter to 1.0, which is essential to classification performance. We use the default values for the other parameters, which the API document says are not critical for performance.

As evaluation measures of these classifier, there are three indicators (precision, recall, and F-measure). First of all, precision is a portion of correct categorization among total classification. And then, the recall means that the number of assigned proper classification is divided by the number of assigned total exact category. Furthermore, F-measure indicates that the combination of precision and recall.

3 Results

3.1 Performance Results of Sentiment Classification

The performance results of sentiment classification are suggested in Table 2. The training data that is used for making model consists of Naver news article, Naver movie review comments' score and Twitter. Three people manually assign sentiment scoring three kinds of sources. Three kinds of feature sets have the highest F-1 in MaxEnt-L1 that is 0.7351, 0.7456, and 0.9296. When we use vector feature set, MaxEnt-L1 classifier indicates the highest accuracy (0.6787). In particular, when we combine feature vector and sequence vector, the values of recall, precision as well as F-1 have the highest in MaxEnt-L1. As a result, MaxEnt-L1 has superior performance among other five classifiers.

3.2 Possibility of Sentiment Prediction for MESI

The Economic Sentiment Indicator (ESI) is a survey-based indicator that synthesis of business survey indices(BSI) and consumer survey indices(CSI). Business and consumer surveys view economic agents' opinions about past, current and future economic

Table 2. The performance results of sentiment classification

Feature set - Vector				
	Accuracy	Recall	Precision	F-1
MaxEnt L1	0.6787 ± 0.0051	0.5000	0.7079	0.7351
Decision-tree	0.5096 ± 0.0036	0.5129	0.5968	0.3765
SVM-kernel	0.4778 ± 0.0099	0.5000	0.4815	0.6222
Ada-boost	0.6695 ± 0.0049	0.2589	0.3234	0.2273
Naïve Bayes	0.6763 ± 0.0052	0.4997	0.4503	0.4071
MaxEnt	0.5129 ± 0.0027	0.5109	0.5163	0.4681
Feature set - Sequence				
	Accuracy	Recall	Precision	F-1
MaxEnt L1	0.8929 ± 0.0168	0.6626	0.7463	0.7456
Decision-tree	0.6834 ± 0.0018	0.6650	0.6991	0.6699
SVM-kernel	0.8942 ± 0.0226	0.6391	0.7189	0.7024
Ada-boost	0.9153 ± 0.0143	0.6436	0.7525	0.6736
Naïve Bayes	0.3789 ± 0.0221	0.5000	0.4479	0.4882
MaxEnt	0.9091 ± 0.0145	0.6460	0.7434	0.6471
Feature set - Combined				
	Accuracy	Recall	Precision	F-1
MaxEnt L1	0.9353 ± 0.0076	0.9305	0.9299	0.9296
Decision-tree	0.6834 ± 0.0018	0.6650	0.6991	0.6699
SVM-kernel	0.8590 ± 0.0188	0.7471	0.8123	0.8002
Ada-boost	0.8942 ± 0.0226	0.6391	0.7189	0.7024
Naïve Bayes	0.8751 ± 0.0122	0.6195	0.7142	0.6245
MaxEnt	0.9556 ± 0.0071	0.5000	0.9033	0.8698

activity. The prime benefits of business and consumer survey data are timeliness and high frequency [13]. It goals to get insights into the opinion of economic agents, both from the business and consumer sides of the economy [14]. The following graph shows the ESI trend during the period of data collection. During the period, the ESI graph shows a large change width. The first plunge is from May to July 2014, and it is believed to have fallen in relation to the 'the Sewol ferry disaster' in April 2015. The second plunge is June 2015, which appears to have been influenced since the end of May 2015, when MERS was first identified (Fig. 1).

3.3 Differences in Sentiment Index by Economic Terms

First, the Pearson correlation coefficient was calculated to see the difference of sentiment value by economic terms. Only the correlation coefficients at the 0.01 or 0.05 level are shown in the following Table 3. Notably, the shaded keyword 'entrepreneurship' showed both positive and negative correlations depending on the relative economic terms. This means that the business itself is a positive signal for the economy as a whole and a negative signal at the same time. For example, the relationship between 'deterioration – entrepreneurship' is positive for entrepreneurship and the relationship between

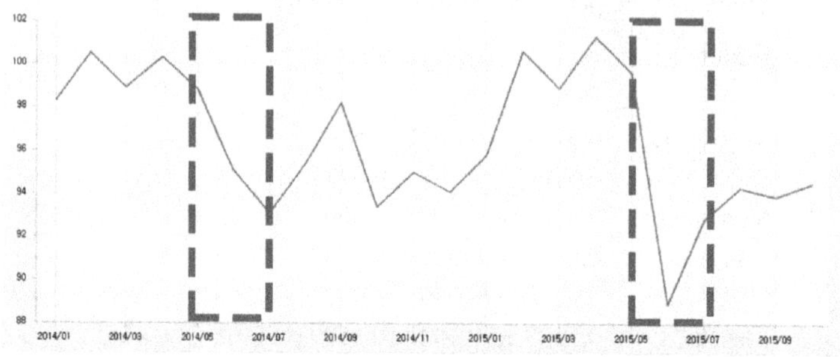

Fig. 1. ESI graph

Table 3. Correlation coefficient by economic terms

Economic term 1	Economic term 2	Correlation coefficient
recession	coffee shops	0.220
universal welfare	liabilities	0.205
liabilities	debt	0.231
liabilities	unemployment	0.237
depression	small business owners	0.268
depression	entrepreneurship	0.218
debt	boom	–0.204
small business owners	entrepreneurship	0.481
delinquent borrower	unemployment	0.242
hardships of life	strike	0.304
deterioration	entrepreneurship	–0.380
owner-operator	traditional markets	0.241
owner-operator	coffee shops	0.272
owner-operator	boom	–0.333
entrepreneurship	chicken	0.230
entrepreneurship	coffee shops	0.421
business sentiment	coffee shops	–0.256

'depression – entrepreneurship' is negative for entrepreneurship. Positive sentiments were entrepreneurship mentors, seminars and policies for business start-ups. On the other hand, the content of seeing entrepreneurship as negative sentiment was that it was difficult to maintain entrepreneurship and that all owner-operator start to start entrepreneurship indiscriminately. In other words, even the same word or issue has different sentiment depending on the context and situation.

3.4 Differences in Sentiment Index by Event Words

As with economic-related terms, the Pearson correlation coefficient was obtained by event-related words sentiment index is as Table 4. Only the significance of the

Table 4. Correlation coefficient by event related words

Event related word 1	Event related word 2	Correlation coefficient
four-river refurbishment project	corruption	0.291
drought	national disaster	0.297
deprivation	pessimistic suicide	0.400
sampo generation	year-end tax adjustment	0.500

correlation coefficient at the 0.01 or 0.05 level is shown in the Table 4. However, the correlation coefficient of event - related words is much smaller than that of economic - related terms - correlation coefficient results. The reason is that the event related words are fewer than the economic related words. In addition, because the words are related to the event, the meanings of the words themselves are different, and the degree of correlation is low. Among the event-related word pairs, 'Sampo generation- year-end tax adjustment' shows a moderate correlation. The Sampo generation is a new term for 2011, which means generation gives up courtship, marriage, and childbirth. The fact that the Sampo generation with this meaning has a moderate positive correlation with the keyword of the year-end tax adjustment shows that those belonging to the Sampo generation who has to live economically well are interested in the year-end tax adjustment.

3.5 Analysis of Sentiment Graph by Keyword

We analyze the sentiment index figures by keyword as shown in Figs. 2 and 3. The horizontal axis and vertical axis indicate time and the sentiment value, respectively. At this time, one of the event keywords 'the Sewol ferry disaster' and 'national disaster' were similar. This is because the Sewol ferry disaster was considered to be one of the national disasters, and it appears that there was a negative value after April 16, 2014 and after the first anniversary of Sewol ferry disaster. These parts are prominently seen in the dotted area. In a similar vein, there is a sentiment graph between the keywords 'MERS' and 'infections'. In the case of MERS, the first MERS confirmed in Korea at the end of May 2015 shows a large fluctuation in the graph as shown in Fig. 3. Also, although not shown in the graph, the emotional graphs of the 'livelihoods', which is one of the economic terms, showed a negative value in June 2015 mainly due to the incapacity of the MERS coping and the welfare breakdown.

3.6 Features that Appear When Integrated as a Whole

Figure 4 shows the overall sentiment graph extracted from the media on a date basis in thick line and the economic sentiment index graph in thin line. As a result of comparing these two graphs to date, the media sentiment and the economic sentiment index show a similar decline at the same time. To put it more precisely, the red economic sentiment index is also showing a declining trend over a period of one month after the decline of the green media sentiment index.

In the first full line, the media sentiment index showed a decline at the end of April 2014, and the economic sentiment index showed a decline in early June of the same

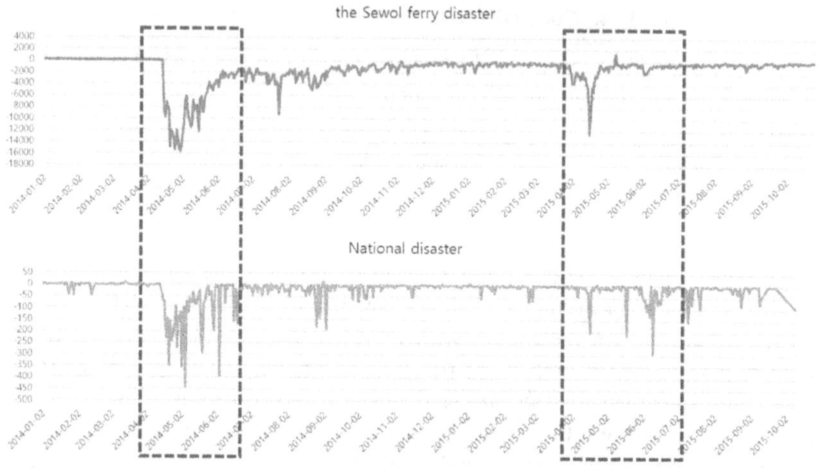

Fig. 2. "The Sewol ferry disaster – National disaster" Keywords sentiment graph

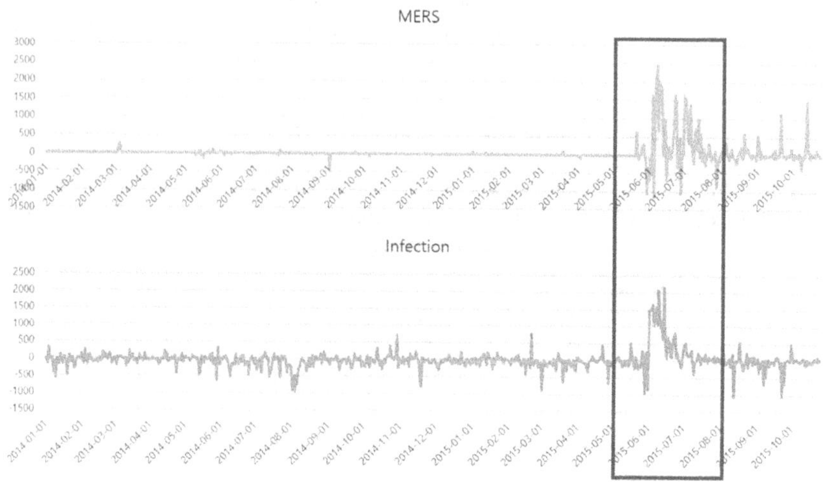

Fig. 3. "MERS – Infection" Keywords sentiment graph

year. In the second dashed section, the media economic sentiment index fell in early November 2014, and the economic media sentiment index plunged in mid-December. Finally, in the double line section, the media sentiment index decreased in early May, and the economic sentiment index fell in mid - June. Again, the media sentiment and the economic sentiment index show a similar decline at the same time, and the economic sentiment index is also showing a declining trend for a month after the fall in the media sentiment index.

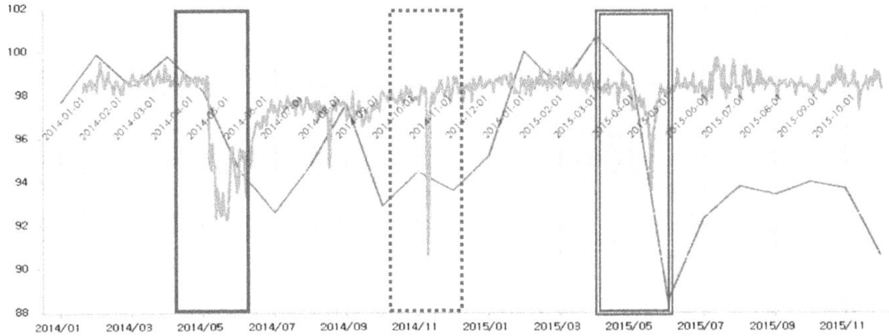

Fig. 4. Comparison of media economic sentiment index and economic sentiment index

4 Conclusion

The rapid growth of usage for social media implies that many individuals and business engage in communications through social media network. [2–4] show the impact of social media on economic performance. In order to investigate whether there is a relationship between his or her personal opinion through social media and economic performance, we employ a newly developed Economic sentiment indicator (MESI) in Korea by following several steps. First, sentiment classifiers based on social media datasets from networks are trained. Then we consider three different feature sets such as feature vector and sequence vector, emotions and lexical properties of sequence of words. We evaluate the performance of 6 classifiers such as MaxEnt-L1, C4.5 Decision Tree, SVM-kernel, Ada-boost, Naïve Bayes and Max Ent. The results show that MaxEnt-L1 outperforms other classifiers that we consider. At the second step, we collect datasets that are related to a number of critical events such as the Sewol ferry disaster, MERS, etc. The final step is to predict sentiment of the collected datasets with the trained classifiers and compare it with economic index.

Acknowledgement. This work was supported by the Ministry of Education of the Republic of Korea and the National Research Foundation of Korea (NRF-2015S1A3A2046711).

References

1. Perrin, B.: Social media usage: 2005–2015. Pew Research Center Surveys (2015)
2. Brown, E.D.: Will Twitter Make You a Better Investor? A Look at Sentiment, User Reputation and Their Effect on the Stock Market (2012)
3. Bollen, J., Mao, H., Zeng, X.: Twitter mood predicts the stock market. J. Comput. Sci. **2**(1), 1–8 (2011)
4. Sitaram, A., Huberman, B.A.: Predicting the future with social media. In: Web Intelligence and Intelligent Agent Technology (WI-IAT), 2010 IEEE/WIC/ACM International Conference on, vol. 1, pp. 492–499. IEEE (2010)
5. Dell'Anno, R., Rayna, T., Solomon, O.H.: Impact of social media on economic growth-evidence from social media. Appl. Econ. Lett. **23**(9), 633–636 (2015)

6. Phillips, S.J., Anderson, R.P., Schapire, R.E.: Maximum entropy modeling of species geographic distributions. Ecol. Model. **190**(3), 231–259 (2006)
7. Sun, C.J., Yao, L., Lin, L., Sha, X.J., Wang, X.L.: Semi-supervised biomedical relation classification using generalized expectation criteria. In: International Conference on Machine Learning and Cybernetics (ICMLC) 2011, vol. 4, pp. 1949–1952. IEEE, July 2011
8. Mann, G.S., McCallum, A.: Generalized expectation criteria for semi-supervised learning with weakly labeled data. J. Mach. Learn. Res. **11**, 955–984 (2010)
9. Polat, K., Güneş, S.: A novel hybrid intelligent method based on C4. 5 decision tree classifier and one-against-all approach for multi-class classification problems. Expert Syst. Appl. **36**(2), 1587–1592 (2009)
10. Schapire, R.E.: The boosting approach to machine learning: an overview. In: Nonlinear Estimation and Classification, vol. 171, pp. 149–171. Springer, New York (2003)
11. Lewis, D.D.: Naive (Bayes) at forty: the independence assumption in information retrieval. In: European Conference on Machine Learning, vol. 1398, pp. 4–15. Springer, Heidelberg, April 1998
12. Vapnik, V.: The Nature of Statistical Learning Theory. Springer, New York (2013)
13. Moon, H.: Construction of an economic sentiment indicator for the korean economy. Korean J. Appl. Stat. **24**(5), 745–758 (2011)
14. Gelper, S., Croux, C.: On the construction of the european economic sentiment indicator*. Oxford Bull. Econ. Stat. **72**(1), 47–62 (2010)

Data and Visual Analytics for Emerging Databases

Carson K. Leung(✉)

University of Manitoba, Winnipeg, MB, Canada
kleung@cs.umanitoba.ca

Abstract. With advances in technology, high volumes of valuable data of different veracity can be generated at a high velocity in wide varieties of data sources in various real-life applications. Examples of these big data include social media data. As a popular data mining tasks, frequent pattern mining discovers implicit, previously unknown and potentially useful knowledge in the form of sets of frequently co-occurring items or events. Many existing data mining algorithms return to users with long textual lists of frequent patterns, which may not be easily comprehensible. Given a picture is worth a thousand words, having a visual means for humans to interact with computers would be beneficial. In this paper, we present a framework for data and visual analytics for emerging databases. In particular, our data and visual analytic framework focuses on mining and analyzing social media data, as well as visualizing the mined 'following' patterns that reveal those groups of frequently followed social entities in a social network.

Keywords: Data analytics · Data mining · Emerging databases · 'following' patterns · Frequent patterns · Knowledge discovery in databases · Social media data · Social network · Visualization · Visual analytics

1 Introduction and Related Works

With advances in technology, high volumes of valuable data of different veracity (e.g., precise data, uncertain or probabilistic data [16,21]) can be generated into emerging databases at a high velocity from wide varieties of data sources. One such emerging database is the social network, which contains rich sets of social media data. Embedded in these social media data are implicit, previously unknown and potentially useful information or knowledge. Hence, *data science*—which applies data mining, database, machine learning, visualization, mathematical, and/or statistical techniques to big data for *data analytics* and/or *visual analytics*—is in demand. Common data mining techniques for data analytics include (i) clustering [10,14], (ii) classification and prediction [3,5,15,32], (iii) anomaly detection [33], and (iv) frequent pattern mining [8,25,28]. Our first *key contribution of this paper* is our data analysis of frequently occurring patterns.

W. Lee et al. (eds.), *Proceedings of the 7th International Conference on Emerging Databases*, Lecture Notes in Electrical Engineering 461,
https://doi.org/10.1007/978-981-10-6520-0_21

As many existing data analytic systems return to users with long textual lists of the data mining results, which may not be easily comprehensible. Given a picture is worth a thousand words, having a visual means for humans to interact with computers would be beneficial. This leads to *visual analytics* [13,17,19,40], which helps users to get a better understanding of the mining results and/or to conduct exploratory analysis of the data. Our second *key contribution of this paper* is our visual analysis of frequently occurring patterns.

As mentioned above, an emerging database is a social network, which contains rich sets of social media data. In general, a social network is made of social entities (e.g., individuals, corporations, collective social units, or organizations) that are linked by some specific types of interdependency (e.g., kinship, friendship, common interest, beliefs, or financial exchange). A social entity is connected to another entity as his next-of-kin, friend, collaborator, co-author, classmate, co-worker, team member, and/or business partner. Nowadays, many popular social networking websites or services (e.g., Facebook, Google+, LinkedIn, Twitter, Weibo [27]) are in use. For instance, a Facebook user can create a personal profile, join common-interest user groups, add other users as friends, categorize theses friends into different customized lists (e.g., classmates, co-workers), and exchange messages. In addition to mutual friendship, Facebook also provides the user with the functionality of 'subscribe', which was relabelled as 'follow' so that a user can subscribe or follow public postings of some other Facebook users without the need of adding them as friends. So, if many friends of a Facebook user u_f followed some individual users or groups of users, then user u_f might also be interested in following the same individual users or groups of users. Similarly, a Twitter user can read the tweets of other users by 'following' them. As such, if many friends of a Twitter user u_t follow some individual users or groups of users, then user u_t might also be interested in following the same individual users or groups of users. Relationships between social entities are mostly defined by following (or subscribing) each other. Each user can be following multiple users and be followed the same or different followers. Note that the follow/subscribe relationship between follower and followee is not the same as the mutual friendship because a user u_{t1} can follow another user u_{t2} who may not even know the user u_{t1}. Our third *key contribution of this paper* is our data and visual analysis of these 'following' patterns.

The remainder of this paper is organized as follows. Next section gives an overview of our framework called SodaVis for social media data and visual analytics. Sections 3 and 4 describe the two components—namely, data analytics and visual analyitcs, respectively—of our framework for the analytics of social media data. Finally, conclusions are presented in Sect. 5.

2 Analytics of Social Media Data

In this section, we start giving an overview of our data and visual analytic framework for emerging databases. Such a framework consists of the following two key components:

1. *Data analytic* component—which focuses on mining and analyzing data—finds implicit, previously unknown and potentially useful information and knowledge (e.g., in the form of frequent patterns) from emerging database such as a social network containing social media data.
2. *Visual analytic* component—which focuses on visualizing and analyzing the data and/or the mining results—represents and displays the mined frequent patterns for easily comprehensible information and discovered knowledge.

Consider the following two tasks and their similarity. In data mining, given a transaction database consists n transactions (each of which contains a set of items chosen from a domain of m merchandise items or events), the task of *mining frequent patterns* aims to find sets of frequently co-occurring items or events. In social computing, given a social media database consists n followers (each of which follows a set of followees chosen from a social network of m users), the task of *mining 'following' patterns* aims to find groups of popular followees (i.e., frequently followed users). Due to the similarity between these two tasks, the task of mining 'following' patterns can be reduced into the task of mining frequent patterns. Consequently, one could adapt frequent pattern mining techniques in mining 'following' patterns.

For an illustrative purpose, let us consider an interesting portion of a social network. Such an interesting portion involves the following $n = 6$ followers and $m = 5$ followees:

1. Alice (user#51) follows Bob (user#132) and Eva (user#614);
2. Bob follows Alice, Carol (user#143) and Eva;
3. Carol follows Alice and Eva;
4. Don (user#608) follows Bob, Carol and Eva;
5. Eva follows Alice, Bob, Carol and Don;
6. Fred (user#620) follows Alice, Bob, Carol and Eva.

Based on the above following relationships, 13 popular groups of frequently followees (who are followed by at least two followers) can be found:

- (1) followee#51, (2) followee#132 and (3) followee#143. These three followees are individually followed by 4 followers.
  (4) followee#614, who is followed by 5 followers.
- (5) followees#51 and 132, as well as (6) followees#132 and 614. Each of these two popular social pairs is followed by 2 followers.
- (7) followees#51 and 143, (8) followees#51 and 614, (9) followees#132 and 143, as well as (10) followees#143 and 614. Each of these four popular social pairs is followed by 3 followers.
- (11) followees#51, 132 and 143, (12) followees#51, 143 and 614, as well as (13) followees#132, 143 and 614. Each of these three popular social triplets is followed by 2 followers.

3 Data Analytics of Social Media Data

Over the past two decades, many *frequent pattern mining* algorithms have been developed for *data analytics*. For instance, many of the early frequent pattern mining algorithms were *Apriori-based algorithms* [2,36], which depend on a generate-and-test paradigm to mine frequent patterns from a transaction database in a level-wise bottom-up fashion by first generating candidates and then checking their actual frequency (i.e., occurrences) against the database. To improve algorithmic efficiency of the original Apriori algorithm [2], several of its variants use techniques like pass bundling [2] (which bundles the generating and checking of candidates of several cardinalities or levels) and partitioning [36] (which divides the original databases into several partitions so that each partition fits into memory) to reduce the number of database scans.

As alternatives to improve algorithmic efficiency of these Apriori-based algorithms (i.e., the original Apriori algorithm and its variants), other serial frequent pattern mining algorithms have also been developed. These include *FP-growth algorithm* [9], which uses an extended prefix-tree structure called Frequent Pattern tree (FP-tree) to capture the content of the transaction database. FP-growth uses a divide-and-conquer paradigm to mine frequent patterns from a transaction database in a bottom-up fashion by recursively extracting relevant paths from the FP-tree to form projected databases (i.e., collection of transactions containing some items), from which subtrees (i.e., smaller FP-trees) capturing the content of relevant transactions are built. To avoid building and keeping multiple FP-trees at the same time during the mining process, algorithms like TD-FP-Growth [39], H-mine [35] and B-mine [12] have been developed. Unlike FP-growth (which mines frequent patterns by traversing the global FP-tree and subtrees in a bottom-up fashion), the *TD-FP-Growth* algorithm [39] mines frequent patterns from a transaction database by traversing only the global FP-tree and in a top-down fashion. During the mining process, instead of recursively building sub-trees, TD-FP-Growth keeps updating the global FP-tree by adjusting tree pointers. Along this direction, the *H-mine* algorithm [35] uses a hyperlinked-array structure called H-struct to capture the content of the transaction database. During the mining process, the H-mine algorithm mines frequent patterns from a transaction database by recursively updating hyperlinks in the H-struct. The *B-mine* algorithm [12] uses a tabular structure called B-table to capture the content of the transaction database. During the mining process, the B-mine algorithm mines frequent patterns from a transaction database by recursively extracting only some relevant portions (i.e., relevant rows and columns) of the B-table.

Besides the aforementioned transaction-centric algorithms which mine the database "horizontally", there are also item-centric algorithms (e.g., Eclat [43], dEclat [41], VIPER [37]) that mine the datasets "vertically" (using a level-wise bottom-up paradigm). To elaborate, *Eclat* [43] treats the database as a collection of item lists. Each list for an itemset keeps a set of IDs of transactions (i.e., tidset) containing all items in the itemset. Frequent patterns can be mined from a transaction database by taking the intersection of lists of frequent itemsets to form itemsets of higher cardinalities. Eclat works well when the database is sparse. In contrast, *dEclat* [41] dEclat uses diffset, which is the set difference between sets of IDs of transactions (i.e., tidsets) of two related itemsets. Frequent patterns can be mined by first taking the complement of the tidset of a singleton itemset and then taking the diffset between a non-singleton itemset and its immediate prefix. As such, dEclat works well when the database is dense. *VIPER* [37] represents the item lists in the form of bit vectors (cf. tidsets in Eclat). As such, VIPER mines frequent patterns by computing the bitwise-AND of bit vectors of frequent itemsets to form itemsets of higher cardinalities.

In addition to the aforementioned serial frequent pattern algorithms, there are also distributed and parallel frequent pattern mining algorithms [38,42]. Many of them mine frequent patterns using distributed or parallel computing, in which frequent patterns are found by multiple processors that either (1) have their own private memory (i.e., distributed memory) with which for information is exchanged via techniques like message passing interface (MPI) among the processors, or (2) have access to a shared memory with which information is exchanged by techniques like Open Multi-Processing (OpenMP) among multiple processors. Among many distributed and parallel frequent pattern mining algorithms, notable Apriori-based algorithms include Count Distribution, Data Distribution, and Candidate Distribution algorithms [1].

In the current era of big data, high volumes of a wide variety of valuable data of different veracities can be generated or collected at a high velocity. To handle big data, MapReduce-based frequent pattern mining algorithms have been developed. As implied by its name, MapReduce [6] involves two key functions: (1) the "map" function and (2) the "reduce" function. Specifically, the input database are read, divided into several partitions (sub-problems), and assigned to different processors. Each processor executes the "map" function on each partition (sub-problem). The "map" function takes a key-value pair of and returns as an intermediate result a list of key-value pairs, which are then shuffled and sorted. Each processor then executes the "reduce" function on every single key from this intermediate result together with the corresponding list of all values. The "reduce" function combines, aggregates, summarizes, filters, and/or transforms the list of values associated with a given key and returns a single (aggregated or summarized) value. The *Single Pass Counting* (*SPC*), *Fixed Passes Combined-counting* (*FPC*) and *Dynamic Passes Combined-counting* (*DPC*) algorithms [30] are some notable Apriori-based algorithms that use MapReduce to mine frequent patterns. Similarly, PFP algorithm [29] is a MapReduce-based parallel FP-growth algorithm. MREclat [44] and Dist-Eclat [34] can be considered as MapReduce versions of the serial vertical frequent pattern mining algorithms Eclat and dEclat respectively. BigFIM [34] can be considered as a MapReduce version of a hybrid of the Apriori and dEclat algorithms. Most of these MapReduce-based algorithms efficiently mine frequent patterns with cloud computing, which makes good use of a network of remote servers hosted on the Internet—such as Amazon Web Services (AWS) or cloud data center—to store, manage, process, and analyze data.

Recently, the concept of using fog computing [31] for frequent pattern mining was proposed. The key ideas of the corresponding algorithms [4] are that, instead of transmitting local data to data centers (as in cloud computing), local data are kept on the local internet of things (IoT) devices from which patterns that are frequent locally on these devices can be found. Then, take the union of these locally frequent patterns to form global candidate patterns. Afterwards, count the frequency of these global candidates on each device, transmit their frequency count to other devices (or a master node), and sum the frequency counts to find globally frequent patterns. As an alternative, in situations where local IoT devices are not powerful, local data are transmitted to their local networking services that lie between the IoT devices and the usual data centers.

Recall that the task of mining 'following' patterns from social media data can be reduced into mining frequent patterns from databases. With these numerous frequent pattern mining algorithms (e.g., serial, distributed or parallel, MapReduce cloud computing-based, and/or fog computing-based algorithms), our data analytic component for mining social media data is designed so flexible in a way that it can adapt and execute any of the aforementioned frequent pattern algorithms to find 'following' patterns from social media data.

Furthermore, our data analytic component for mining social media data also provides users with direct algorithms for mining social media data. Specifically, we incorporate *FoP-miner* [11], *ParFoP-miner* [26], *BigFoP* [23] and *WAH-miner* [20] into our framework. Note that FoP-miner is a serial algorithm designed for mining 'following' patterns. As extensions of FoP-miner, the ParFoP-miner and BigFoP algorithms are parallel version and MapReduce-based version of FoP-miner respectively. As another extension of FoP-miner, the WAH-miner algorithm handles big social media data.

4 Visual Analytics of Social Media Data

Recall that the data analytic component of our flexible framework allows users to efficiently find from social media data the 'following' patterns. These data analytic results may be returned in long textual lists, which may not be easily comprehensible. Given a picture is worth a thousand words, having a visual means for users to interact with computers would be beneficial because a visual representation matches the power of the human visual and cognitive system and enables human to interact with computers. In response, the *visual analytic component* of our flexible framework is designed and developed in such a way that it allows users to visualize the original data and analyze the mined results (i.e., frequent patterns). In addition, it also resolves the challenges of (1) showing patterns with their prefix-extension relationships and (2) indicating the frequency of each pattern. Note that, patterns $\{a\}$ and $\{a, b\}$ are said to be *prefixes* of $\{a, b, c\}$, whereas $\{a, b, c, d\}$ and $\{a, b, c, e\}$ are said to be *extensions* of $\{a, b, c\}$.

Like the data analytic component, the visual analytic component of our framework also adapts and takes advantages of some existing frequent pattern visualizers. For instance, our framework calls *FIsViz* [21] to visualize 'following' patterns (i.e., groups of k popular followees being frequently followed by at least $m = 2$ followers) as polylines connecting k nodes in a two-dimensional space with (x, y)-coordinates, in which followees (e.g., users#614, 51, 132, 143, 608 and 620) are listed on the x-axis and the number of followers are indicated by the y-axis. The x-locations of all nodes in the polyline indicate the individual followee contained in a 'following' pattern Z, and the y-location of the rightmost node of a polyline for Z indicates the number of followers for Z. Hence, prefix-extension relationships can be observed by traversing along the polylines. See Fig. 1(a). In addition, to facilitate exploration of data and mining results, FIsViz also provides users with interactive detail-on-demand features. When the mouse hooves on a polyline connecting two nodes u and v, FIsViz shows a list of followees being followed both u and v. Similarly, when the mouse hovers over a node, FIsViz shows a list of all patterns contained in all polylines starting or ending at this node.

To avoid the bending and crossing-over of polylines, our framework provides users with alternative visualizers adapted from WiFIsViz [22] and FpVAT [18]. As shown in Fig. 1(c), *WiFIsViz* uses two half-screens to visualize 'following' patterns. Both half-screens are wiring-type diagrams (i.e., orthogonal graphs), which represent 'following' patterns as horizontal lines connecting k nodes in a two-dimensional space (where the x-axis lists all the followees). The left half-screen provides the follower information by using the y-location of the horizontal line to indicate the number of followers who follow the popular groups of followees. The right half-screen lists all 'following' patterns in the form of a trie.

Similarly, *FpVAT* [18] also uses wiring-type diagrams to visualize 'following' patterns. However, FpVAT shows all the 'following' patterns (i.e., groups of popular followees) and the number of their corresponding followers on the same full-screen. See Fig. 1(b).

Instead of polylines or wiring-type diagrams (i.e., orthogonal graphs), both *RadialOutViz* [24] and *RadialInViz* [7] use alternative design with emphasis on showing the prefix-extension relationships among the 'following' patterns. For example, both RadialInViz and RadialOutViz visualizes groups of popular followees in a radial layout, which leads to a benefit of being *orientation-free*. As such, the legibility of the represented 'following' patterns is not be impacted by the orientation. Hence, they are ideal for the collaborative environment (cf. traditional two-dimensional rectangular space, which favors the viewer who visualizes data or mining results at the up-right position but not favors those on the opposite side or the left/right sides). Moreover, both RadialInViz and RadialOutViz represent 'following' patterns in a hierarchical fashion so that extensions of a 'following' pattern Z are embedded within sectors representing the prefixes of Z. The number of fellowers of Z is represented by the radius of the sector representing Z. A key difference between RadialInViz and RadialOutViz is that the former radiating in and the latter radiating out. Between the two, RadialInViz represents the 'following' patterns in a clearer matter because information is not crowded near the center of the circular space. See Fig. 2.

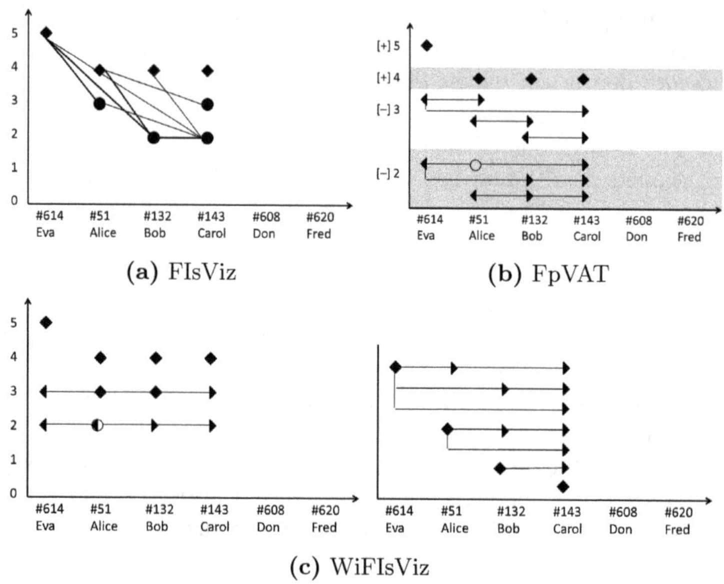

(a) FIsViz (b) FpVAT

(c) WiFIsViz

Fig. 1. Visual analytics of social media data via orthogonal graphs

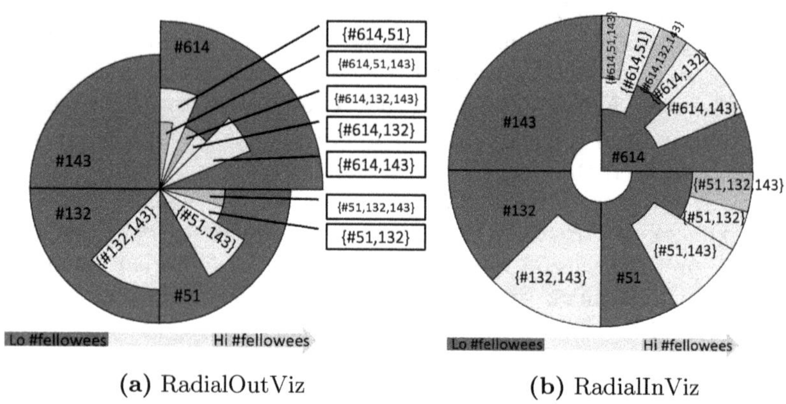

(a) RadialOutViz (b) RadialInViz

Fig. 2. Visual analytics of social media data via circular graphs

5 Conclusions

This paper presents a data and visual analytic framework for emerging databases. In particular, the data analytic component of our framework allows users to apply either frequent pattern mining algorithms (ranging from serial, distributed and parallel, MapReduce cloud computing-based, to fog computing-based algorithms) or direct 'following' pattern miners to mine and analyze social media data. The visual analytic component of our framework allows users to apply either orthogonal graph-based visualizers or circular graph-based visualizers to visualize and analyze the mined 'following' patterns (i.e., groups of popular or frequently followed followees in social networks). As ongoing work in the current era of growing volumes of social media data and beyond, we are extending our framework to accept more efficient algorithms for data analytics and/or more effective visualizers for visual analytics of social media data.

Acknowledgement. This project is partially supported by NSERC (Canada) and University of Manitoba.

References

1. Agrawal, R., Shafer, J.C.: Parallel mining of association rules. IEEE TKDE **8**(6), 962–969 (1996)
2. Agrawal, R., Srikant, R.: Fast algorithms for mining association rules in large databases. In: VLDB 1994, pp. 487–499 (1994)
3. Bernhard, S.D., Leung, C.K., Reimer, V.J., Westlake, J.: Clickstream prediction using sequential stream mining techniques with Markov chains. In: IDEAS 2016, pp. 24–33 (2016)
4. Braun, P., Cuzzocrea, A., Leung, C.K., Pazdor, A.G.M., Tanbeer, S.K.: Mining frequent patterns from IoT devices with fog computing. In: HPCS 2017, pp. 691–698. IEEE (2017)
5. Choudhery, D., Leung, C.K.: Social media mining: prediction of box office revenue. In: IDEAS 2017, pp. 20–29 (2017). doi:10.1145/3105831.3105854
6. Dean, J., Ghemawat, S.: MapReduce: simplified data processing on large clusters. In: OSDI 2004, pp. 137–150 (2004)
7. Dubois, P.M.J., Han, Z., Jiang, F., Leung, C.K.: An interactive circular visual analytic tool for visualization of web data. In: IEEE/WIC/ACM WI 2016, pp. 709–712 (2016)
8. Duong, V.T.T., Khan, K.-U., Jeong, B.-S., Lee, Y.-K.: Top-k frequent induced subgraph mining using sampling. In: EDB 2016, pp. 110–113 (2016)
9. Han, J., Pei, J., Yin, Y.: Mining frequent patterns without candidate generation. In: ACM SIGMOD 2000, pp. 1–12 (2000)
10. Hoque, M.N., Ahmed, C.F., Lachiche, N., Leung, C.K., Zhang, H.: Reframing in clustering. In: IEEE ICTAI 2016, pp. 350–354 (2016)
11. Jiang, F., Leung, C.K.: Mining interesting "following" patterns from social networks. In: Bellatreche, L., Mohania, M.K. (eds.) DaWaK 2014. LNCS, vol. 8646, pp. 308–319. Springer, Heidelberg (2014). doi:10.1007/978-3-319-10160-6_28
12. Jiang, F., Leung, C.K., Zhang, H.: B-mine: frequent pattern mining and its application to knowledge discovery from social networks. In: Li, F., Shim, K., Zheng, K., Liu, G. (eds.) APWeb 2016, Part I. LNCS, vol. 9931, pp. 316–328. Springer, Heidelberg (2016). doi:10.1007/978-3-319-45814-4_26
13. Lee, J.H., Kim, J.M., Choi, Y.S.: SNS data visualization for analyzing spatial-temporal distribution of social anxiety. In: EDB 2016, pp. 1106–1109 (2016)
14. Lee, R.C., Cuzzocrea, A., Lee, W., Leung, C.K.: Majority voting mechanism in interactive social network clustering. In: ACM WISM 2017, Article no. 14 (2017). doi:10.1145/3102254.3102268

15. Lee, W., Song, J.J.S., Leung, C.K.-S.: Categorical data skyline using classification tree. In: Du, X., Fan, W., Wang, J., Peng, Z., Sharaf, M.A. (eds.) APWeb 2011. LNCS, vol. 6612, pp. 181–187. Springer, Heidelberg (2011). doi:10.1007/978-3-642-20291-9_19
16. Leung, C.K.: Mining frequent itemsets from probabilistic datasets. In: EDB 2013, pp. 137–148 (2013)
17. Leung, C.K., Carmichael, C.L., Hayduk, Y., Jiang, F., Kononov, V.V., Pazdor, A.G.M.: Data mining meets HCI: data and visual analytics of frequent patterns. In: Frasconi, P., Landwehr, N., Manco, G., Vreeken, J. (eds.) ECML-PKDD 2016, Part III. LNCS (LNAI), vol. 9853, pp. 289–293. Springer, Heidelberg (2016). doi:10.1007/978-3-319-46131-1_37
18. Leung, C.K., Carmichael, C.L.: FpVAT: a visual analytic tool for supporting frequent pattern mining. ACM SIGKDD Explor. **11**(2), 39–48 (2009)
19. Leung, C.K., Carmichael, C.L., Johnstone, P., Xing, R.R., Yuen, D.S.H.: Interactive visual analytics of big data. In: Ontologies and Big Data Considerations for Effective Intelligence, pp. 1–26 (2017)
20. Leung, C.K., Dela Cruz, E.M., Cook, T.L., Jiang, F.: Mining 'following' patterns from big sparse social networks. In: IEEE/ACM ASONAM 2016, pp. 923–930 (2016)
21. Leung, C.K.-S., Irani, P.P., Carmichael, C.L.: FIsViz: a frequent itemset visualizer. In: Washio, T., Suzuki, E., Ting, K.M., Inokuchi, A. (eds.) PAKDD 2008. LNCS (LNAI), vol. 5012, pp. 644–652. Springer, Heidelberg (2008). doi:10.1007/978-3-540-68125-0_60
22. Leung, C.K., Irani, P.P., Carmichael, C.L.: WiFIsViz: effective visualization of frequent itemsets. In: IEEE ICDM 2008, pp. 875–880 (2008)
23. Leung, C.K., Jiang, F.: Big data analytics of social networks for the discovery of "following" patterns. In: Madria, S., Hara, T. (eds.) DaWaK 2015. LNCS, vol. 9263, pp. 123–135. Springer, Heidelberg (2015). doi:10.1007/978-3-319-22729-0_10
24. Leung, C.K.-S., Jiang, F.: RadialViz: an orientation-free frequent pattern visualizer. In: Tan, P.-N., Chawla, S., Ho, C.K., Bailey, J. (eds.) PAKDD 2012, Part II. LNCS (LNAI), vol. 7302, pp. 322–334. Springer, Heidelberg (2012). doi:10.1007/978-3-642-30220-6_27
25. Leung, C.K., Jiang, F., Dela Cruz, E.M., Elango, V.S.: Association rule mining in collaborative filtering. In: Collaborative Filtering Using Data Mining and Analysis, pp. 159–179 (2017)
26. Leung, C.K., Jiang, F., Pazdor, A.G.M., Peddle, A.M.: Parallel social network mining for interesting 'following' patterns. Concurrency Comput. Pract. Exper. **28**(15), 3994–4012 (2016)
27. Leung, C.K., Jiang, F., Poon, T.W., Crevier, P.-E.: Big data analytics of social network data: who cares most about you on Facebook? In: Highlighting the Importance of Big Data Management and Analysis for Various Applications, pp. 1–15 (2018). doi:10.1007/978-3-319-60255-4_1
28. Leung, C.K., MacKinnon, R.K., Jiang, F.: Finding efficiencies in frequent pattern mining from big uncertain data. World Wide Web **20**(3), 571–594 (2017)
29. Li, H., Wang, Y., Zhang, D., Zhang, M., Chang, E.Y.: PFP: parallel FP-growth for query recommendation. In: ACM RecSys 2008, pp. 107–114 (2008)
30. Lin, M., Lee, P., Hsueh, S.: Apriori-based frequent itemset mining algorithms on MapReduce. In: ICUIMC 2012, Article no. 76 (2012)
31. Linthicum, D.S.: Connecting fog and cloud computing. IEEE Cloud Comput. **4**(2), 18–20 (2017)

32. MacKinnon, R.K., Leung, C.K.: Stock price prediction in undirected graphs using a structural support vector machine. In: IEEE/WIC/ACM WI-IAT 2015, vol. 1, pp. 548–555 (2015)
33. Mateo, M.A.F., Leung, C.K.: Design and development of a prototype system for detecting abnormal weather observations. In: C3S2E 2008, pp. 45–59 (2008)
34. Moens, S., Aksehirli, E., Goethals, B.: Frequent itemset mining for big data. In: IEEE BigData 2013, pp. 111–118 (2013)
35. Pei, J., Han, J., Lu, H., Nishio, S., Tang, S., Yang, D.: H-mine: hyper-structure mining of frequent patterns in large databases. In: IEEE ICDM 2001, pp. 441–448 (2001)
36. Savasere, A., Omiecinski, E., Navathe, S.: An efficient algorithm for mining association rules in large databases. In: VLDB 1995, pp. 432–444 (1995)
37. Shenoy, P., Bhalotia, J.R., Bawa, M., Shah, D.: Turbo-charging vertical mining of large databases. In: ACM SIGMOD 2000, pp. 22–33 (2000)
38. Tanbeer, S.K., Ahmed, C.F., Jeong, B.-S.: Parallel and distributed frequent pattern mining in large databases. In: IEEE HPCC 2009, pp. 407–414 (2009)
39. Wang, K., Tang, L., Han, J., Liu, J.: Top down FP-growth for association rule mining. In: Chen, M.-S., Yu, P.S., Liu, B. (eds.) PAKDD 2002. LNCS (LNAI), vol. 2336, pp. 334–340. Springer, Heidelberg (2002). doi:10.1007/3-540-47887-6_34
40. You, Y.S., Lee, S., Kim, J.: Design and development of visualization tool for movie review and sentiment analysis. In: EDB 2016, pp. 117–123 (2016)
41. Zaki, M.J.: Fast vertical mining using diffsets. In: ACM KDD 2003, pp. 326–335 (2003)
42. Zaki, M.J.: Parallel and distributed association mining: a survey. IEEE Concurrency **7**(4), 14–25 (1999)
43. Zaki, M.J.: Scalable algorithms for association mining. IEEE TKDE **12**(3), 372–390 (2000)
44. Zhang, Z., Ji, G., Tang, M.: MREclat: an algorithm for parallel mining frequent itemsets. In: CBD 2013, pp. 177–180 (2013)

A Method to Maintain Item Recommendation Equality Among Equivalent Items in Recommender Systems

Yeo-jin Hong, Shineun Lee, and Young-ho Park(✉)

Department of IT Engineering, Sookmyung Women's University, Cheongpa-ro 47-gil 100, Yongsan-gu, Seoul 04310, South Korea
yhpark@sm.ac.kr

Abstract. Collaborative Filtering is a useful algorithm to offer personalized recommendations for users. However, there are several technical challenges in collaborative filtering, including the first-rater problem, where an item not yet evaluated cannot be recommended until it has been rated. In the paper, the presenting method deals with the first-rater problem that is similar to the process starvation is operating systems. The method reduces the score gap between items and makes it possible for a new item or an item with no user preference to be recommended automatically. Thus, the system can recommend items in the same group without bias. Finally, we present an analysis of an example of the algorithm.

Keywords: Recommender systems · Collaborative filtering · First-Rater problem · Data sparsity

1 Introduction

As the amount of web content increases, users find themselves spending a lot of time and energy trying to find content that is useful to them. Recommender systems offer a solution to this problem by identifying and displaying those items that the user is likely to select. There are many types of recommender systems, and each system produces predictions or recommendations for users through its own framework. Collaborative Filtering (CF) is a type of recommender system which has been widely adopted in many applications that offer personalized recommendations for products, services, locations, learning contents, etc. CF algorithms can be classified into two approaches. One is memory-based CF, which uses similarities between users or items to find the best matches. It also is referred to as neighborhood-based CF. The other is model-based CF, which estimates user ratings using statistical techniques, such as latent factor models and matrix factorization models, to provide recommendations [1].

Although CF algorithms are suitable for personalized recommendations, there are several challenges related to data sparsity. One of the data sparsity issues, which is termed *cold start*, occurs when a new user is added to the system. Due to lack of preference information, the user is unable to receive accurate recommendations from the system. This problem applies not only to new users but also to new items entered

W. Lee et al. (eds.), *Proceedings of the 7th International Conference on Emerging Databases*, Lecture Notes in Electrical Engineering 461,
https://doi.org/10.1007/978-981-10-6520-0_22

into the system, which cannot be recommended until they are rated by a user. This problem is referred to as the *first-rater* problem or *early-rater* problem, where an item with no rating cannot be recommended until it has received evaluation from users.

Thus, the number of items which can be recommended in the recommender system is reduced. Consequently, the same items are recommended repeatedly to the user, which may result in the decrease in user satisfaction. In this paper, we focus on the first-rater problem. To alleviate the problem, we designed a method using inverse recommend frequency of an item. Our contribution in this work is twofold: First, the method improves recommendation coverage of the system. Existing methods hardly recommend items which have lower ratings than the other items or have no ratings even if the items are enough to be recommended by the system. Second, this method can be easily applied to recommender systems. Recent recommender systems usually combine two or more methods in order to improve its performance, but some latest techniques are hard to apply to a recommender system which already combined with other techniques due to its characteristics. This method can be useful to apply existing recommender systems since it has a simple scoring process and characteristic.

The paper proceeds as follows. Section 2 describes the entire process of user-based CF.

Section 3 discusses related work focused on the first-rater problem. Section 4 explains the basic process of the proposed method. In Sect. 4, we describe how the method works with an example of English sentence recommender that was designed for a language learning application. Section 5 concludes this paper and describes our future work.

2 Preliminaries

In this section, we describe the entire process of user-based CF. User-based CF algorithm, which is a type of memory-based collaborative filtering, recommends for a user items which are selected by similar users. The basic assumption of CF is that users with similar preferences would like similar items as well. When a request from a user reaches the system, CF recommends items chosen by other similar users (Fig. 1).

Usually these approaches consist of three steps to recommend items. First, a recommender system calculates similarities of all the users to group the similar users. Second, it selects users who have a higher similarity within the group. Third, it computes the values of the similar users' ratings weighted by their similarity [1].

2.1 Similarity Measure

In user-based CF, a recommender system calculates user similarity to offer personalized recommendations. There are several methods to calculate the similarity between users, such as Pearson correlation coefficient, cosine similarity, Spearman correlation coefficient, etc. In this paper, we use Pearson correlation coefficient to compare similarity between users.

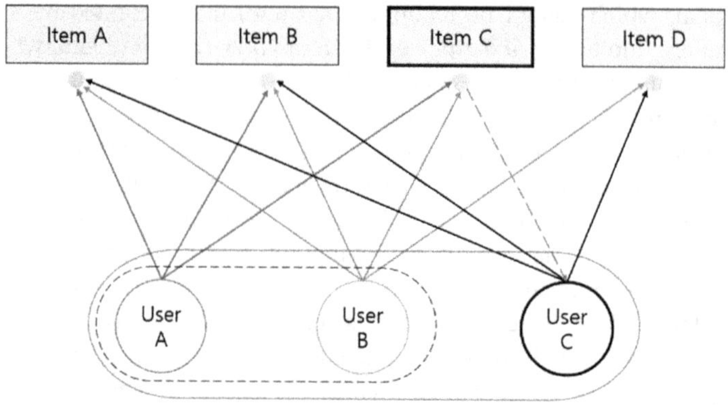

Fig. 1. User-based collaborative filtering

$$sim(u,v) = \frac{\sum_{i \in I} (r_{u,i} - \bar{r}_u)(r_{v,i} - \bar{r}_v)}{\sqrt{\sum_{i \in I} (r_{u,i} - \bar{r}_u)^2} \sqrt{\sum_{i \in I} (r_{v,i} - \bar{r}_v)^2}}. \quad (1)$$

where sim represents similarity between user u and v, I stands for items rated by user u and v, $r_{u,i}$ is the rating for item i from user u, and $\bar{r}_u$ is the average rating from user u [2–4]. Similarity calculated using Pearson Correlation has the output value between –1 to 1. If the value is close to –1, the users u and v are dissimilar from each other. After calculating similarity using Pearson correlation coefficient, a recommender system with user-based CF selects top N users with the highest similarity.

2.2 Item Rating

A set of similar users is used to predict items to recommend. Ratings for the items are calculated based on the selected users' purchase history. The ratings can be computed in many different ways. Generally, a recommendation system uses the weighted sum of others' rating according to the following formula [1].

$$P_{a,i} = \bar{r}_a + \frac{\sum_{u \in U} (r_{u,i} - \bar{r}_u) w_{a,u}}{\sum_{u \in U} |w_{a,u}|}. \quad (2)$$

where $\bar{r}_a$ and $\bar{r}_u$ are the average ratings for the user a and user u on all rated items, and $w_{a,u}$ is the similarity between the user a and user u, which is usually calculated using Pearson correlation coefficient that is explained above. When the rating process is done, the system recommends top N items which have higher rating than the other items. This process can avoid the *first-rater* problem to some degree, but there still remains the problem that items with no ratings cannot be recommended.

In the paper, we modify this rating process to avoid the *first-rater* problem by including recommend frequency of each item into a scoring process of a recommender system. Section 4 describes details of the method.

3 Related Work

In this section, we discuss previous studies addressing the *first-rater* problem. In [4], the authors proposed a Community Experts based Recommendation system, or CER, to solve the *first-rater* problem by building an interest similarity group in a virtual community. The authors presented a system in which the users teamed up based on similar interests and then identified the experts from the field of interest. The system offers recommendations in the form of positive, negative or neutral opinions based on the reviews in the experts' blog. [5] developed a hybrid recommender system for sports news. The recommender system combines content-based approach with CF to overcome the challenges of CF. The content-based algorithm in the study considers every keyword in an article as an element of a vector space and calculates the relative dimension of the keyword weight. It then shows 50 articles which have the highest score based on the keyword weight. [6] proposed a method using naive bayes classification which is used to classify documents. The basic idea of naive bayes classification is calculating probability to classify given documents into one of many categories. The proposed method finds keywords in a paper and computes similarity based on keyword frequency and weight based on the author's type, co-authors or related authors. [7] developed a hybrid recommender system, which collects users' opinions in the form of user-item rating matrix using a proposed Centering-Bunching based Clustering algorithm, or CBBC. This algorithm fixed number clusters and stored it in its database for future recommendation. The algorithm then generates new recommendations using the clusters. [8] developed a system that makes predictions using item information which is already in the user purchase history. This method makes a set of preference words from the purchase history and predicts preference for items. Then, the system recommends an item which has highest score by comparing each set of preference words. Most of the studies worked with hybrid systems since the *first-rater* problem can be solved when a recommender system has another source of information that can be used as grounds for recommendation. However, hybrid systems have a limitation in application scope since it should consider each algorithm's characteristics. Moreover, it cannot be used for different type of recommendations since it is usually constrained to a specific field.

4 Inverse Recommend Frequency

In this section, we introduce an algorithm concerned with the *first-rater* problem in user-based CF. To alleviate the *first-rater* problem, we propose a modified scoring process for CF algorithm using inverse recommend frequency of each item. This method is based on the *TF-IDF* technique, short for term frequency–inverse document frequency, which is used in the information retrieval field. It is a statistical method to measure the significance of a word in a document.

The whole recommendation process is applied as follows. First, a recommender system with user-based CF approach computes similarity between users based on their purchase history. And the system selects top K users who have higher similarities than the others. Second, the system creates a rating matrix $\mathbf{R}$ including the selected users,

	i_1	i_2	i_3	…	i_n
a_1	1	4	5	…	3
a_2	2	2		…	4
a_3		2	3	…	4
…				…	
a_k			1	…	

Fig. 2. An *example* of a rating matrix R

items the users bought, and ratings represented by $r_{a,i}$. $r_{a,i}$ is the rating given to item i by user a. Figure 2 shows an example of a rating matrix **R**.

The matrix **R** is used to calculate recommendation score S_i. S_i is a score for an item i in the matrix **R**, which is criteria for recommendations. The Recommendation score S_i is calculated by the formula in below.

$$S_i = \sum\nolimits_{u \in U} \overline{r}_a + r_{u,i} w_{a,u} (\lambda + F)^{-1}. \tag{3}$$

Where **U** is the set of selected users, and w is similarities between user a and user u. F is recommend frequency for each item and λ represents *Base Quantifier* to increase

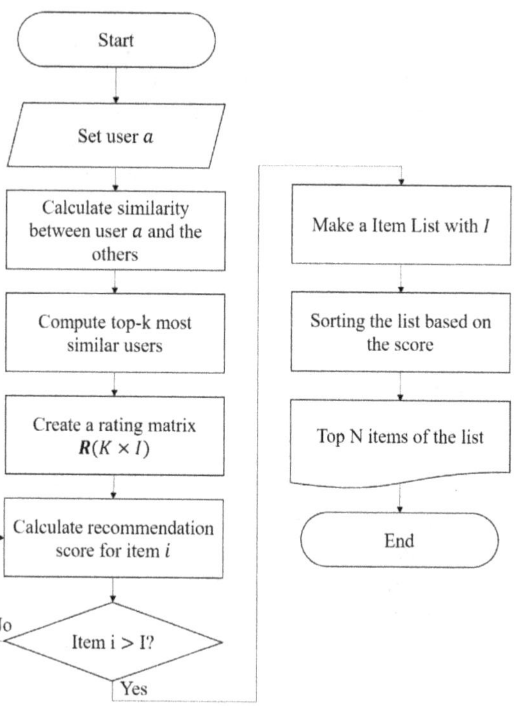

Fig. 3. Recommend process with inverse recommend frequency method

the recommend frequency of each item by λ. It avoids division by zero since if an item has not been recommended by the system before, its recommend frequency is zero. When the rating process is done over all the items in the matrix **R** in this way, the system makes a list of recommendation based on the ratings. Finally, after sorting the list in the descending order, the recommender system recommends the top N-items to the user. Figure 3 shows a flowchart of this process.

This method reduces the score gap between items and allow users to get new recommendations. Moreover, this method can be applied in many different algorithms for recommender systems since it has a simple recommendation process.

5 Analysis of an Example of the Inverse Recommend Frequency

In this section, we demonstrate the proposed method using an example. Let's suppose there is a set of items for a recommender system. This group has items which have different recommend frequency but their ratings and similarities are the same. Item C in Fig. 4 has no recommend frequency.

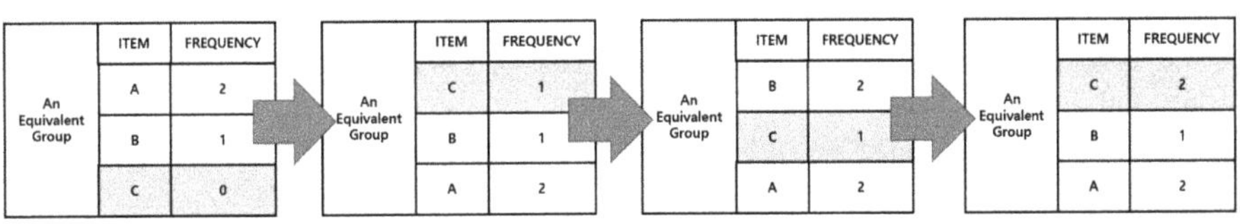

Fig. 4. An *example* of inverse recommend frequency method

The recommender system with the proposed method first calculates scores for every item in the set. A score for each item is each item's preference multiplied by the users' similarity and with its base quantifier added. In this case, the system will first recommend item C since it has the highest score as the proposed method. The system recommends item B or C since both have same score in the set when it is required to offer a new recommendation at the set. In this example, the system recommends item B. Again, item C can be recommended when the system finds a proper item to recommend since the item has the lowest score in the set. Therefore, the system maintains balance of recommend frequency without bias as long as the system follows the proposed method.

6 Conclusion

In this paper, we propose inverse recommend frequency solving the *first-rater* problem. The *first-rater* problem is that an item which has no rating cannot be recommended until it has received evaluation from users. The proposed method calculates recommendation scores with recommend frequency of each item when they make a recommendation to users. The method can recommend items with no rating in the same group by reducing the score gap between items, which can improve the probability that an item is recommended by the system. Finally, this algorithm can be applied in any recommender system and modified based on the system's objective.

However, we did not conduct experiments to evaluate our method. And there is a remaining issue where an item with a high score cannot be distinguished clearly from others. Our future work in this field is to verify the performance of the proposed method and evaluate the results. Also, we will develop a method that can detect which items are more valuable for users among ordinary items in order to improve the proposed method.

Acknowledgement. This work was supported by Institute for Information & communications Technology Promotion (IITP) grant funded by the Korea government (MSIP) (No. R0115-16-1009, Development of smart learning interaction contents for acquiring foreign languages through experiential awareness).

References

1. Melville, P., Vikas, S.: Recommender Systems. Encyclopedia of Machine Learning, pp. 829–838. Springer, US (2011)
2. Son, J., Kim, S.B., Kim, H., Cho, S.: Review and analysis of recommender systems. J. Korean Inst. Ind. Eng. **41**(2), 185–208 (2015)
3. Su, X., Taghi, M.K.: A Survey of Collaborative Filtering Techniques. Advances in Artificial Intelligence, pp. 1–19
4. Kumar, A., Bhatia, M.: Community expert based recommendation. J. Com. App. **37**, 7–13 (2012)
5. Lenhart, P., Herzog, D.: Combining Content-based and Collaborative Filtering for Personalized Sports News Recommendations. In: CBRecSys, pp. 3–10 (2016)
6. Lee, S.G., Lee, B.S., Bak, B.Y., Hwang, H.K.: A study of intelligent recommendation system based on naive bayes text classification and collaborative filtering. J. Inf. Manage. **41**(4), 227–249 (2010)
7. Shinde, K.S., Kulkarni, U.: Hybrid personalized recommender system using centering-bunching based clustering algorithm. Expert Syst. Appl. **39**, 1381–1387 (2012)
8. Choi, Y.S., Moon, B.R.: A prediction system of user preferences for newly released items based on words. J. Korean Inst. Intell. Syst. **16**(2), 156–163 (2006)

Time-Series Analysis for Price Prediction of Opportunistic Cloud Computing Resources

Sarah Alkharif, Kyungyong Lee(✉), and Hyeokman Kim

Kookmin University, Seoul, South Korea
{saraalkharif,leeky,hmkim}@kookmin.ac.kr

Abstract. Cloud computing resources are offered in various forms, and surplus of computing resources are provided at cheaper price. A leading cloud computing vendor, Amazon Web Services, provides such opportunistic resources as EC2 spot instance whose price changes dynamically based on the resource demand from users. We analyze the spot instance price logs and apply various predictive analysis algorithms to better predict future spot instance price. By applying various train dataset modeling heuristics, we uncover that the SARIMA algorithm achieves the best prediction accuracy in spot price prediction; it shows 17% more accuracy than other algorithms that are widely used for spot instance applications. By applying contributions in this paper, we expect that spot instance users can decrease monetary cost while improving system stability.

1 Introduction

Cloud computing provides various types of compute resources to serve diverse application scenarios. The cloud computing frees the burden of system administration overheads without incurring prohibitive initial hardware purchase cost. From the service provider's perspective, fully utilizing the already established hardware resources and services is crucial to maximize monetary gain. As the users' resource demand can vary from time to time, some cloud computing providers offer services at cheaper price than the regular price to maximize hardware/service utilization. For instance, Amazon Web Services (AWS), a leading cloud computing vendor, provides its surplus of EC2 computing resources at a cheaper price in the form of *spot instance*. A user who wants to use a spot instance bids for a price that one is willing to pay, and if the bid price is higher than the spot price that is decided by the service provider, one can get the resource allocated and pays for the spot price in the hourly basis. Other than AWS, Google Cloud Engine provides such opportunistic resources in the form of *preemptive instances*, and Microsoft Azure provides its excess compute capacity as *low-priority VM*.

Though users can utilize the opportunistic resources at a cheaper price, sudden service termination can happen at anytime as the demand for the computing resource changes. To mitigate the chance of sudden service interruption,

W. Lee et al. (eds.), *Proceedings of the 7th International Conference on Emerging Databases*, Lecture Notes in Electrical Engineering 461,
https://doi.org/10.1007/978-981-10-6520-0_23

few works were conducted to better predict and model the price change of EC2 spot instance in literature. Ben-Yehuda et al. [1] and Zhao et al. [10] tried to predict the future spot instance price using various predictive analysis algorithms, but they all concluded that the spot price is rather random and hard to make meaningful prediction for future price changes. Since then, most of studies that are related to spot instance focus on the handling sudden service interruption [2,5,6,9] or spot instance bid strategy [7,11].

In this paper, we apply various time-series analysis algorithms to predict price change pattern of AWS EC2 spot instances. By carefully designing the period of train datasets, we could uncover that applying seasonal-arima (SARIMA) can improve the accuracy of price change prediction error by 17% on average comparing to the naive method that references the most recent price to predict future price [5]. In addition to the contribution of improved price prediction accuracy, we could also discover that the configuration values to get the best prediction accuracy differs significantly across different availability zones and instance types. Based on the extensive experiments and promising results, we bring up an opportunity of improving spot instance prediction accuracy that can result in significant cost gain for cloud computing users with increased system stability.

2 Time-Series Analysis for AWS EC2 Spot Instance Price

In this work, we use various time-series analysis algorithms to predict future spot instance price. First, we apply a simple *mean* method that uses the average of previous price to make prediction. In the *mean* method, assuming the price of instance in time $t-1$ is x_{t-1}, to predict the price of an instance at time t, we use previous instance price until lag that is notated as n (Eq. 1).

$$\hat{y_t} = \frac{x_{t-1} + x_{t-2} + ... + x_{t-n}}{n} \tag{1}$$

Naive method is another simple model that references only the most recent spot instance price to make prediction for future time windows 2. For naive and mean methods, the expected value can be used to make prediction for longer period of time later than t.

$$\hat{y_t} = x_{t-1} \tag{2}$$

Seasonal naive method is similar to naive method that uses the most recent observation, but it utilizes seasonal information in making prediction. Assuming the seasonal period is s, the predicted value at time t is the observed value at time $t-s$ [3].

$$\hat{y_t} = x_{t-s} \tag{3}$$

Auto Regressive Integrated Moving Average (ARIMA) is a popular statistical model and widely applied for time series datasets. The algorithm is a combination of auto-regressive (Eq. 4) and moving-average model (Eq. 5), with three parameters (p, d, q). p is the number of auto-regressive terms that indicates dependencies on past values, d is the degree of differencing to make input dataset stationary, and q is the number of lagged forecast errors in the prediction equation that depends only on random error terms.

$$\hat{y_t} = w_0 + \beta_1 y_{t-1} + \beta_2 y_{t-2} + ...\beta_n y_{t-n} + \epsilon_t \tag{4}$$

$$\hat{y_t} = w_0 + \epsilon_t + \delta_1 \epsilon_{t-1} + \delta_2 \epsilon_{t-2} + ... + \delta_n \epsilon_{t-n} \tag{5}$$

If we set $d = n$, the equation becomes

$$\hat{y_t} = w_0 + \beta_1 y_{t-1} + \beta_2 y_{t-2} + ...\beta_n y_{t-n} + \delta_1 \epsilon_{t-1} + \delta_2 \epsilon_{t-2} + ... + \delta_n \epsilon_{t-n} \tag{6}$$

The term β_i is the weight that is applied to prior values in the time series, δ_i is auto-correlation coefficients at lags, and ϵ_i is the residual error term.

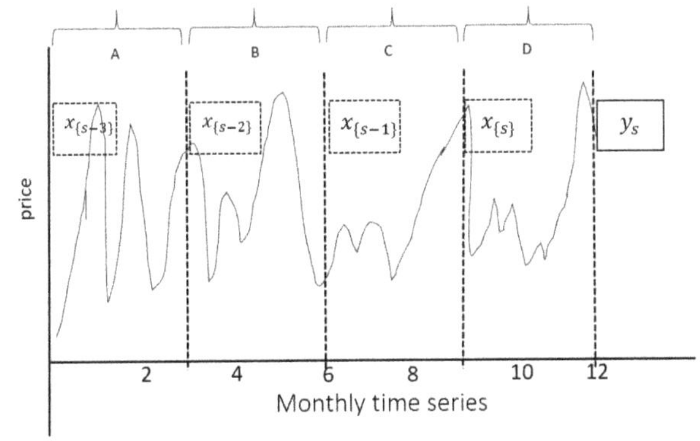

Fig. 1. SARIMA example

SARIMA stands for Seasonal ARIMA that allows ARIMA model to take into account seasonal characteristics that some seasonal patterns repeat along many periods in data sets. For example, in Fig. 1, taking historical time series data X for duration of one year, we will predict y_s with a seasonal period, $s = 4$, then the model splits the length of time series by the number of seasonal period. We get 4 periods, and each period contains 3 months of data. In the figure, assuming (A) as 4th lag, (B) 3rd lag, (C) 2nd lag, and (D) as the first lag, the seasonal lag time for X would be $(x_0, x_{s-1}, x_{s-2}, x_{s-3})$ and the number of periods will be chosen based on number of lags given by AR (Eq. 4) and MA (Eq. 5). If we set AR and MR coefficient as 2 then the AR would be β_s and β_{s-1} and the MA δ_s and δ_{s-1}.

Prophet model is developed by Facebook's Core Data Science team [8]. Prophet model forecasts time series dataset based on an additive and non-linear model fit with seasonality and holidays. The prophet algorithm predicts y_t by computing growth, seasonality, and holidays:

$$\hat{y_t} = g_t + s_t + h_t + \epsilon_t \tag{7}$$

The term g_t is the growth function to compute how the series has grown and the expected values for continuing growth, and s_t seasonality change based on series behaviors, h_t the effects of holidays, ϵ_t the error term.

3 Evaluation

To evaluate the effectiveness of applying various algorithms to predict spot instance price, we fetch 11 months (from March. 2016 to Feb. 2017) of spot price log files from the AWS public API service. From the log files, we extracted the timestamp, spot price, availability zone, and instance type. The price of each instance is stored hourly-basis using *time-series* data format in R with the period of 24 h. The on-demand instance price is different for different instance types in different regions, we normalize the spot instance price to that of on-demand instance. The normalized value indicates the cost gain that one can expect while using spot instances; smaller value indicates more gain.

At the time of writing, there are over 60 s of AWS EC2 instance types that are served in over 30 availability zones. It becomes prohibitive to present the experiment results from all instance types, and we select representative instances in General, GPU, Compute-, Memory-, and Storage-optimized types that are m4.2xlarge, g2.2xlarge, c3.2xlarge, r3.2xlarge, and i2.2xlarge, respectively. The instances are not served in all availability zones, and we choose 18 zones that provide the aforementioned instance types for experiments.

We evaluate naive, seasonal naive, mean, seasonal ARIMA, linear regression, and Prophet algorithms using packages in R 3.2.4. Among them, linear regression and Prophet always perform worse than Arima, and we do not show the result from them. Different algorithms have distinct heuristics to choose the train dataset window. Naive and seasonal naive methods simply reference values from the previous observations. For mean method, we use previous 1, 3, 7, 15, 30, 60, 90, and 120 days of prices to get mean value for prediction. However, using only the most recent data (1 day) shows the best result, and we exclude other results. For seasonal Arima, we differentiate the training dataset period as 30, 60, 90, 120, and 150 days. In the prediction step, we use a model built by *auto.arima* method of R. After building a model, we use the model for the next 1, 4, 8, 15, and 30 days. Overall, seasonal Arima has two configurations in the modeling data, *previous days used in modeling, the number of days for a model to be used*, and we notate the value using parenthesis. For all algorithms, we predict the normalized spot instance price for the next 24 h and calculate root-mean-squared-error to evaluate each model.

Figure 2 shows the test error of different algorithms. For seasonal Arima and mean, we select the best performing configuration values. Each algorithm is executed in all 18 availability zones, and the mean test error is presented. Regardless of instance types, Arima algorithm shows the best prediction accuracy among other methods. Previous works on predicting spot instance price insisted that using predictive analysis algorithms did not help to improve the prediction accuracy. With that, most systems using spot instances usually rely only on the very

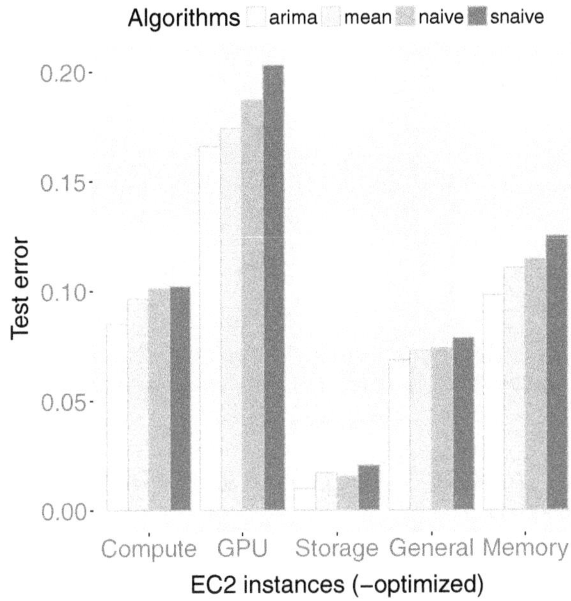

Fig. 2. Test error rates of different predictive algorithms. Regardless of the instance types, Arima shows the least error rate (lower is better).

recent price (naive method) only. However, with thorough experiments and train data configuration, we could uncover the effectiveness of using Arima model to predict spot instance price for the first time. In the figure, we can observe that different instance types show different test error rate, and we expect the different hardware specifications, such as GPU cards, can result in distinct supply and demand pattern.

From Fig. 2, we observed that the Arima algorithm shows the best performance. As noted earlier, we use various combinations of modeling data period and model use days. To see the effect from the different parameters, we show the test errors of GPU instances for different train data configurations in Fig. 3. A group of bars in the left two clusters show the test error value of us-west-1a and us-west-2c that show the least impact from the distinct parameters. The right two bars show the test error of ap-northeast-1a and ap-southeast-1a that show the most impact from the parameters. Other availability zones that are not shown in the chart show the pattern in-between. In the legend, the first value separated by comma means the number of days used in the SARIMA modeling step, and the second value after the comma indicates the number of days a model is used after it is built. In ap-southeast-1a zone, the worst configuration shows 50% more error rate than the best configuration. Furthermore, the worst configuration in ap-southeast-1a (150, 1) is the best configuration for ap-northeast-1a. We suspect that such diversity was not considered in the previous works that

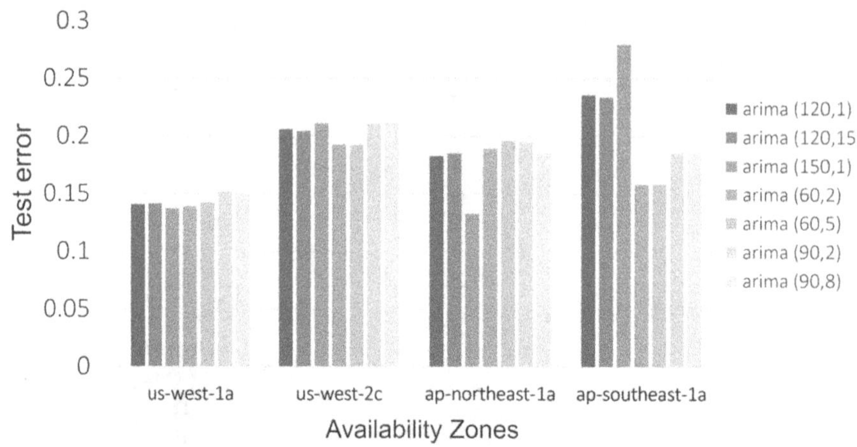

Fig. 3. The impact of train data configurations to the overall test error rate of GPU instance (lower is better).

try to predict spot instance price, and they could not eventually find a model to make better prediction.

To check if there is optimal Arima train data configuration, we list the best performing train data parameters in Table 1. From the table, we can see that there is no globally optimal configurations. Contrary to general belief, building a model with train dataset with shorter time window sometimes performs better than longer train dataset. Furthermore, using a model longer period of time occasionally perform better. From the table, we conclude that predicting spot instance price needs careful consideration in building train dataset, and the configuration needs to be dynamically updated.

4 Discussion and Future Work

With thorough analysis about spot instance price prediction algorithms, we uncover the improved prediction accuracy as well as challenges in making better prediction. Based on the observation, we are going to further improve the algorithm in the following way.

The Good: Spot Price Change Prediction Most of previous works that tried to predict spot instance price concluded that the price is random, and applying predictive analysis algorithms does not really help to improve prediction quality [1,10]. In this work, we applied multiple time-series analysis algorithms by carefully designing the period of modeling data and parameters. With extensive evaluation, we could uncover that applying predictive analysis algorithms improves the price prediction accuracy over 17% on average comparing to a method that uses only the most recent price [5,7].

The Challenge: No Globally Optimal Model Despite of increasing prediction accuracy by applying various techniques, we could not find the globally

Table 1. Best Arima model configuration for different instance types in distinct availability zones

Availability zones	General	Compute	Memory	Storage
ap-northeast-1a	(60,4)	(60,4)	(60,4)	(120,15)
ap-northeast-1c	(60,2)	(60,2)	(120,15)	(120,15)
ap-southeast-1a	(150,1)	(150,1)	(150,1)	(150,1)
ap-southeast-1b	(150,1)	(60,2)	(150,1)	(150,1)
ap-southeast-2a	(60,4)	(60,4)	(150,1)	(150,1)
ap-southeast-2b	(60,4)	(60,2)	(150,1)	(120,15)
eu-west-1a	(60,4)	(90,8)	(30,2)	(150,1)
eu-west-1b	(60,4)	(60,2)	(30,1)	(120,1)
eu-west-1c	(60,2)	(60,4)	(30,1)	(150,1)
us-east-1a	(50,1)	(150,1)	(30,1)	(60,2)
us-east-1c	(150,1)	(150,1)	(30,1)	(150,1)
us-east-1d	(60,2)	(150,1)	(30,2)	(150,1)
us-east-1e	(60,4)	(150,1)	(30,1)	(60,4)
us-west-1a	(60,2)	(150,1)	(30,1)	(120,15)
us-west-1b	(60,4)	(150,1)	(30,1)	(120,15)
us-west-2a	(60,2)	(150,1)	(150,1)	(150,1)
us-west-2b	(60,4)	(150,1)	(30,1)	(150,1)
us-west-2c	(60,2)	(60,2)	(60,2)	(150,1)

optimal algorithm and training data specification for different availability zones and distinct instance types. It makes challenging to apply the algorithms for real applications that can be deployed in any environments.

The Promising: Applying Hybrid Models Even with the diversity of prediction accuracy for different algorithms and train data configuration, it is observed that the train error and test error show high correlation. Pearson product-moment correlation coefficient of train and test error is 0.904 - note that the coefficient has a value from −1.0 to 1.0, and the value of 1.0 means a perfect positive linear correlation, −1.0 means a negative correlation, while 0.0 means no correlation. We currently work on referencing the train error to better choose the algorithms and train data period. We are going to apply the heuristic to an application that utilizes GPU-based AWS EC2 spot instances to execute deep learning tasks in a cost efficient way [5].

The Benefit: Lower Cost while Using Spot Instances With the improvement in the prediction accuracy, we expect it will result in the cost gain by cherry-picking few availability zones and instance types with lower prices. We are working on a theoretical model that specifies correlation between the prediction accuracy and the real cost gain. We are also working to utilize the predicted outcome to anticipate instances that are likely to incur unexpected service interruption that is the crucial factor of making users reluctant to use spot instances. In the prediction step, we try to anticipate the spot instance price of the next 24 h. In an ideal case, if the task migration cost among different availability zones and instance types are negligible, we can issue more frequent migrations by relying on prediction module that has lower prediction error rate as the prediction window becomes shorter. By applying task migration heuristics that are proposed in literature [5,6], we expect to decrease the prediction time window to increase the accuracy.

5 Related Work

Since the introduction of spot instance from AWS EC2 service, many attempts were made to predict price change pattern in the near future. After thorough investigation of spot instance price logs, Ben-Yehuda et al. [1] concluded that applying predictive analysis algorithms in the prediction of spot instance price is meaningless as it changes randomly. Javadi et al. [4] tries to apply statistical model to the price logs by applying MLE method. Though the approach helps to understand the spot price distribution, it does not help to predict future price. Similar to our work, Zhao et al. [10] applied ARIMA model to make spot instance price prediction, but they could not uncover the findings as we do in this paper. The authors performed experiments only for one instance type in a single region. As shown in this paper, the price change pattern differs significant among different environments, and we expect the authors missed the characteristics.

As it was widely known that the spot instance price is hard to be predicted, most of recent work focused on increasing stability of applications that run on spot instances. DeepSpotCloud [5] and Flint [6] proposed fast task migration mechanisms when a service interruption event happens. As the spot price prediction is challenging, both approaches used the naive method [7,11], and we believe that the finding in this paper can significantly improve the cost gain and system stability of the previous approaches.

6 Conclusion

In this work, we try to predict future price for diverse instance types of AWS EC2 spot instances in 18 availability zones using various predictive analysis algorithms (naive, seasonal naive, mean, SARIMA, linear regression, and Prophet) to see if they can help to better predict the future spot instance price. Different from what is generally known, we uncover that SARIMA model performs better than simple methods. To the authors' best knowledge, this is the first work that uncovers applying predictive analysis helps to better predict the future spot instance price. To get better result, we need to tune the train dataset by differentiating the modeling period and model use time, and the tuning steps needs to optimized to further improve the accuracy.

Acknowledgements. This work is supported by the National Research Foundation of Korea (NRF) Grant funded by the Korean Government (MSIP) (No. NRF-2015R1A5A7037615 and NRF-2016R1C1B2015135), the ICT R&D program of IITP (2017-0-00396), and the AWS Cloud Credits for Research program.

References

1. Agmon Ben-Yehuda, O., Ben-Yehuda, M., Schuster, A., Tsafrir, D.: Deconstructing amazon ec2 spot instance pricing. ACM Trans. Econ. Comput. **1**(3), 16:1–16:20 (2013). http://doi.acm.org/10.1145/2509413.2509416
2. Gong, Y., He, B., Zhou, A.C.: Monetary cost optimizations for MPI-based HPC applications on amazon clouds: checkpoints and replicated execution. In: Proceedings of the International Conference for High Performance Computing, Networking, Storage and Analysis, SC 2015, pp. 32:1–32:12. ACM, New York (2015). http://doi.acm.org/10.1145/2807591.2807612
3. Hyndman, R.J., Athanasopoulos, G.: Forecasting: Principles and Practice (2012). https://www.otexts.org/book/fpp
4. Javadi, B., Kondo, D., Vincent, J.M., Anderson, D.: Discovering statistical models of availability in large distributed systems: an empirical study of seti@home. IEEE Trans. Parallel Distrib. Syst. **22**(11), 1896–1903 (2011). doi:10.1109/TPDS.2011.50
5. Lee, K., Son, M.: Deepspotcloud: leveraging cross-region GPU spot instances for deep learning. In: 2017 IEEE 10th International Conference on Cloud Computing (CLOUD) (2017)
6. Sharma, P., Guo, T., He, X., Irwin, D., Shenoy, P.: Flint: Batch-interactive data-intensive processing on transient servers. In: Proceedings of the Eleventh European Conference on Computer Systems, EuroSys 2016, pp. 6:1–6:15, NY, USA. ACM, New York (2016). doi:10.1145/2901318.2901319
7. Sharma, P., Irwin, D., Shenoy, P.: How not to bid the cloud. In: 8th USENIX Workshop on Hot Topics in Cloud Computing (HotCloud 2016). USENIX Association, Denver, CO (2016)
8. Taylor, S.J., Benjamin, L.: Forecasting at scale. In: Facebook Technical Report (2017)
9. Yan, Y., Gao, Y., Chen, Y., Guo, Z., Chen, B., Moscibroda, T.: Tr-spark: Transient computing for big data analytics. In: Proceedings of the Seventh ACM Symposium on Cloud Computing, SoCC 2016, pp. 484–496. ACM, New York (2016). http://doi.acm.org/10.1145/2987550.2987576
10. Zhao, H., Pan, M., Liu, X., Li, X., Fang, Y.: Optimal resource rental planning for elastic applications in cloud market. In: Proceedings of the 2012 IEEE 26th International Parallel and Distributed Processing Symposium, IPDPS 2012, pp. 808–819. IEEE Computer Society, Washington, DC (2012). http://dx.doi.org/10.1109/IPDPS.2012.77
11. Zheng, L., Joe-Wong, C., Tan, C.W., Chiang, M., Wang, X.: How to bid the cloud. In: Proceedings of the 2015 ACM Conference on Special Interest Group on Data Communication, SIGCOMM 2015, pp. 71–84. ACM, New York (2015). http://doi.acm.org/10.1145/2785956.2787473

Block-Incremental Deep Learning Models for Timely Up-to-Date Learning Results

GinKyeng Lee, SeoYoun Ryu, and Chulyun Kim(✉)

Sookmyung Women's University, Cheongpa-ro 47-gil 100, Yongsan-gu, Seoul 04310, South Korea
jinkyung908@naver.com, sunnyryu95@gmail.com, cykim@sm.ac.kr

Abstract. As mobile devices and personal computers have been more frequently used through the Internet, data generated by not only people but also devices have been continuously piling up. The data growing endlessly is called Big Data and Deep Learning algorithms with the Big Data have been introducing the next level of artificial intelligence. It is generally applicable that the more data deep learning algorithms train, the more accurate the deep learning algorithms are. Then, an important problem is which size of data is enough for deep learning algorithms to train the data. In many cases, it is not practical that we wait for the data to grow bigger enough, and thus we need a new learning model that can reduce this latency time and timely derive learning results with useful accuracy. In this paper, we propose novel block-incremental learning models for deep learning and experimentally show that the proposed model can timely derive learning results with useful accuracy and the final accuracy is even better than the traditional deep learning algorithms with the same size of training data.

Keywords: Incremental learning · Learning model · Machine learning

1 Introduction

Machine learning techniques that learn data from a computer, select meaningful results and apply it [1]. Therefore, machine learning becomes a hot topic in many fields simultaneously with the emergence of Big Data [2]. The most important factor for a computer to learn and judge for itself is the amount of computer learning, or amount of data [3].

However, the data accumulates sequentially. Therefore, we must consider the enormous waiting time of the huge training data for computer to learn. That is because if the waiting time to obtain accurate and meaningful results is long, the time to use the results will be delayed. Therefore, when a certain amount of data is input, there is a need to utilize the data to derive the learning results timely, and to derive accurate results and utilize the results.

For example, suppose that you want to learn the purchasing information of a user in a newly opened shopping mall and recommend related products, and let's assume that it is necessary to construct learning data for about one year for meaningful learning.

W. Lee et al. (eds.), *Proceedings of the 7th International Conference on Emerging Databases*, Lecture Notes in Electrical Engineering 461,
https://doi.org/10.1007/978-981-10-6520-0_24

In that case, the service will be available at least one year later. However, failing to provide any service until the customer's information is accumulated for a year is a huge loss to both the company and the customer. In this situation, anyone can feel the need to learn the purchasing information of the user and present the service as timely as possible to the customer. Namely, this timely learning result should not be less accurate than the result of learning by building a year of learning data. Therefore, we propose a model to incrementally learn data and derive timely results as an alternative to this need.

The contribution of this paper is as follows.

- Suggest a new learning model based on incremental learning.
- Confirmed through experiments that the model learned using incremental learning eventually showed a higher accuracy than the existing learning model.
- Using incremental learning to derive timely accuracy results to reduce the application time of the learning model.

The composition of this paper is as follows.

- In Sect. 2, we briefly introduce related work of this paper.
- In Sect. 3, we introduce several approaches to algorithms that produce timely results.
- In Sect. 4, we explain and compares the algorithm introduced in Sect. 2 and the experimental results of the existing algorithm.
- Finally, in Sect. 5, shows conclusions, limitations, and plans obtained from the experimental results in Sect. 4.

2 Related Work

2.1 Basic Machine Learning Method

Computers cannot distinguish dogs and cats with photographs alone, but people can be easily distinguished. To this end, a method called "machine learning" was devised. It is a technique to input a lot of data into a computer and classify similar things. When a photo similar to a stored dog photo is entered, the computer classifies it as a dog photo [4]. There have already been many machine learning algorithms for how to classify data. 'Decision tree', 'Bayesian', 'support vector machine (SVM)' and 'artificial neural network' are representative. Of these, deep running is a descendant of artificial neural networks. Recently, several researches on artificial neural networks, in particular deep neural networks, have been conducted [5]. Among them, the Recurrent Neural Network (RNN) applied a deep neural network structure to the language modeling field [6], and Convolutional Neural Network (CNN) is well applied in the field of computer vision [7].

2.2 Transfer Learning

The model we are proposing is inspired by transfer learning and is based on getting the results by learning the data to be discriminated [8]. Moreover, Data learning with transfer learning shows a one-off. Unlike transfer learning, the model proposed in this paper has a more efficient and lasting effect because it learns repeatedly by gradually accumulating data in order to increase learning accuracy by learning by incremental learning method [9]. That is, in the existing learning method, when there are n pieces of data to be learned, it is possible to see the result after learning all the n pieces of data to be learned, and data can be learned only once.

3 Algorithm

In this chapter, we will introduce and compare 4 algorithms which create timely conclusion with and without incremental learning.

3.1 Definition of n, block, subset and finalT

We use the term of n, block, subset and finalT in this paper. First, n represents the number of training data to reach goal accuracy in original learning model. Block means a fixed quantity of data produced in time T, and the sum of block will get to n as time T increases, and we define correspond T as finalT. Lastly, subset in this paper refers the training dataset of each algorithm for learning. To make this clear, we will use the example from introduction. Suppose newly opened mall has to build 60,000 training data to offer service to their customer and 5,000 data will be gathered in every month. Then, n will be 60,000 and block will be 5000. At the same time, finalT will be 12 as it will take 12 months to get 60,000 of data from 5000 of monthly data. And subset will be 60,000 of data in this learning model.

3.2 Iteration Algorithm

Iteration Algorithm is learning method that use block generated at time T as subset and train this with model newly produced in every T. It can be said that we use finalT existing learning models to train each block. In Fig. 1, it trains the data block (1) generated at T = 1 with Model (1) produced at T = 1, and Block (2) is trained with Model (2) made at T = 2. Finally, it will train block (FinalT) with model (finalT) newly created at T = finalT.

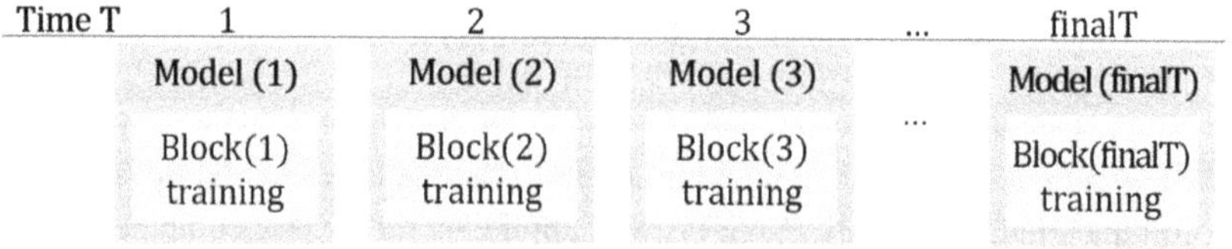

Fig. 1. Iteration learning model

```
iteration( )
{
    for t ←1 to finalT {
        getSubset(subset, Block, t);
        model = train(subset);
    }
}

getSubset(subset, Block, t)
{
    subset.clear()

    i ← Block*t to Block*(t+1)
    subset.add(i);
}
```

Code. 1. iteration algorithm pseudo code

Iteration algorithm uses subset from getSubset ( ). As T updated from 1 to final T, subset will be also renewed with new subset that correspond to current time T for training. And this subset will be trained with model generated in every T.

3.3 iterationNext Algorithm

IterationNext algorithm is very similar to the Iteration Algorithm in that it trains data timely generated in every T. However, in contrast with Iteration Algorithm, IterationNext algorithm trains the block(T + 1) maintaining the model that was used to train the previous block(T). In other words, the result of the previous data block is reflected in the current block training. In Fig. 2, it learns the data block (1) generated at T = 1 with Model (1), and at T = 2, Block (2) is trained with $\text{Model}(1)^2$ reflecting the result of training Block(1). It iterates training in this way until T = FinalT using a model that reflects the training result block (final (T − 1)).

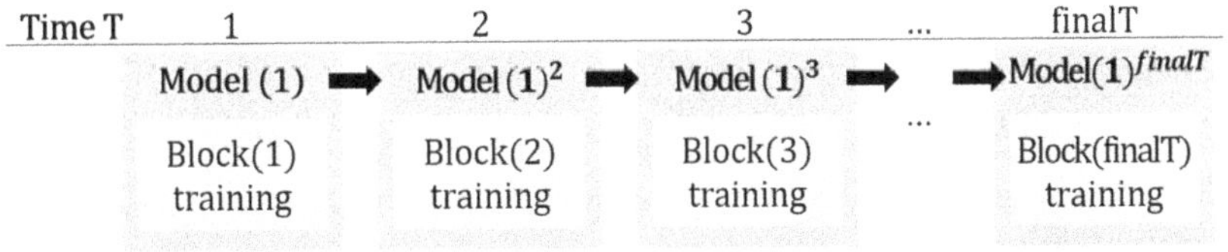

Fig. 2. iterationNext learning model

```
IterationNext()
{
    for t ←1 to finalT {
        getSubset(subset, Block, t);
        if (t == 1)
          model = train(subset);
        else
          model = nextTrain(subset, model); ▷training subset with model
                                              reflecting previous subset.
    }
}
getSubset(subset, Block, t)
{
    subset.clear()

    i ← Block*t to Block*(t+1)
    subset.add(i);
}
```

Code. 2. iterationNext algorithm pseudo code

iterationNext algorithm uses same way to get subset like iteration algorithm. It gets new subset from getSubset ( ) in every T from 1 to final T. But iterationNext algorithm generate new model at T = 1 and it uses model received from time (T − 1) to train block (T) in T > 1.

3.4 BlockIncrement Algorithm

Unlike iteration and iterationNext algorithm which subset is temporarily used, BlockIncrement algorithm train with cumulative subset. But it does not pass the model like iteration algorithm. In Fig. 3, it trains the data block (1) generated at T = 1 with Model (1), and Block (2) is trained with Model (2) made at T = 2. Eventually, it will train with Block (1) +Block (2) +…+Block (FinalT) which is equal to n at time (T).

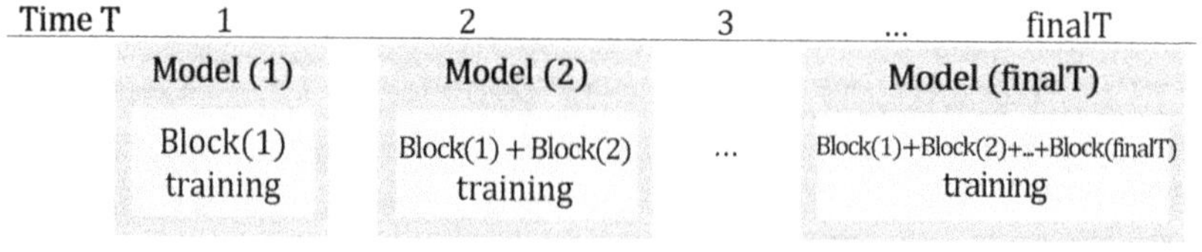

Fig. 3. BlockIncrement learning model

```
BlockIncrement()
{
   for t ←1 to finalT {
      getBlockIncrementSubset(subset, Block, t);
      model = train(subset);
}

getBlockIncrementSubset(subset, Block, t)
{
   i ← Block * t to Block * (t + 1)
   subset.add(i);
}
```

Code. 3. BlockIncrement algorithm pseudo code

BlockIncrement algorithm uses getBlockIncrementdSubset ( ) to get subset for training. Subset will be accumulated as T updated from 1 to final T. In addition, this subset will be trained with model generated in every T.

3.5 BlockIncrementNext Algorithm

BlockIncrementNext is similar to BlockIncrement algorithm in the way that trains with cumulative subset. Although they are very alike, there is a difference between two in that BlockIncrementNext algorithm employ model inherited from former block training. In Fig. 4, after block (1) is trained with model (1), it passes this model to next training for block (1) +block (2). Finally, it will train block(1)+block(2)+……+block (finalT) which is n with model reflecting the result of previous training block(1)+block (2)+……+block(finalT – 1).

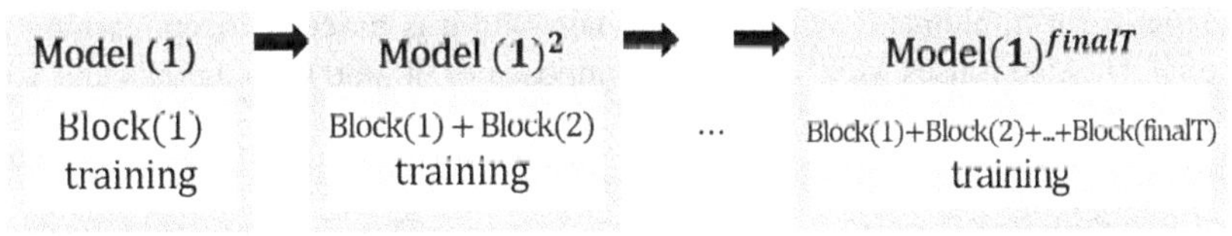

Fig. 4. BlockIncrementNext learning model

```
BlockIncrement_next()
{
   for t ←1 to  finalT {
      getBlockIncrementSubset(subset, Block, t);

      if (t == 1)
          model = train(subset);
      else
          model = nextTrain(subset, model); ▷training subset with model
                                                  reflecting previous subset.
}

   getBlockIncrementSubset(subset, Block, t)
{
   i ← Block * t to Block * (t + 1)
   subset.add(i);
}
```

Code. 4. BlockIncrementNext algorithm pseudo code algorithm

BlockIncrementNext algorithm makes accumulated subset with getBlockIncrementdSubset ( ) for training in every T from 1 to final T. And it will generate model for training subset at T = 1, then this model will be handed over to (T + 1) from T in T > 1.

4 Experimental Result

4.1 Experiment Environment

This experiment implements with java language and it is based on deep learning open source DL4J. Also it uses MNIST dataset composed of 60,000 training data and 10,000 test data [10].

4.2 Experimental Result

The comparison algorithm in this experiment is as follows.

- oriTrain: Existing Learning algorithm.
- iteration: (3.2) Block learning without passing the model configuration.
- iterationNext: (3.3) Block learning with passing the model configuration.
- BlockIncrement: (3.4) Cumulative block learning without passing the model.
- BlockIncrementNext: (3.5) Cumulative block learning with passing the model.

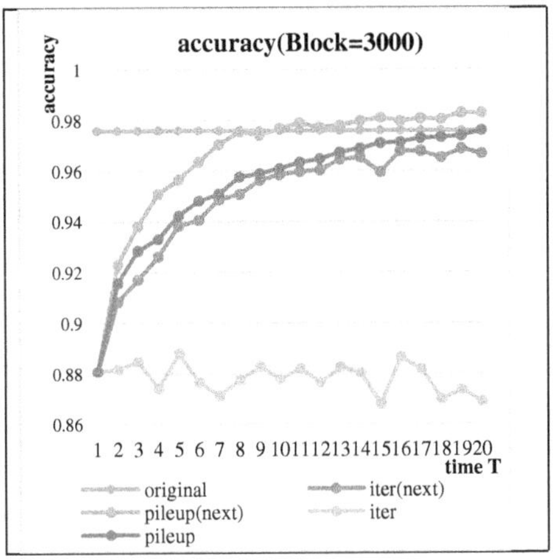

Graph. 1. Comparison of accuracy in Block = 3000

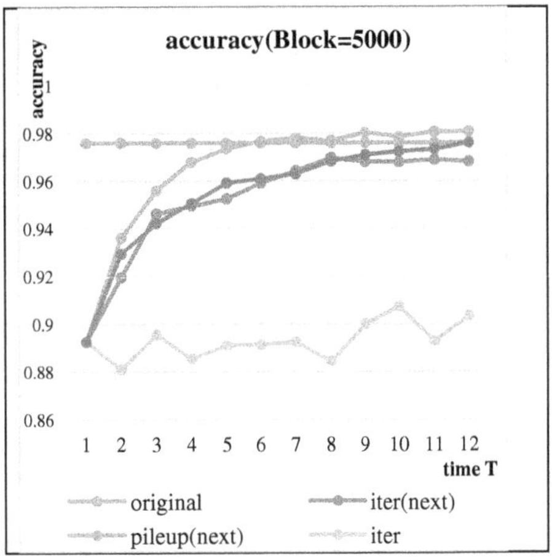

Graph. 2. Comparison of accuracy in Block = 5000

These five algorithms were performed for each of 4 experiments with block = 3000 and block = 5000, and the experimental results were the average of 4 experimental results.

Graph 1 shows the accuracy comparison graph of five algorithms when Block = 3000 and Graph 2 is the graph of five algorithm accuracy comparison when Block = 5000. Both of them show that as the block increases, the detail of the graph is the same even though the accuracy is different.

The accuracy of iteration algorithm and iterationNext algorithm, which derive the results for each block, are much lower than the accuracy of oriTrain. It is because each block generated for each T is trained to produce a timely result, the number of training data is too small for the accuracy to reach the oriTrain's in both training models. Unlike these two, BlockIncrement learns by accumulating blocks. However, since the results of the previous training are not used, the accuracy of the oriTrain is reached only when the final time T = FinalT. But the learning method using the BlockIncrementNext algorithm, which accumulates the blocks, proceeds through training, passes the model, and reflects the previous results. In the early stage of the graph, the accuracy of oriTrain, which is a model that collectively learns n data, is superior to our proposed BlockIncrementNext model. However, blockIncrementNext model makes accuracy increasing rapidly as learning progresses and beyond this accuracy from T * (T * <FinalT). In block 5000, we can confirm that the BlockIncrementNext's accuarcy will be exceeded the accuracy of oriTrain from T = 6 and in block 3000 the accuracy of oriTrain will be exceeded from T = 10. In addition, the accuracy of the blockIncrementNext is maintained higher than the oriTrain continuously from the corresponding T. Accordingly, BlockIncrementNext can be used without training after T * point beyond the accuracy of existing oriTrain. Thus, the model we proposed learns through timely accumulation of data, and the results are timely derived, so that it can be used

more quickly than the existing model. Therefore, we ultimately present the BlockIncrementNext model as a reasonable way to reduce the time loss.

4.3 Differences Between Existing oriTrain and BlockIncrementNext

The total time taken to derive the result from oriTrain is the (waiting time until **n data** accumulates) + (actual training) and the available time from result is above T = FinalT. Whereas, the total time required to obtain the results from the BlockIncrementNext method is (waiting time until **block data** accumulates) + (actual training time). Also available time is above T * (T * <FinalT). We will explain it with the shopping mall example that we mention it earlier. If you have accumulated customer data from January in the shopping mall, results of using the existing learning model will be reliable in February of the next year, the results using BlockIncrementNext learning model will be around July according to the experimental results In addition, you can check whether the accuracy from learning is sufficient to get usable results and at the same time, you can stop the learning if the accuracy is adequate. To check the results of January data and use if accuracy is enough, it will take {(1 month) + (January data training time)}. Then, you realize that subset is not enough yet. However, this process is necessary. This is because the model used in this process will help to improve accuracy in the next train. Training will be continued until you get June data block as accuracy is high enough to use when you check.

To sum up, the BlockIncrementNext model has advantages such that there is no long waiting time to accumulate data compared with the existing model and that it can be used quickly with high accuracy at T* (T* <finalT).

5 Conclusion

We present a BlockIncrementNext model as a rational approach to the need to timely derive learning outcomes, and to derive accurate results and utilize the results. This model will reduce the time lag between data accumulation and reduce the time loss and ensure a faster time to use than the existing model with high accuracy results.

However, the BlockIncrementNext model has the limitation that the block time is accumulated and the learning time is longer than the origin because BlockIncrementNext model accumulates blocks and learns the accumulated data. When learning in this way, there is a limit point that has longer learning time than existing learning model. We will suggest new models to complement these problems in future work. Despite these limitations, this model can have positive effects such as high accuracy, timely response, and quick availability of meaningful results compared to existing learning models. Therefore, in this paper, we propose a BlockIncrementNext model with such meaningful results.

Acknowledgments. This research was supported by Support Program for Women in Science, Engineering and Technology through the National Research Foundation of Korea (NRF) funded by the Ministry of Science, ICT and future Planning (No. 2016H1C3A1903202).

References

1. Frank, E.: Machine Learning Techniques for Data Mining. Lecture Notes. University of Waikato, New Zealand, 25 October 2000
2. 정용찬. 『빅데이터 혁명과 미디어 정책 이슈』 (KISDI Premium Report 12-02). 정보통신정책연구원 (2012)
3. Domingos, P.: A few useful things to know about machine learning. Commun. ACM **55**(10), 78–87 (2012)
4. Vedaldi, A.: Cats and dogs. In: Proceedings of the 2012 IEEE Conference on Computer Vision and Pattern Recognition (CVPR), pp. 3498–3505. IEEE Computer Society, June 2012
5. Schmidhuber, J.: Deep learning in neural networks: an overview. Neural Netw. **61**, 85–117 (2015)
6. Mikolov, T., Karafiát, M., Burget, L., Cernocký, J., Khudanpur, S.: Recurrent neural network based language model. In: Interspeech, vol. 2, p. 3, September 2010
7. Kim, Y.: Convolutional neural networks for sentence classification. In: Interspeech, vol. 2, p. 3 (2014). arXiv preprint: arXiv:1408.5882
8. Raina, R., Battle, A., Lee, H., Packer, B., Ng, A.Y.: Self-taught learning: transfer learning from unlabeled data. In: Proceedings of the 24th International Conference on Machine Learning, pp. 759–766. ACM, June 2007
9. Cauwenberghs, G., Poggio, T.: Incremental and decremental support vector machine learning. In: NIPS, vol. 13, December 2000
10. https://deeplearning4j.org

Harmonic Mean Based Soccer Team Formation Problem

Jafar Afshar[1(✉)], Arousha Haghighian Roudsari[1], Charles CheolGi Lee[1], Chris Soo-Hyun Eom[1], Wookey Lee[1], and Nidhi Arora[2]

[1] Industrial Engineering, INHA University, 100, Inha-ro, Nam-gu, Incheon 22212, Republic of Korea
{jafar.afshar, arousha.haghighian, rich, sunching, trinity}@inha.edu

[2] GD Goenka World Institute, Lancaster University, Haryana, India
rustagi.nidhi@gmail.com

Abstract. This paper provides a review of the literature on team formation problem. We illustrate the review based on two different classifications; operations research and data mining. The papers in each class are then described according to their contributions to the literature. We aim to facilitate the chasing of conducted works in relevant area of interest and to identify trends and fields which should be topic to future studies. Also, we present the problem of the existing Team Formation method and showed our approach outperformed the conventional approaches, where we explained not only the abilities of the players but also the harmony has been one of the key factors that has been worked in the real world soccer datasets.

Keywords: Team formation problem · Operations research · Data mining

1 Introduction

Most of the tasks, that people usually perform, demand some degree of collaboration and communication which can be named as teamwork [3, 12]. The importance of project management and teamwork in engineering and computer science is undeniable [9, 11]. This importance manifests itself in the attraction of many researchers to conduct research on the problem called Team Formation.

A set of interrelated tasks which have been planned to be executed within certain budget, time, and other limitations can be accomplished with a group (team) of experts who have a full set of complementary skills. However, the fulfillment of a project is not only associated with the level of expertise that experts required, but also an effective collaboration and communication among team members. Thus, the emerging problem to tackle is selecting a team of experts with a minimum cost, who cover the required skills of the project and able to have an effective communication and collaboration with each other.

Concerning the literature of the team formation problem, the research papers published in this topic can be classified into two communities; Operations Research

W. Lee et al. (eds.), *Proceedings of the 7th International Conference on Emerging Databases*, Lecture Notes in Electrical Engineering 461,
https://doi.org/10.1007/978-981-10-6520-0_25

(OR) and data mining [6, 10]. The difference between these two communities lies in the existence of social network among individuals. Unlike the research papers published in the operation research community, the communication and collaboration among team members have been considered in the data mining related publications.

In this paper, we aim to provide a review of the literature on team formation. The relevant manuscripts have been extracted from different databases such as DBLP, Web of Science, etc. Some additional publications have been also found from the cited papers in the references of the obtained manuscripts. We ended up with a total number of 36 research papers. A diagram to illustrate the distribution of the manuscripts according to their publication date is shown in Fig. 1.

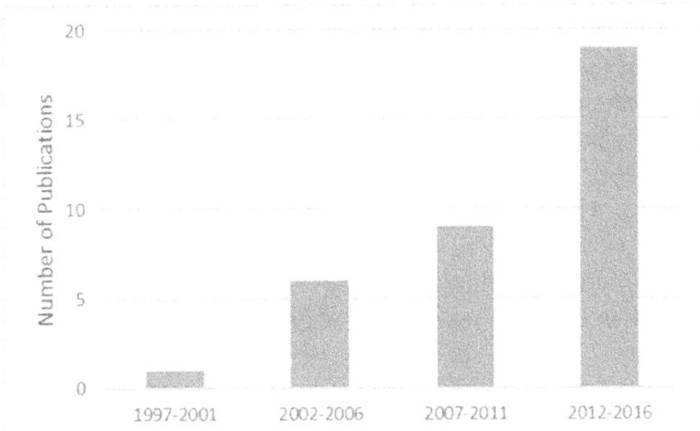

Fig. 1. Number of papers according to Publication date

We organize the literature of team formation using two perspectives; 1. Studies conducted without considering social network which are in the field of Operations Research (OR) class. 2. Studies conducted with considering social network in data mining category. In such a way, a researcher can probe for a list of papers according to his or her specific needs.

2 Related Works on Team Formation Problem

Traditionally, team formation has been discussed in the field of operational research. As mentioned earlier, most of the research conducted in operations research community ignored the network structure of the team members and the focus is limited on only their expertise level. Generally, team formation studies in OR are based on a trend in which the problem is formulated as an integer linear program (ILP) to obtain an optimal match between individuals and the desired functional requirements [8]. In these research studies, the problem is often tackled applying techniques such as genetic algorithm [13], branch-and-cut [15], and simulated annealing [2].

In [15], Zakarian and Kusiak proposed a methodology for team formation which is based on the Analytical Hierarchy Process (AHP) and the Quality Function Deployment

(QFD) method. They also developed an integer programming model to form the multi-functional teams in which the team members are prioritized based on engineering characteristics. The authors asserted that their research is the first work which uses an analytical approach for building teams. A similar model to the Zakarian's one was proposed by Boon and Sierksma to boost [2] the quality of a team. The authors formulated linear optimization models to calculate the optimal teams for assigning the right candidates to the right team.

Wi et al. [13] analyzed the knowledge of the individuals to form new teams with managers and used a genetic algorithm to select team managers and members. In this research, a fuzzy inference system was used to assess the knowledge of the candidates and their familiarity. Using similar approach, Baykasoglu et al. [1] developed a fuzzy optimization model and solved it using a simulated annealing algorithm to build a team for a given project. To do so, the budget and time limitations of a project and interpersonal relations of team members were brought into consideration.

Unlike the research studies in OR community, the network structure between candidates has been considered in data mining papers for team formation problem. To take the social network into account, it is often modeled as a graph, where the nodes represent candidates, who possess one or more than one skills, and the edge between pair of candidates is weighted as either the communication cost (distance) or the similarity of the two corresponding candidates.

For the first time in the literature, Lappas et al. [8] introduced the problem of finding a team of candidates from a social network. They tried to build a team which encompasses the required candidates for the project accomplishment so that the candidates can collaborate and communicate with each other effectively. In this study, the authors estimated the communication cost of the team by proposing two functions; diameter communication cost (Cc-R) and minimum spanning tree communication cost (Cc-MST). However, Kargar and An [5] asserted that the communication cost by the functions proposed in [8] might not be measured well when the communication of two skills is brought into consideration. To support their assertion, they stated that the diameter function only calculates the communication cost of two candidates that are far from each other, and the communication costs of all required candidates are not measured by MST function. Furthermore, they pointed out that both functions are sensible to even a slight change to the graph due to the instability essence of them. Hence, the authors in [5] proposed two new communication cost functions (the sum of distances and the leader distance) to overcome the limitations of [8] ones. They tackled the problem and the procedure of finding top-k teams of candidates with/without a leader. They also proved that one of the functions proposed is NP-hard and designed an approximation algorithm to solve it.

Later, Kargar et al. [6] developed a bi-objective function to consider both node attributes and arc values. In the function proposed, each node represented a candidate who is associated with a weight as the personnel cost, and the weight on the edges represented the communication costs between pairs of candidates. The authors used the new combined cost function to select a team of candidates that covers all the required skills and minimizes the communication and personnel costs of the project. They proved that the problem with new cost function is NP-hard and proposed four algorithms (an approximation algorithm and three heuristic algorithms) to solve the problem.

Similar objective to [5], Juang et al. [4] proposed two efficient algorithms (BCPruning and SSPruning algorithms) to find a team of candidates for covering the required skills as well as minimizing the communication cost. By proposing BCPruning algorithm, the better initial leader candidates, for obtaining a lower communication cost, were selected and this results in an effective pruning of candidates. On the other hand, a lower bound of communication cost for each leader was found by SSPruning algorithm to prune some candidates without any computation. In [14], Yang et al. proposed approximation algorithms for team formation problem to find near optimum teams. The problem was presented on a novel heterogeneous network structure which form the team by prioritizing the skills of the candidates. The information used in the paper was extracted from research news for constructing a co-occurrence network.

Surprisingly, in the last decades, majority of the manuscripts related to team formation problem have been emerged in data mining community [7, 10, 12]. This issue manifests itself in the papers such as [5, 6, 8, 14] so that their complementary role in this specific topic is undeniable. This research is led to the discovery of the theoretical background on the team formation problem and represented the problem definition with formal notation as following section.

3 Problem Definition and Solution Framework

The existing methods constitute a team by considering simple harmony among players or player ability, but we construct a team by incorporating both of them. In order to see our method Harmonic mean, let's see the followings.

Let's assume that a candidate of team, C, and a set of position or skill, $S = \{s_1, s_2, \ldots, s_m\}$. For a given candidate $c_i \in C$ and his position ability $c_i = \{s_1^i, s_2^i, \ldots, s_m^i\}$, a team T is denoted by $T = \{c_1, c_2, \ldots, c_m\}$. Then, we define the harmonic mean for the team as:

$$Harmonic\, mean = \alpha \times T_{node} + (1 - \alpha) \times T_{arc}$$

where T_{node} is the sum of the personal abilities of the team members, and T_{arc} is the sum of the team members' harmony between the team members as follows:

$$T_{arc} = \sum_{c_i, c_j \in T \text{ and } i \neq j} \frac{\{s_1^i(s_2^j + s_3^j + \ldots + s_m^j) + \ldots + s_m^i(s_1^j + s_2^j + \ldots + s_{m-1}^j)\}}{m}$$

For example, given two team members $c_p = \{s_1^p, s_2^p, s_3^p\}$ and $c_q = \{s_1^q, s_2^q, s_3^q\}$ and $m = 3$, $T_{arc}(c_p, c_q) = s_1^p(s_2^q + s_3^q) + s_2^p(s_1^q + s_3^q) + s_3^p(s_1^q + s_2^q)$ or $T_{arc}(c_p, c_q) = s_1^q(s_2^p + s_3^p) + s_2^q(s_1^p + s_3^p) + s_3^q(s_1^p + s_2^p)$.

In this paper, the algorithm and other related details will be removed for the size limitation and the explanation also explained in the conference. In short, this kind approach is the first approach for the team formation problem to the best of our knowledge, and in the experiment our approach outperformed the conventional approaches in terms of time and efficiency that represented in the next section.

4 Experiment and Discussion

We applied existing method and our approach a real world FIFA data set of the year 2010–2011 where some of the interesting data has been selected and exploited in this paper such as FC Barcelona, Manchester United, and Real Madrid players in FIFA 11 and compared it with other tactics and other teams in the same team.

Figure 2 shows the experimental results for our method and the existing methods, where we can see that the players in the three teams are excellent in the personal abilities (see the last bars of the teams) that is easy to guess. The team harmony, however, of the tem FC Barcelona (year 2010–2011) was much better than the other two teams, which can the reason why the team won UEFA champions league at that time.

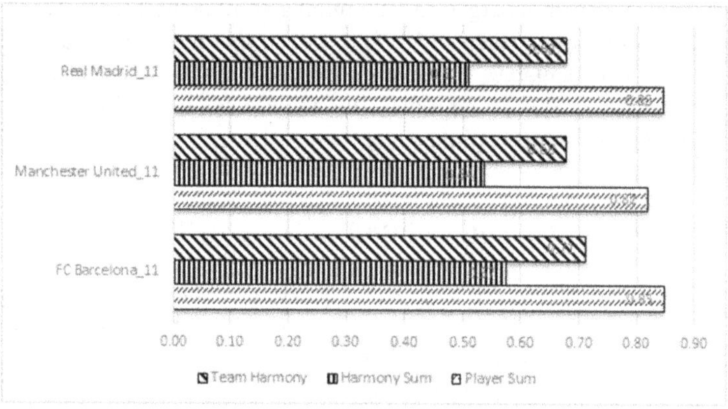

Fig. 2. Result for existing methods and our method

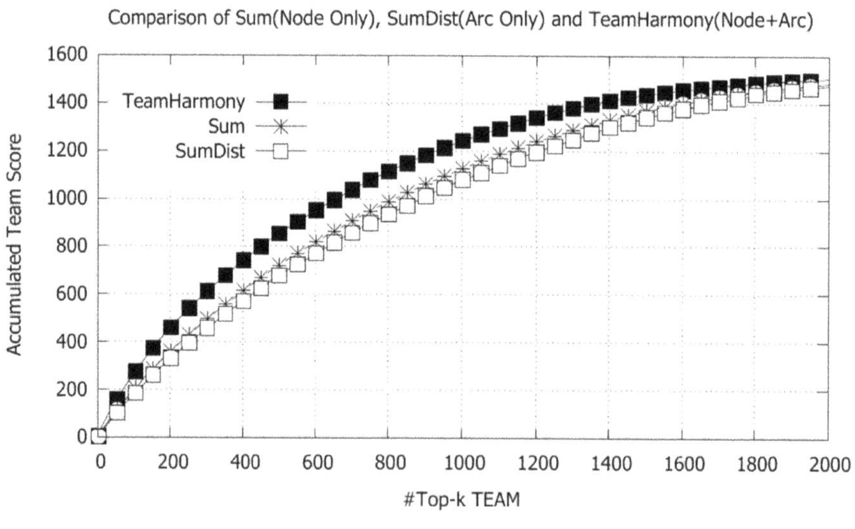

Fig. 3. Comparison with Team Harmony Method and Sum of the distance method

Figure 3 shows the experimental results for our method and the existing method, the sum of the distance approach and the simple sum, where the accumulated score of the team formation of our approach is superior to the other two approaches with respect to the top-k increasing. Note that the node sum approach considers only the abilities of the players that has been the conventional way of research, and the arc sum approach considers only the relationship between the players of the team. The our team harmony approach represented always the best among all the three approaches up to the numbers to the last combination possibility. We can claim that our team harmony approach has been the best of all the team formation methods for any number of team members.

5 Conclusion

In this paper, we provided a review of the literature on team formation problem. We illustrated the literature from two different communities such as operations research community where the social network among team members is not considered in the manuscripts. And the data mining community where the team members collaborate and communicate with each other on team formation problem. The review is to facilitate the chasing of conducted works in relevant area of interest and to identify trends and fields which should be topic to future studies. Also, we present the problem of the existing Team Formation method and showed our approach outperformed the conventional approaches, where we explained not only the abilities of the players but also the harmony worked, and we showed that it was why FC Barcelona won in UEFA champions league in year 2010–2011.

Acknowledgement. "This work was supported by the National Research Foundation of Korea (NRF) grant funded by the Korea government(MSIP) (No. NRF-2016R1A2B4014245)."

References

1. Baykasoglu, A., Dereli, T., Das, S.: Project team selection using fuzzy optimization approach. J. Cybern. Syst. **38**(2), 155–185 (2007)
2. Boon, B.H., Sierksma, G.: Team formation: Matching quality supply and quality demand. Eur. J. Oper. Res. **148**(2), 277–292 (2003)
3. Galegher, J., Kraut, R.E., Egido, C.: Intellectual Teamwork: Social and Technological Foundations of Cooperative Work. Psychology press, New York (2014)
4. Juang, M.C., Huang, C.C., Huang, J.L.: Efficient algorithms for team formation with a leader in social networks. J. Supercomput. **66**(2), 721–737 (2013)
5. Kargar, M., An, A.: Discovering top-k teams of experts with/without a leader in social networks. In: Proceedings of International Conference on Information and Knowledge Management, pp. 985–994 (2011)
6. Kargar, M., An, A., Zihayat, M.: Efficient bi-objective team formation in social networks. In: Machine Learning and Knowledge Discovery in Databases, pp. 483–498 (2012)
7. Kim, J., Lee, W., Song, J., Lee, S.: Optimized combinatorial clustering for stochastic processes. Cluster Comput. **20**(2), 1135–1148 (2017)

8. Lappas, T., Liu, K., Terzi, E.: Finding a team of experts in social networks. In: Proceedings of ACM SIGKDD, pp. 467–476 (2009)
9. Lingard, R., Barkataki, S.: Teaching teamwork in engineering and computer science. In: Proceedings of FIE (2011)
10. Park, S., Song, J., Lee, J., Lee, W., Ree, S.: How to measure similarity for multiple categorical data sets? Multimedia Tools Appl. **74**(10), 3489–3505 (2015)
11. Smith, K.A., Imbrie, P.K.: Teamwork and Project Management. McGraw-Hill, New York (2005)
12. Lee, W., Leung, C., Lee, J.: Mobile web navigation in digital ecosystems using rooted directed trees. IEEE Trans. Ind. Electron. **58**(6), 2154–2162 (2011)
13. Wi, H., Oh, S., Mun, J., Jung, M.: A team formation model based on knowledge and collaboration. Expert Syst. Appl. **36**(5), 9121–9134 (2009)
14. Yang, J., Li, M., Wu, B., Xu, C.: Forming a research team of experts in expert-skill co-occurrence network of research news. In: Proceedings of ASONAM, pp. 1143–1150 (2016)
15. Zakarian, A., Kusiak, A.: Forming teams: an analytical approach. IIE Trans. **31**(1), 85–97 (1999)

Generating a New Dataset for Korean Scene Text Recognition with Augmentation Techniques

Mincheol Kim and Wonik Choi(✉)

Inha University, Incheon 22212, Korea
mckim8686@gmail.com, wichoi@inha.ac.kr

Abstract. Korean text recognition in a natural scene is a challenging task due to the complexity of character shapes and the lack of dataset comparing to English or other languages. In this paper, we present a new dataset with the goal of improving the recognition of Korean natural scene text. Our dataset is generated by data augmentation techniques without losing a reality. The number of augmented images is 3 million and these images are made up of about 30 non-commercial fonts and 511,000 words from a standard Korean language dictionary. This enormous amount of data offers new possibilities for training deeper neural networks. In our extensive experiments, results show that our dataset effectively trains convolutional recurrent neural networks that achieve state-of-the-art performance on the Korea Advanced Institute of Science & Technology (KAIST) scene text database with very few data-acquisition costs.

Keywords: Scene text recognition · Data augmentation · Neural network

1 Introduction

Scene text recognition, which recognizes text in natural images, is a challenging problem but widely useful for automatically identifying signboards and guideboards, etc. While the recognition of English text within natural images has been well studied in [1–4], Korean text recognition is still a challenging problem due to the complexity of character shapes and the lack of a dataset. There have been many efforts [5–10] to recognize or detect Korean text in natural scenes or in handwritten text. Handwritten text recognitions [5, 7, 8] and detection [6] methods have been proposed, however, the maximum number of images per class used in these papers is only 2,187 that is relatively smaller than those in the Modified National Institute of Standards and Technology (MNIST) dataset [11], which has 6,000 instances per class. This lack of dataset gets worse in case of natural scene text detection and recognition. The dataset used in experiments has only 113 images in [10] and 445 images in [9]. To solve the lack of a dataset on a text recognition in English natural scene text images, there have been many efforts [2, 3, 12] to generate an augmented scene text dataset.

In this paper, we present a new dataset that contains 511,000 words from a standard Korean language dictionary with more than 30 non-commercial fonts. In total, the dataset has 3 million images with 3269 characters. In contrast to the latest Korean scene

W. Lee et al. (eds.), *Proceedings of the 7th International Conference on Emerging Databases*, Lecture Notes in Electrical Engineering 461,
https://doi.org/10.1007/978-981-10-6520-0_26

text database [13, 14], our dataset has more classes, instances per classes, and images. Convolutional recurrent neural networks trained with our augmented dataset are almost the same accuracy as the state-of-the-art technique.

Our scene text synthesis system is drawn in Sect. 2, our experimental results with state-of-art method, convolutional recurrent neural networks (CRNNs) [4], is described in Sect. 3, and Sect. 4 concludes the paper.

2 Scene Text Synthesis System

Supervised learning of deep neural networks such as CRNNs, which have millions of trainable parameters, needs a huge training dataset with labels. While datasets [14–19] are publicly available, the datasets have a very limited number of image, and even text in the dataset [15, 16] is a list of characters, not words. In the real world, it is a very demanding task that is almost impossible to collect millions of a dataset, especially natural text scene. To cope with this problem, many researchers have proposed methods [2, 3, 20] that synthesize a dataset.

Ku-1 dataset has 1,500 characters and 1,000 images. PE92, PHD08, CBNU DB and KAIST DB have 2,350 characters and the number of images in datasets ranges from 100 to 2,470. Ku-1, PE92 and PHD08 consist of characters, not words. CBNU DB and KAIST DB contains words. Ku-1 and PE92 are handwritten text, PHD08 is printed matters, CBNU DB and KAIST DB are natural scene images. Our dataset consists of 3,269 characters, has 3 million images and is synthetic natural scene images. Characteristics of each dataset are shown in Table 1.

Table 1. Characteristics of Datasets

	Type	# of characters	# of images	# of fonts
Ku-1 [18]	Handwritten	1,500	1,000	–
PE92 [17]	Handwritten	2,350	100	–
PHD08 [15]	Printed	2,350	2,187	9
CBNU DB [19]	Natural scene	2,350	2,250	–
KAIST DB [14]	Natural scene	2,350	2,470	–
Ours	Synthetic	3,269	3,000,000	30

In this section, we describe our scene text synthesis system, which helps deeper networks trained with an augmented dataset. Our system generates realistic scene text images under hypothesis that text in the wild is generally rendered by printing devices with various fonts, styles and backgrounds, so that a dataset that is generated by our system gives the trained models generality to real images. The system consists of four steps: text and shadow rendering, random curve and rotating in 3D, blending text and background with natural images and merging layers. The pipeline of our system can be illustrated as follows.

1. Text and shadow rendering – our system renders text and its shadow for randomly selected font and each word in dictionary. Font size is varied from uniform distribution, text and shadow color are sampled from color clusters that are generated from natural images [21] and gradation is randomly applied to text.
2. Random curve and rotating in 3D – a rendered text follows a random curve that has various amplitude and frequency and rotates in x- and z-axis according to uniformly distributed random probabilities.
3. Blending text/background with natural images – text and colored background are blended with randomly selected natural images. Blended method is arbitrarily selected from addition, multiply, subtract, difference, etc.
4. Merge all layers – merges all layers with background, shadow and text.

The whole process of our system is depicted in Fig. 1. The number of total images is 3 million and our dataset contains 3,269 character classes and 511,000 words. Each class has 5,000 images in average and it is relatively huge size comparing to KAIST scene text dataset [14] that is made up of about 2,470 images, including Korean and English.

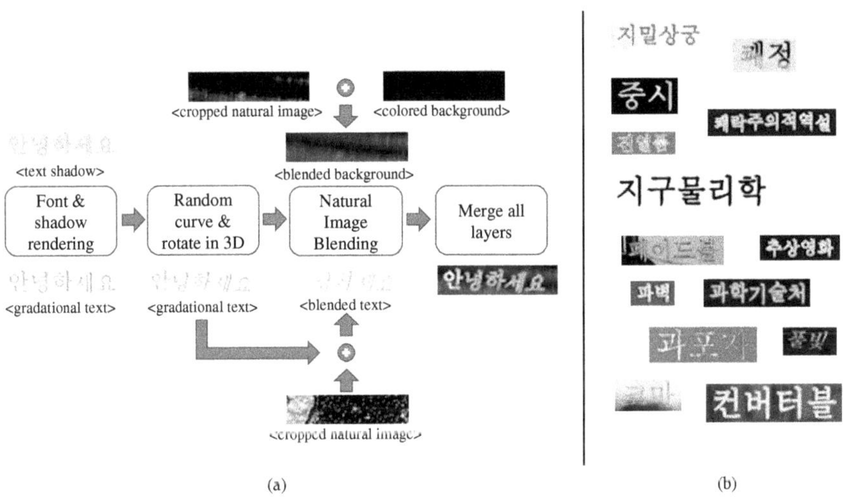

Fig. 1. (a) The pipeline of a scene text synthesis system. (b) samples from the dataset generated by the scene text synthesis system.

3 Experimental Results

In this section, we show the validity of dataset that is generated using an augmented approach. To verify our dataset, we train a convolutional recurrent neural network [4] that shows the most effective recognition technique for English and we modified the number of output classes to 3269. The structure of the convolutional recurrent neural networks used to validate our dataset is illustrated in Fig. 2. Our CRNN networks is similar to that of Shi et al. [4], but outputs for recurrent layers are extended to 3269 as

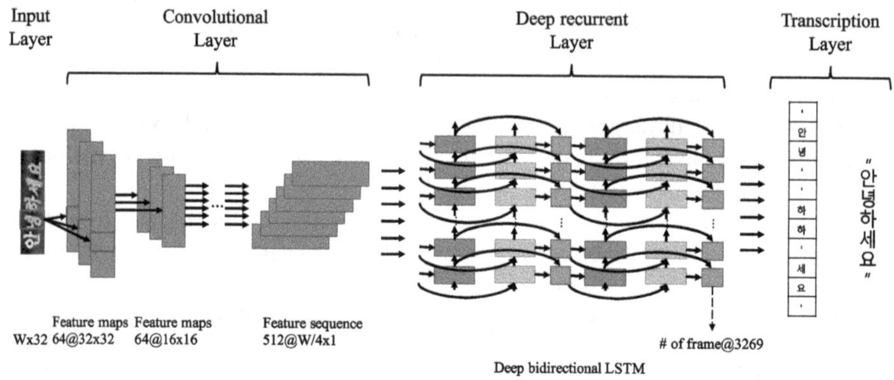

Fig. 2. Illustration of convolutional recurrent neural networks for Korean scene text recognition.

depicting in Fig. 2. Experiments were conducted on a machine with an Intel Xeon E5-2620 v3 CPU, 32 GB of RAM, which is equipped with an NVIDIA Tesla K40m GPU. An experimental environment is depicted in Table 2. We use ADA-DELTA [22] to train the network, setting the parameter ρ to 0.9. The training process takes about three days to reach convergence. To test the network, we use a real dataset [14] as a test dataset. The result shows that the trained network with our dataset achieves almost same accuracy of 74.8%, as state-of-the-arts. Figure 3 shows results of recognition of three samples from our synthetic dataset in epoch 10, 45 and 180. For an image in the first row, the first two characters are incorrectly in epoch 10. In epoch 45, second character is improperly recognized and, in epoch 180, whole characters are perfectly recognized. The recognition accuracy of the others in Fig. 3 is also improved similar to the first case.

Table 2. Experimental Environment

Device	Specification	
CPU	Xeon E5-2620 v3 @ 2.4 GHz	
Main memory	32 GB	
GPU	NVIDIA Tesla K40m	
	# of Cores	2880
	Memory	12 GB
	Compute Capability	3.5

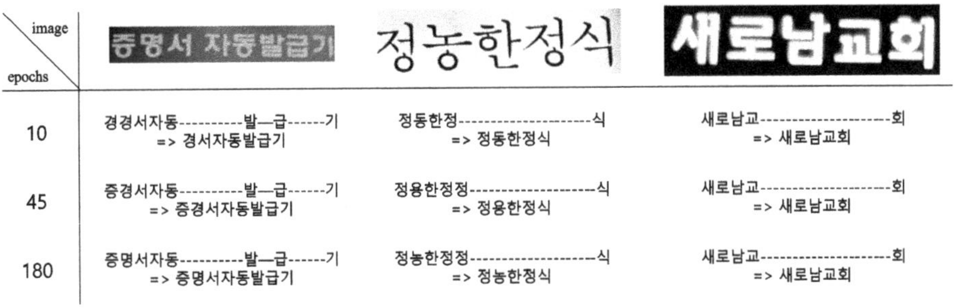

Fig. 3. Recognition results in epoch 10, 45 and 180

4 Conclusion

In this paper, we have introduced a new dataset with the goal of improving the recognition of Korean natural scene text. The dataset was generated by data-augmentation techniques that consist of four steps including text/shadow rendering, perspective distortion, blending text with natural images and merging layers without data-collection effort. Using 511,000 words from a standard Korean language dictionary and 30 Korean fonts, 3 million images were generated without loss of reality. Then, as we trained CRNNs with our dataset, we showed the validity of the augmented dataset and achieved adequate accuracy of 74.8%.

In this paper, we introduced cropped images that contain only Korean text without a background region. However, in the future works, we will generate a new dataset depicting a whole scene, including not only characters but also real-world backgrounds. And we will also propose fully integrated networks to detect a location of text and recognize characters in a natural scene text. Then we will publicly open our dataset on our project website.

Acknowledgement. This work was supported by the Korea Institute of Energy Technology Evaluation and Planning(KETEP) and the Ministry of Trade, Industry & Energy(MOTIE) of the Republic of Korea (No. 20161210200610).

References

1. Marial, A., Jibrael, J.: Feature extraction of optical character recognition: survey. Int. J. Appl. Eng. Res. **12**(7), 1129–1137 (2017). ISSN:0973-4562
2. Wang, T., Wu, D.J., Coates, A., Ng, A.Y.: End-to-end text recognition with convolutional neural networks. In: 2012 21st International Conference on Pattern Recognition (ICPR), pp. 3304–3308 (2012)
3. Jaderberg, M., Simonyan, K., Vedaldi, A., Zisserman, A.: Synthetic data and artificial neural networks for natural scene text recognition, CoRR, vol. abs/1406.2 (2014)
4. Shi, B., Bai, X., Yao, C.: An end-to-end trainable neural network for image-based sequence recognition and its application to scene text recognition. CoRR, vol. abs/1507.0 (2015)

5. Park, K.-W., Kim, B.-H., Han, D.-S., Son, S., Kang, W.-Y., Zhang, B.-T.: Handwritten hangul recognition using multi-column deep neural networks. J. Korean Inst. Intell. Syst. **26** (1), 159–160 (2016)
6. Kim, S., Jeong, S., Lee, G.-S., Suen, C.Y.: Word segmentation in handwritten Korean text lines based on gap clustering techniques. In: Proceedings of Sixth International Conference on Document Analysis and Recognition, pp. 189–193 (2001)
7. Sung Kim, W., Park, R.-H.: Off-line recognition of handwritten Korean and alphanumeric characters using hidden Markov models. Pattern Recognit. **29**(5), 845–858 (1996)
8. Kim, Y., Cha, E.: Streamlined GoogLeNet algorithm based on CNN for Korean character recognition. J. Korea Inst. Inf. Commun. Eng. **20**(9), 1657–1665 (2016)
9. Park, J., et al.: Automatic detection and recognition of Korean text in outdoor signboard images. Pattern Recognit. Lett. **31**(12), 1728–1739 (2010)
10. Lee, J., Park, J.-S., Hong, C.-P., Seo, Y.-H.: Illumination-robust foreground extraction for text area detection in outdoor environment. KSII Trans. Internet Inf. Syst. **11**(1), 345–359 (2017)
11. Lecun, Y., Bottou, L., Bengio, Y., Haffner, P.: Gradient-based learning applied to document recognition. Proc. IEEE **86**(11), 2278–2324 (1998)
12. Gupta, A., Vedaldi, A., Zisserman, A.: Synthetic data for text localisation in natural images. CoRR, vol. abs/1604.0 (2016)
13. Jung, J., Lee, S.H., Cho, M.S., Kim, J.H.: Touch TT: Scene text extractor using touchscreen interface. ETRI J. **33**(1), 78–88 (2011)
14. KAIST Scene Text Database - TC11, http://www.iapr-tc11.org/mediawiki/index.php?title=KAIST_Scene_Text_Database. Accessed 13 June 2017
15. Ham, D.-S., Lee, D.-R., Jung, I., Oh, I.-S.: Construction of printed hangul character database PHD08 TT - construction of printed hangul character database PHD08. J. Korea Contents Assoc. **8**(11), 33–40 (2008)
16. PHD08 Dataset, http://cv.jbnu.ac.kr/index.php?mid=notice&document_srl=189. Accessed 17 Mar 2017
17. Kim, D.H., Bang, S.Y.: An overview of hangul handwritten image database PE92 TT - an overview of hangul handwritten image database PE92. In: Proceedings of the 4th Annual Conference on Human and Cognitive Language Technology, pp. 567–575 (1992)
18. Kim, D.-I., Kim, S.-Y., Lee, S.-W.: Design and construction of a large-set off-line handwritten hangul character image database Ku-1 TT - design and construction of a large-set off-line handwritten hangul character image database Ku-1. In: Proceedings of the 9th Annual Conference on Human and Cognitive Language Technology, pp. 152–159 (1997)
19. Heo, G.-S., Oh, I.-S.: Sign image database collected at Jeonju Hanok Village TT - sign image database collected at Jeonju Hanok Village. J. Korea Contents Assoc. **6**(11), 243–248 (2006)
20. Goodfellow, I.J., et al.: Generative adversarial nets. Adv. Neural. Inf. Process. Syst. **27**, 2672–2680 (2014)
21. Lucas, S.M., Panaretos, A., Sosa, L., Tang, A., Wong, S., Young, R.: ICDAR 2003 robust reading competitions. In: Proceedings of the Seventh International Conference on Document Analysis and Recognition, p. 682 (2003)
22. Zeiler, M.D.: ADADELTA: An Adaptive Learning Rate Method. CoRR, vol. abs/1212.5, December 2012

Markov Regime-Switching Models for Stock Returns Along with Exchange Rates and Interest Rates in Korea

Suyi Kim, So-Yeun Kim, and Kyungmee Choi(✉)

Hongik University, Sejong-ro 2639, Sejong 30016, Republic of Korea
Kmchoi@hongik.ac.kr

Abstract. We apply the Hamilton 2-regime Markov Switching model to the stock returns along with exchange rates and interest rates from January 1993 to December 2016 in Korea. Two regimes are distinct in the Korean stock market. In regime 1 with low-volatility, the stock returns of Korea are significantly affected first by their exchange rates and secondly by their interest rates. More precisely, both exchange rates and interest rates negatively influence the stock returns during relatively stable periods in Korea. In regime 2 with high-volatility, the Korean stock market is explained by none of the two explanatory variables.

Keywords: Markov regime switching model · Stock returns · Exchange rates · Interest rates

1 Introduction

Since the Asian financial crisis and the global financial crisis, the regime shift behavior has been more often recognized in the stock markets. Before the two crises, many empirical studies have already focused on the regime shift behavior or structural breaks in stock market volatility. Hamilton [1] developed Markov regime switching models (MRSM). Cai [2] developed Markov Switching Autoregressive Conditional Heteroscedasticity (MS-ARCH) model. Gray [3] developed Markov Switching Generalized Autoregressive Conditional Heteroscedasticity (MS-GARCH) model and Henry [4] extended it to a 2-regime Markov Switching Exponential Generalized Autoregressive Conditional Heteroscedasticity (MS-EGARCH) model. Recent researches associated stock returns with only either interest rates or foreign exchange rates. Henry [4] evaluated the relationship between stock returns and short-run interest rates employing MS-EGARCH in the US. Regime switching behavior of stock returns with foreign exchange rates was examined by Diamandis and Drakos [5], Lin [6], and Chkili and Nguyen [7].

We examine the regime shift behavior of stock returns associated with both interest rates and foreign exchange rates in the Korean stock market based on the 2-regime MRSM (Hamilton [1]) from January 1993 to December 2016.

W. Lee et al. (eds.), *Proceedings of the 7th International Conference on Emerging Databases*, Lecture Notes in Electrical Engineering 461,
https://doi.org/10.1007/978-981-10-6520-0_27

2 Methods

The data used in this paper is monthly total return indexes of Korean KOSPI from January 1993 to December 2016, along with the monthly exchange rates and the monthly interest rates on the same period. The monthly stock prices and exchange rates are originated from ECOS of Bank of Korea and the monthly interest rates are from monthly monetary and financial statistics of OECD.

For estimating the correlations among the stock returns, the exchange rates, and the interest rates in the Korean stock market, the 2-regime Markov Switching model by Hamilton [1] is an adequate approach. In terms of methodology, the autoregressive terms are not necessary in our models due to the logarithmic transformation of stock returns, while Hamilton models [1] mostly use the autoregressive terms. The Markov Regime Switching Model (MRSM) following Hamilton [1] assumes that there are multiple regimes with different volatilities and the processes switch between the two regimes. The regime 1 consists of the low-volatility periods and the regime 2 consists of the high-volatility periods. In the empirical studies, the process is said to be in regime 2 with high-volatility if the estimated probability of regime 2 is close to 1. The regime 2 is used to identify unstable economic situation.

The logarithms of the stock prices are considered as a Brownian motion (Osborne [8]). As an analogy, at time t, let Y_t be the logarithm of monthly stock returns as follows:

$$Y_t = \log\left(\frac{Stock_t}{Stock_{t-1}}\right) \tag{1}$$

Then Y_t lies in one of the two regimes s_t, where $s_t = 1, 2$. We similarly transform the two exogenous variables, exchange rates and interest rates to define E_t and I_t as follows:

$$E_t = \log\left(\frac{Exchange\ Rate_t}{Exchange\ Rate_{t-1}}\right) \tag{2}$$

$$I_t = \log\left(\frac{Interest\ rate_t}{Interest\ rate_{t-1}}\right) \tag{3}$$

The MRSM can be written as follows:

$$Y_t|s_t = \beta_{0s_t} + \beta_{1s_t}E_t + \beta_{2s_t}I_t + \varepsilon_t|s_t \tag{4}$$

where $\varepsilon_t|s_t \sim N\left(0, \sigma^2_{s_t}\right)$. Our model assumes that the stock returns switch between the two regimes based on the Markov transition probabilities which are denoted by

$$p_{ij} = \Pr[s_{t+1} = j|s_t = i],\ i = 1, 2,\ j = 1, 2 \tag{5}$$

where $p_{i1} + p_{i2} = 1$ for i = 1, 2. Note that the exchange rates and the interest rates do not switch. We assume that $\sigma_1^2 < \sigma_2^2$.

The most widely used criteria for model selection is the Maximum Log-likelihood (MLL) function. If the sample size is large, the MLL difference between general model and restricted model is asymptotically a chi-square distribution with the degrees of freedom as dimension difference of the two parameter spaces when the restricted model is correct. As additional criteria, the Akaike information criterion (AIC) and the Bayesian Information Criterion (BIC) are considered and "smaller is better" for both of them (Akaike [9]; Akaike [10]; Pinheiro and Bates [11]; Rice [12]).

We used msmFit function (Sanchez-Espigares and Jose [13]) and stepAIC function (Venables and Ripley [14]) in R 3.3.1. Throughout the paper, the p-values less than 0.10 are considered to be statistically significant.

3 Result

The two plots in Fig. 1 show co-movements of processes from January 1993 to December 2016 in the Korean stock market: (1) stock prices (KOSPI) vs. exchange rates and (2) stock prices (KOSPI) vs. interest rates. In the first plot of stock prices vs. exchange rates, there seem to be clear multiple clusters of connected processes. Within each cluster, stock prices and exchange rates are negatively associated. In the second plot of stock prices vs. interest rates, there seem to be two clusters. Within the low-interest cluster, negative relation is observed and within the high-interest cluster, no linear association between stock prices and interest rates is observed.

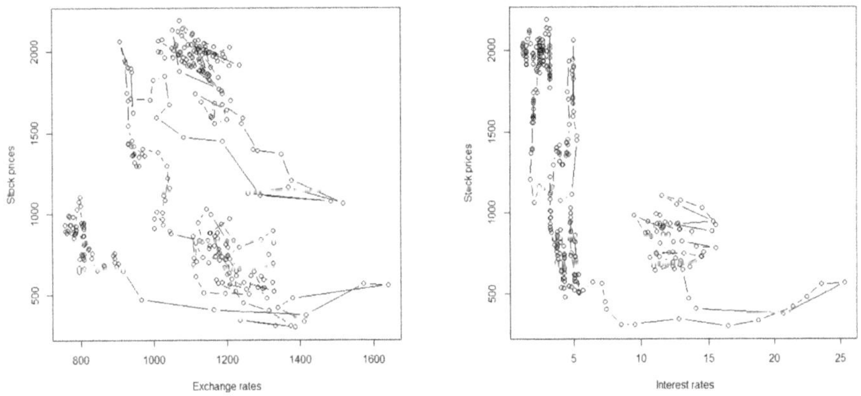

Fig. 1. The left plot is stock prices (KOSPI) vs. exchange rates and the right plot is stock prices (KOSPI) vs. interest rates from January 1993 to December 2016 in the Korean stock market.

For the stock returns, we fit the following four 2-regime switching models: (1) intercept only (2) exchange rates (3) interest rates (4) both exchange rates and interest rates. According to the criteria, MLL, AIC, and BIC, model 4 is the best (Table 1). The most significant exogenous variable is exchange rates and the second significant one is

Table 1. Model selection criteria for the Markov Regime Switching Models for stock returns along with exchange rates and interest rates.

	Exogenous variables	AIC	BIC	MLL	Regime 1 R^2	Regime 2R^2	Selected
(1)	None	−740.1	−721.5	372.1	0.000	0.000	BIC
(2)	E_t	−775.6	−738.3	391.8	0.183	0.051	
(3)	I_t	−746.3	−709.0	377.1	0.052	0.000	AIC, MLL
(4)	E_t and I_t	−783.3	−727.3	397.6	0.233	0.057	

Note: R^2 is the coefficient of determination. The response variable is stock returns. Regime 1 consists of low-volatility periods and regime 2 consists of high-volatility periods.

interest rates. Table 2 presents the parameter estimates for both regimes of the model 4. The volatility of regime 2 ($\sigma_2 = 0.1169$) is about 3 times the volatility of regime 1 ($\sigma_1 = 0.0424$). In regime 1 with low-volatility, the exchange rates negatively influence on the stock returns ($P < 0.01$) and similarly the interest rates negatively influence the stock returns ($P < 0.1$). It means that the stock returns increase when interest rates or exchange rates decrease. The transition probability from regime 1 with low-volatility to regime 2 with high-volatility is more than twice the transition probability from regime 2 to regime 1. In regime 2 with high-volatility, Korean stock returns are not influenced by either its exchange rates or interest rates.

The coefficient of determinant R^2 in Table 1 means the explained variability compared to the total variability of the process. In regime 1 with low-volatility of the model 2, we can see that the exchange rates alone can explain 18.3% of the stock returns (R^2 = 0.183). In regime 1 with low-volatility of the model 3, we can see that the interest rates alone can explain 5.2% of the stock returns (R^2 = 0.052). In regime 1 with low-volatility of the model 4, the exchange rates and interest rates together can explain 23.3% of the stock returns. All the coefficients of determinant in regime 2 with high-volatility are quite small. The coefficient of determinant R^2 are increased up to near 0.6 in 3-regime switching model maybe because each regime gets shorter and is composed of relatively homogeneous processes. The 2-regime model, however, is fitted in this study since it looks more intuitive than the 3-regime model.

In view of the methodology, the parameter estimates for the MRSM in Table 2 were slightly inconsistent which was already pointed out by Campbell [15], Kim et al. [16], and Yuan [17]. "The unfiltered two-state Markov-switching model suffers estimation instability while the filtered model turns out to be temporally consistent" or "MLE is inconsistent for regime-switching models in general". However, its overall significance turned out to be consistent throughout the analysis. Above all, the plots of probabilities for the two regimes well capture the historic Asian financial crisis and also the global financial crisis.

In Fig. 2, on the left are time series plots of stock returns, exchange rates, and interest rates. On the right are plots of the probabilities of being in two regimes estimated from the MRSM for the stock returns along with both exchange rates and interest rates. The upper plot corresponds to the estimated probabilities of being in regime 1 with low-volatility and the lower plot corresponds to the estimated probabilities of being in

Table 2. Parameter estimates of the selected Markov Regime Switching Model in Korea.

	Parameters	Estimates (Standard error)
Regime 1 with low-volatility	β_{01}	9.6104 (0.1857)
	β_{11}	−0.8984 (0.0859)***
	β_{21}	−0.1872 (0.1138)*
	σ_1	0.0424
Regime 2 with high-volatility	β_{02}	6.5934 (16.2391)
	β_{12}	−0.5526 (3.0434)
	β_{22}	0.1214 (0.5098)
	σ_2	0.1169
Transition probabilities	p_{12}	0.0275
	p_{21}	0.0122

Note: 1 The response variable is stock returns. The exogenous variables are E_t and I_t for Korea. β_{01} and β_{02} are intercepts for regime 1 with low-volatility and regime 2 with high-volatility. β_{11} and β_{12} are the coefficients of exchange rates, β_{21} and β_{22} are the coefficients of interest rates, and σ_1 and σ_2 are the standard deviations of the two regimes. ***: significant at 0.01 level. *: significant at 0.1 level.

regime 2 with high-volatility. At time t, sum of the two probabilities is equal to 1. The process is practically said to be in regime 2 with high-volatility if its corresponding estimated probability is close to 1. In regime 2 with high-volatility of Korea, there are sets of periods with estimated probabilities close 1, which include Asian financial crisis of 1997 and the Global financial crisis of 2008. Korean economy suffered great instability during both the Asian financial crisis and the global financial crisis.

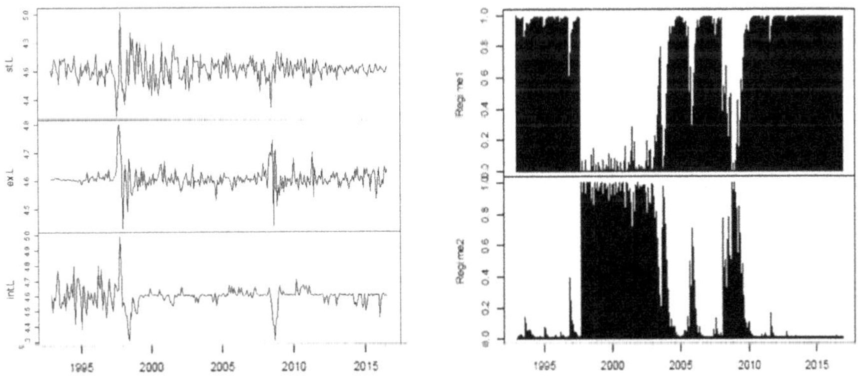

Fig. 2. The left plots are stock returns, exchange rates, and interest rates. The right plots are the probabilities of being in regime 1 and regime 2 estimated from the MRSM for the stock returns along with both exchange rates and interest rates. The upper plot corresponds to regime 1 with low-volatility and the lower plot corresponds to regime 2 with high-volatility.

4 Conclusion and Discussion

We examine the regime shift behavior of stock returns associated with both interest rates and foreign exchange rates in the Korean stock market based on the 2-regime MRSM (Hamilton [1]) from January 1993 to December 2016. The best fitted model for this analysis is 2-regime Markov regime switching models with both exchange rates and interest rates.

In this model, we have found the evidence to support the existence of two distinct regimes in the Korean stock market: regime 1 with low-volatility and regime 2 with high-volatility. Regime 2 with high-volatility includes the Asian financial crisis of 1997 and the Global financial crisis of 2008 in the Korean stock markets. The duration of Regime 2 with high-volatility after the Asian financial crisis of 1997 is much longer than its duration after the Global financial crisis of 2008.

In regime 1 with low-volatility, the stock returns of Korea are closely related to their exchange rates and interest rates. The exchange rates are more influential to the Korean stock returns than the interest rates in this regime. In regime 2 with high-volatility, Korean stock returns are not influenced by either its exchange rates or interest rates. It is because that the policy relating to exchange rates and interest rates is not instantly reflected on the stock markets in unstable economic periods.

The future study will include the analysis of the regime switching dynamics for the four stock markets, Korea, Japan, USA, and China. The latter three countries are closely related to the Korean economy. In addition, we examine the simultaneous cross-correlations among these stock markets using the MRSM for the same periods.

Acknowledgments. This work was supported by the 2017 Hongik University Academic Research Support Fund.

References

1. Hamilton, J.D.: A new approach to the economic analysis of nonstationary time series and the business cycle. Econometrica **57**, 357–384 (1989)
2. Cai, J.: A Markov model of unconditional variance in ARCH. J. Bus. Econ. Stat. **12**, 309–316 (1994)
3. Gray, S.F.: An analysis of conditional regime-switching models. Working Paper, Fuqua School of Business, Duke University (1995)
4. Henry, O.T.: Regime switching in the relationship between equity returns and short-term interest rates in the UK. J. Bank. Finance **33**(2), 405–414 (2009)
5. Diamandis, P., Drakos, A.: Financial liberalization, exchange rates and stock prices: exogenous shocks in four Latin America countries. J. Policy Model. **33**, 381–394 (2011)
6. Lin, C.H.: The co-movement between exchange rates and stock prices in the Asian emerging markets. Int. Rev. Econ. Finance **22**, 161–172 (2012)
7. Chkili, W., Nguyen, D.K.: Exchange rate movements and stock market returns in a regime-switching environment: evidence for BRICS countries. Res. Int. Bus. Finance **31**, 46–56 (2014)
8. Osborne, M.F.M.: Brownian motion in the stock market. Operations Research (1959)

9. Akaike, H.: A new look at statistical model identification. IEEE Trans. Autom. Control **19**, 716–723 (1974)
10. Akaike, H.: A Bayesian extension of the minimum AIC procedure. Biometrika **66**, 237–242 (1979)
11. Pinheiro, J.C., Bate, D.M.: Mixed-Effects Models in S and S-Plus. Springer, New York (2004)
12. Rice, J.A.: Mathematical Statistics and Data Analysis. Thomson, Brooks/Cole (2007)
13. Sanchez-Espigares, J.A., Jose, A.L.: MSwM: Fitting Markov Switching Models. R package version 1.2. https://CRAN.R-project.org/package=MSwM. Last Accessed 30 June 2017
14. Venables, W.N., Ripley, B.D.: Modern Applied Statistics with S, 4th edn. Springer, New York (2002)
15. Campbell, S.D.: Specification Testing and Semiparametric Estimation of Regime Switching Models: An Examination of the US Short Term Interest Rate. Brown University Department of Economics Working Paper #2002-26 (2002)
16. Kim, C.J., Piger, J., Startz, R.: Estimation of Markov regime-switching regression models with endogenous switching. J. Econom. **143**(2), 263–273 (2008)
17. Yuan, C.: Forecasting Exchange Rates: The Multi-State Markov-Switching Model with Smoothing. International Review of Economics & Finance http://economics.umbc.edu/files/2014/09/wp_09_115.pdf. Last Accessed 30 June 2017

A New Method for Portfolio Construction Using a Deep Predictive Model

Sang Il Lee and Seong Joon Yoo([✉])

Department of Computer Engineering,
Sejong University, Seoul 05006, Republic of Korea
sjyoo@sejong.ac.kr

Abstract. A well-designed portfolio plays a key role in achieving investment goal. In this paper, we use Recurrent Neural Networks with Long Short Term Memory (LSTM) Units, to predict potential returns of a collection of investments. Then, we construct diversified portfolios by giving thresholds for the potential returns and examine the return and risk levels of the portfolios. These results show conclusively that it is possible to build a portfolio given a desired degree of return and risk by adjusting the thresholds, which is promising in asset allocations reflected investors' risk preference.

1 Introduction

Machine and Deep learning are rapidly becoming one of the most important components of financial field. Prediction of financial time series is regarded as a challenging problem since the financial market is essentially nonlinear, complex and evolutionary in nature. Machine learning technique have been successfully used for modelling and forecasting financial time series. Application of machine learning algorithms to the financial market has attract a lot of attention since they provide flexibility in modelling high-dimensional financial data sets (Atsalakis and Valavanis 2009; Dixon et al. 2015; Huck 2009, 2010; Krauss 2017; Moritz and Zimmermann 2014; Sermpinis et al. 2013; Takeuchi and Lee 2013; Cavalcante 2016).

It has recently been proposed that more sophisticated models for stock prediction: two neural network architectures and generalized regression neural networks to predict the stock market (Mostafa 2010), a model combined wavelet transform and artificial neural networks (ANNs) (Chandar et al. 2016), particle swarm optimization of ensemble neural networks with fuzzy aggregation (Pulido et al. 2014), ANNs integrated with an improved bacterial chemotaxis optimization (Zhang and Wu 2009), and a hybrid neural networks and multiple regression model (Averkin et al. 2016).

Most previous studies on application of machine (deep) learning to financial sector focus on the prediction accuracy of financial time series or automatic trading rules. In this paper, we deal with a portfolio construction and asset allocation methodology which is considered to be the most important of investment

W. Lee et al. (eds.), *Proceedings of the 7th International Conference on Emerging Databases*, Lecture Notes in Electrical Engineering 461,
https://doi.org/10.1007/978-981-10-6520-0_28

principles. In the 1950s Markowitz developed the Modern Portfolio Theory (or mean-variance analysis) which is a mathematical framework for assembling a portfolio of assets such that the expected return is maximized for a given level of risk (Markowitz 1952). His work demonstrated that investors should analyze the combined expected returns and risks of different combinations of asset classes, not considered individually. However, his work is now considered unrealistic since it is based on normally distributed asset returns. That is, today, it is well-established that equity returns are not Normally distributed.

In this paper, we propose a simple way for building a diversified portfolio of a specific target risk-return without any financial assumptions using a long short-term neural (LSTM) network model.

The rest of this paper is organized as follows: Sect. 2 presents an asset universe, features and proprecessing. Section 3 describes the LSTM Neural Network for stock return prediction. Section 4 introduces thresholds for filtering the assets and observes the risk and return of deep predictive portfolios. Section 5 concludes this paper.

2 Data

The asset universe consists of 10 top stocks in terms of market value from the Standard and Poor's 500 index (S&P500),

- Apple, Amazon, Bank of America Corporation, Berkshire Hathaway Inc. Class B, General Electric Company, Johnson&Johnson, JPMorgan Chase & Co., AT&T Inc., Wells Fargo & Company.

Daily stock dataset, containing the five attributes (e.g. Open, High, Low, Adjust Close, and Volume) is taken for January 2004 to December 2016 from Yahoo Finance. Then, we carry out preprocessing as follows:

$$x_p^{(t)} = (x^{(t)} - x^{(t-1)})/x^{(t-1)}, \tag{2.1}$$

where $x^{(t)}$ is a sample data at time t, and $x_p^{(t)}$ is the percentage change of $x^{(t)}$. We divide the entire data into two parts, the training data set for setting the parameters of models and the test set for out-of-sample evaluation of the models.

3 Prediction Models

We use LSTM to capture a dynamic interdependence among the features and their highly non-linear behaviour. LSTM originally introduced by Hochreiter and Sulovsk (1997), has ability to learn long term patterns in sequential data, and it has recently been applied to diverse area such as sound, handwriting recognition, machine translation and many others. The equations for memory block illustrated in Fig. 1 is as follows:

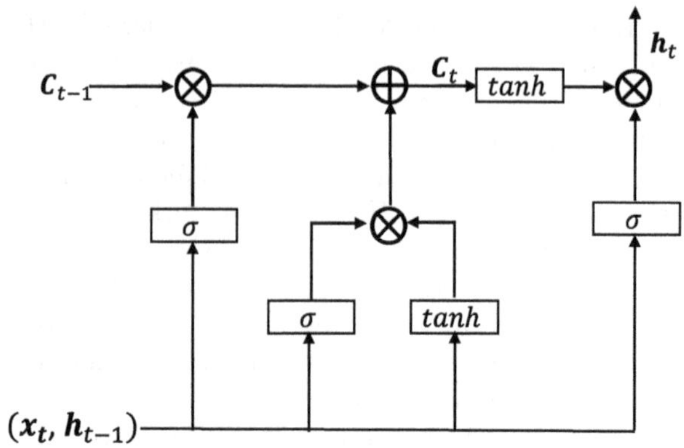

Fig. 1. The structure of LSTM cell

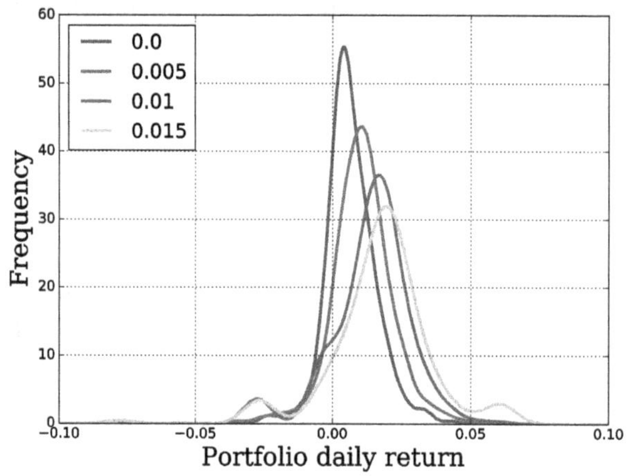

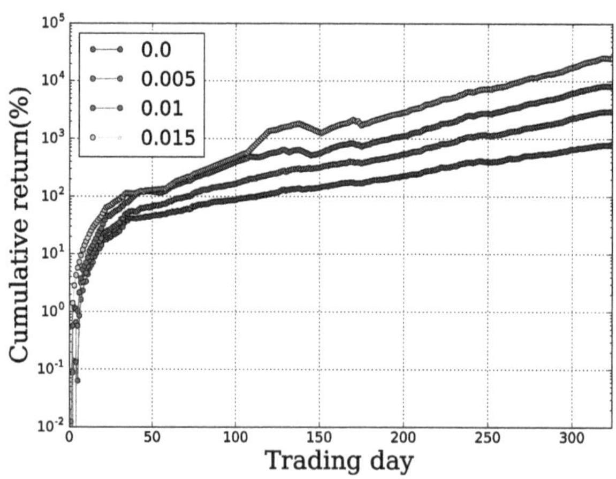

Fig. 2. Experiment 1. (Top) Cumulative return of portfolios and (bottom) histogram of portfolio daily returns for different values of $\theta^+ = 0.0, 0.005, 0.01$, and 0.015.

$$i_t = \sigma(W^{ix}x_t + W_{ih}h_{t-1}) \tag{3.1}$$
$$f_t = \sigma(W^{fx}x_t + W_{fh}f_{t-1}) \tag{3.2}$$
$$o_t = \sigma(W^{ox}x_t + W^{oh}h_{t-1}) \tag{3.3}$$
$$c_t = f_t \odot c_{t-1} + i_t \odot H(W^{cx}x_t + W^{ch}h_{t-1}) \tag{3.4}$$
$$h_t = o_t \odot c_t, \tag{3.5}$$

where i_t, f_t, o_t and c_t are input gate, forget gate, output gate and memory cell, respectively. W^{ix}, W^{fx}, W^{ox} and W^{cx} denotes the weight matrices connecting x_t to the three gates. $\odot$ is the element-wise product operation with gate value, $\sigma(\cdot)$ represents the logistic sigmoid function, and $H(\cdot)$ denotes the hyperbolic tangent.

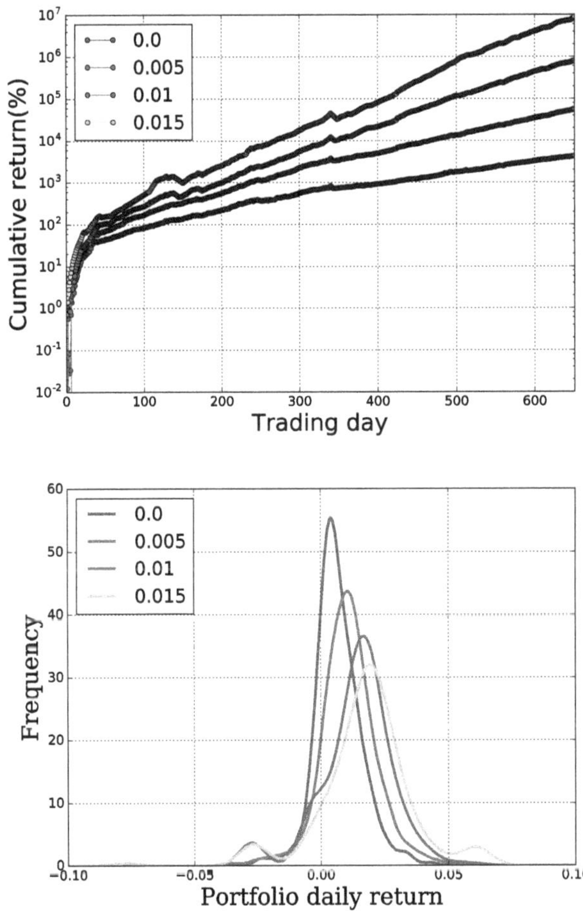

Fig. 3. Experiment 2. (Top) Cumulative return of portfolios and (bottom) histogram of portfolio daily returns for different values of $\theta^+ = 0.0, 0.005, 0.01$, and 0.015.

We build LSTM models for one-step ahead forecasts of daily stock return. They has one layer that contains 100 hidden layer units and recurrent dropout with 0.5 strength, and are fitted using the efficient ADAM optimization algorithm (Kingma and Ba 2014) and the mean squared error loss function.

4 Experimental Results

We define threshold parameters, θ^+ and θ^-, to construct a portfolio from the asset universe. The parameters are used for classifying assets to be long and short. That is, a long (short) equity portfolio consists of the asset of which the predicted return is higher (lower) than θ^+ (θ^-).

It is generally observed that short and long-term trends in stock movement, and thus we use different three in-sample and out-of-sample periods to achieve robustness of our results and conclusion:

- Experiment 1:90% of entire for training, 10% of entire for testing
- Experiment 2: 80% of entire for training, 20% of entire for testing
- Experiment 3: 70% of entire for training, 30% of entire for testing

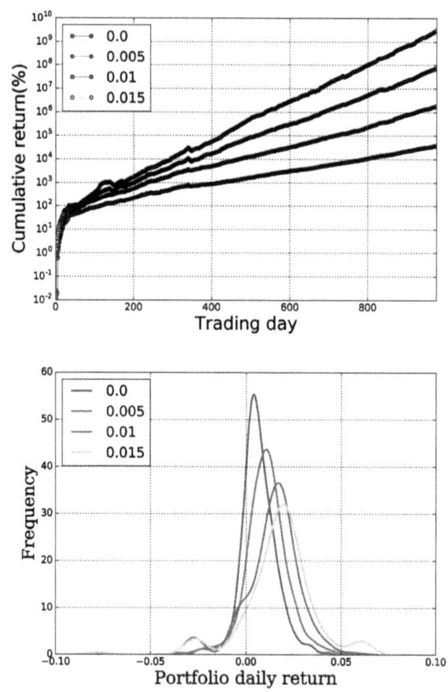

Fig. 4. Experiment 3. (Top) Cumulative return of portfolios and (bottom) histogram of portfolio daily returns for different values of $\theta^+ = 0.0, 0.005, 0.01$, and 0.015.

Figures 2, 3 and 4 graphically shows the cumulative returns (top) and histograms of daily returns (bottom) of portfolios at the threshold levels of $(\theta^+, \theta^-) = (0.0, 0.0), (0.005, 0.0), (0.01, 0.0)$, and $(0.015, 0.0)$. The increase of θ^+ causes the histogram to shift to the right and to increase the standard deviation, which means that the higher the potential return of the portfolios, the higher the risk. This is intuitively reasonable since the increase of θ^+ reduces the effect of diversification on asset allocation by selecting a few of assets with a high expected return. These results show that we are able to optimize investor's portfolios according to their individual risk preference by adjusting θ^+: portfolios with high and low θ^+ are suitable for a risk seeking investor and a risk tolerant investor, respectively. Table 1 provides statistical measurements for the portfolios.

Table 1. Mean and standard deviation (SD) of the portfolios

Exp.	Statistics	θ^+			
		0.0	0.005	0.01	0.015
Exp. 1	Mean	0.0064	0.0103	0.0137	0.0169
	SD	0.0088	0.0103	0.0137	0.0160
Exp. 2	Mean	0.0058	0.0098	0.0140	0.0178
	SD	0.0093	0.0097	0.0129	0.0159
Exp. 3	Mean	0.0061	0.0101	0.0140	0.0179
	SD	0.0081	0.0094	0.0125	0.0150

5 Conclusion

We have proposed a new method for constructing portfolios of a specific risk-return, based on predicted returns produced by LSTM networks. Our method enable to build target portfolios at specific levels of risk and return by adjusting the thresholds classifying assets. It is useful for providing investors with portfolios matching individuals' risk preference, which helps them make appropriate investment decisions. In this paper, we does not present the effect of the performance of portfolios produced from combinations of the thresholds and optimality for a better risk-return profile. In future work, we will further investigate the optimality of deep predictive portfolios.

Acknowledgement. This work was supported by the ICT R&D program of MSIP/IITP. [2017-0-00302, Development of Self Evolutionary AI Investing Technology]

References

Atsalakis, G.S., Valavanis, K.P.: Surveying stock market forecasting techniques-Part II: soft computing methods. Expert Syst. Appl. **36**(3), 5932–5941 (2009)

Dixon, M., Klabjan, D., and Bang, J. H.: Implementing deep neural networks for financial market prediction on the Intel Xeon Phi. In: Proceedings of the 8th Workshop on High Performance, pp. 1–6 (2015)

Hochreiter, S., Schmidhuber, J.: Long short-term memory, case. Neural Comput. **9**(8), 1735–1780 (1997)

Kingma, D., Ba, J.: Adam: a method for stochastic optimization, arXiv preprint. arXiv:1412.6980 (2014)

Huck, N.: Pairs selection and outranking: an application to the S&P 100 index. J. Oper. Res. **196**(2), 819–825 (2009)

Huck, N.: Pairs trading and outranking: the multi-step-ahead forecasting case. Eur. J. Oper. Res. **207**(3), 1702–1716 (2010)

Krauss, C., Do, X.A., Huck, N.: Deep neural networks, gradient-boosted trees, random forests: statistical arbitrage on the S&P 500. Europ. J. Oper. Res. **259**(2), 689–702 (2017)

Moritz, B., Zimmermann, T.: The relation between past and future stock returns. Working paper, LMU Munich and Harvard University (2014)

Sermpinis, G., Theofilatos, K., Karathanasopoulos, A., Georgopoulos, E.F., Dunis, C.: Forecasting foreign exchange rates with adaptive neural networks using radial-basis functions and particle swarm optimization. Eur. J. Oper. Res. **225**(3), 528–540 (2013)

Takeuchi, L., Lee, Y.-Y.: Applying deep learning to enhance momentum trading strategies in stocks. Working paper, Stanford University (2013)

Cavalcante, R.C., Brasileirob, R.C., Souzab, V.L.F., Nobregab, J.P., Oliveirab, A.L.I.: Computational intelligence and financial markets: a survey and future directions. Expert Syst. Appl. **55**(15), 194–211 (2016)

Mostafa, M.M.: Forecasting stock exchange movements using neural networks: empirical evidence from Kuwait. Expert Syst. Appl. **37**(9), 6302–6309 (2010)

Chandar, S.K., Sumathi, M., Sivanandam, S.N.: Prediction of stock market price using hybrid of wavelet transform and artificial neural network. Indian J. Sci. Technol. **9**(8) (2016)

Pulido, M., Melin, P., Castillo, O.: Particle swarm optimization of ensemble neural networks with fuzzy aggregation for time series prediction of the Mexican Stock Exchange. Inf. Sci. **280**, 188–204 (2014)

Zhang, Y., Wu, L.: Stock market prediction of S&P 500 via combination of improved BCO approach and BP neural network. Expert Syst. Appl. **36**(5), 8849–8854 (2009)

Averkin, A., Yarushev, S., Dolgy, I., Sukhanov, A.: Time series forecasting based on hybrid neural networks and multiple regression. In: Proceedings of the First International Scientific Conference on Intelligent Information Technologies for Industry (IITI 2016), pp. 111–121. Springer International Publishing (2016)

Markowitz, H.M.: Portfolio selection. J. Finance **7**(1), 77–91 (1952)

Personalized Information Visualization of Online Product Reviews

Jooyoung Kim and Dongsoo Kim(✉)

Soongsil University, 369 Sangdoro, Dongjak-Gu, Seoul, Korea
{whytimesgone,dskim}@ssu.ac.kr

Abstract. This paper presents a new method for visualizing online product reviews considering customer profiles. Typically, product review data are unstructured and have no fixed format or structure. The review data can be used by customers and also an e-business company. Potential consumers can acquire useful information on product characteristics and decide whether to buy or not depending on the review data. Also, the company can understand customers' experiences or opinions on the product and reflect them in developing marketing strategies. In order to provide valuable information to the customers from enormous and unstructured review data, the process of collecting, storing, and preprocessing of review data should be performed firstly. And then text mining and personalization techniques can be integrated to extract properly visualized data. Thus, customers can utilize review data conveniently with the assistance of the proposed system.

Keywords: Text mining · Information visualization · Product review · Personalization

1 Introduction

In the past, companies collected opinions or evaluations of their products by conducting surveys online or offline. However, this method of obtaining information is inefficient because of the time, expense and truthfulness of the data depending on the customer's willingness to respond [1]. Recently, online reviews created by consumers in a voluntary fashion can provide meaningful information to a company by describing their experience or evaluation of the product, and can also provide useful information to potential customers interested in the product [2]. For potential customers, online reviews are one of the important factors influencing their purchasing, which is supported by previous research that word-of-mouth marketing through reviews has a significant impact on others [3]. However, since online reviews are in the form of unstructured data, in order to be able to extract meaningful information, it is necessary to convert them into regular data form [4].

This study proposes a new method to analyze online product review data by applying text mining method based on R program. In the proposed method, natural language and document processing technology are applied to refine unstructured online reviews into regular data and provide analysis results in a form that best matches the purpose of use. Specifically, a review table composed of data generated through text

W. Lee et al. (eds.), *Proceedings of the 7th International Conference on Emerging Databases*, Lecture Notes in Electrical Engineering 461,
https://doi.org/10.1007/978-981-10-6520-0_29

mining is constructed, and data mining is used to derive customized review information for each purpose.

The rest of this paper is organized as follows. Section 2 reviews related work on review data analysis. Section 3 presents an overall framework of the proposed method. Section 4 presents the analysis results, followed by Sect. 5 concluding the paper and suggesting future research directions.

2 Related Work

In the age of Web 2.0, Internet users can produce information easily. Therefore, online review data on which an individual evaluates a particular product are being produced as fast as the increase in the number of informal data on the Internet [5]. Many Internet users look for product reviews through search engines and find opinions on products from real users [6]. Opinions in review data with actual users' evaluations serve as a factor of judgment in the purchase decision making of Internet users. This review data recognized as a way of word of mouth marketing has become one of the decision-making factors of potential customers who are interested in a specific product. In addition, it has become a means to obtain customer feedback from the perspective of the company that produces the product [7].

However, without a proper processing method, meaningful information cannot be obtained from a large amount of informal online review data. Due to the massive information overflow, potential customers tend to read popular or recent reviews and make purchasing decisions. It is inefficient for consumers to perform purchasing activities based on such biased information because they lack comprehensive judgment. Therefore, it is necessary to construct meaningful information by finding meaning and rule information by processing online reviews into valuable information, rather than data listed in simple collective intelligence [8]. In order to process review data, it is necessary to apply a text mining technique to unstructured data.

Most of the existing studies on review data analysis based on text mining are opinion mining and social mining researches, which collect the opinions of customers. Opinion mining allows us to understand the distribution of positive and negative responses from customers [9]. Woolley et al. proposed a new graphical model that represents a comparison between the products for which the opinions of the customers are presented for use as an indicator for identifying potential threats to the company, new product design and marketing activities [10]. Social mining is a method of performing text mining using SNS data. Although it is a technique belonging to the opinion mining category, there are many studies that interpreted meaning by using social data in text mining.

Most of the existing studies have focused on the classification of the emotional aspects of words or the study of natural language processing. However, in this study, it is meaningful to summarize the review by extracting the keywords, preprocessing each review, classifying it according to the characteristics or purpose of the user, and analyzing the keywords again.

3 Proposed Method

3.1 Overall Framework

Figure 1 shows an overall research framework of review data analysis based on text mining.

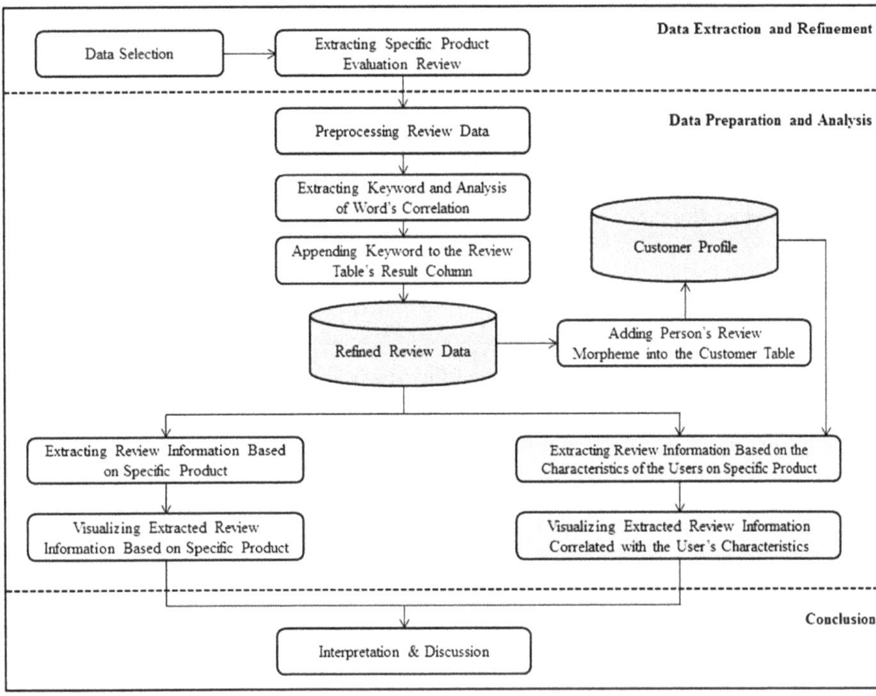

Fig. 1. Research framework of review data analysis based on text mining

The framework is an improved version from the authors' previous work [12]. The proposed framework includes a process of extracting specific product review data, which is one of the unstructured data existing on the Internet, and collecting a comprehensive review of the product and a customized review according to the characteristics of an individual. The whole process of analysis composed of data extraction and refinement, data preparation and analysis, and conclusion is shown in the figure. The steps differentiating from existing studies are shaded.

3.2 Keyword Extraction and Word Correlation Analysis

The TF (Term Frequency) is a value that indicates how often a particular word appears in a document. There are two ways to calculate the TF value. First, the Boolean Frequency is a way of entering 1 for the value of the matrix if the word occurs once in the document (or sentence), otherwise 0. Since the value is a Boolean expression indicating true or false, the frequency of how the word appears in the document is not considered.

Second, the increasing frequency is a method of adjusting the frequency value of the word according to the length of the document. In the formula below, t means the extracted term and d means the collected document. The value of x means a weight, and if the emphasis is on the appearance of a word, the value of x may be set small.

$$tf(t,d) = x + \frac{(1-x) \times f(t,d)}{\max\{f(w,d) : w \in d\}} \tag{1}$$

The IDF (Inverse Document Frequency) is a value indicating how common a word appears in the entire document set. As shown in the formula below, it can be obtained by taking the log value to the value obtained by dividing the total number of documents by the number of documents containing the word.

$$idf(t,D) = \log\left(\frac{|D|}{1 + |\{d \in D : t \in d\}|}\right) \tag{2}$$

The TF-IDF value is expressed by the following equation. The higher the frequency of words in a particular document, and the fewer the documents containing the word, the higher the TF-IDF value. Using this value, it is possible to filter words that are common in all documents and reveal the unique characteristics of each document.

$$tfidf(t,d,D) = tf(t,d) \times idf(t,D) \tag{3}$$

LSA (Latent Semantic Analysis) is a technique for analyzing the potential meaning [11]. A term-document matrix is obtained by obtaining frequency of words included in each document, and the relationship between the documents is obtained by measuring the degree of similarity of words. Because it measures the similarity of words, even individual words marked differently can be grouped together if they are similar in meaning. It is a method of analyzing the meaning of each document by reducing the dimension in a way that minimizes the loss of information.

The steps of deriving meaningful information using LSA are as follows. First, we extract words to create a word-document matrix. A row of the word-document matrix means an extracted word, and each column means a document (or sentence). Each element can take a value of 0 or 1. The next step is to count the frequency of occurrences of words in the word-document matrix. To get the result, the values of the elements are added in rows and ranked. The ranked words are sorted in descending order and the words are extracted to reduce the dimension. The third step is to reduce the dimension to create a new matrix. Although the previous word-document matrix represents all the words extracted from the review data in rows, each row of the matrix created by reducing the dimension consists of the top 20 words. The fourth step is to create a co-occurrence matrix. By multiplying the matrix created in the third step by its transpose matrix, we can obtain the co-occurrence matrix that forms rows and columns with the top x keywords. Through the co-occurrence matrix, simultaneous word analysis can be used to determine the association between words. The final step is to list the selected words, that is, other words associated with the words that represent the dimension.

3.3 Results Storage in Review Tables

To provide customized review information for each purpose, the data to be used when presenting comprehensive reviews and the data to be used when presenting personalized reviews are separated and stored in each review table. In order to present the comprehensive reviews, we use the results obtained using the LSA algorithm described in the previous section. When presenting the personalized reviews, review keywords derived by the Boolean TF value from the TF-IDF are used.

3.4 Comprehensive Review Information Extraction and Visualization

In this step, comprehensive review information about a specific product is extracted using the extracted review data, stored in a product table, and visualized. Data visualization represents information extracted in the form of a graph, making it easy to find patterns or relationships that are difficult to find with just numbers or letters. Visualization helps understanding by those who refer to it, and contributes to decision-making by making it possible to find meaning. Therefore, it aims to convey the meaning more effectively than the complex data listing.

3.5 Personalized Review Information Extraction and Visualization

There are two types of customized and personalized review information extraction. The first method is to filter the data based on the characteristics of the individual in the user table, and then analyze the association with the given data. The second method extracts a keyword from the review data created by the user and displays the information of the review data of the other person including other words having a strong relation with the keyword. After performing the conformance checking between the two reviews, it lists highly relevant review information to the customer.

4 Application Case and Analysis Results

4.1 Application Case Description

The data of this study were obtained by selecting a specific product on a website selling cosmetic products and then extracting the review data created by the customers for the product using the web crawling technique. Figure 2 illustrates the composition of the review data used in the case study. Through the crawling technique, the product table includes attributes such as product ID, product name, product type, and the price. In the customer table, customer ID and the other attributes indicating the characteristics of the customer are inserted. The review data used in the case study is divided into two parts because the good and bad points of the product are classified, and the analysis is performed assuming two parts as individual parts.

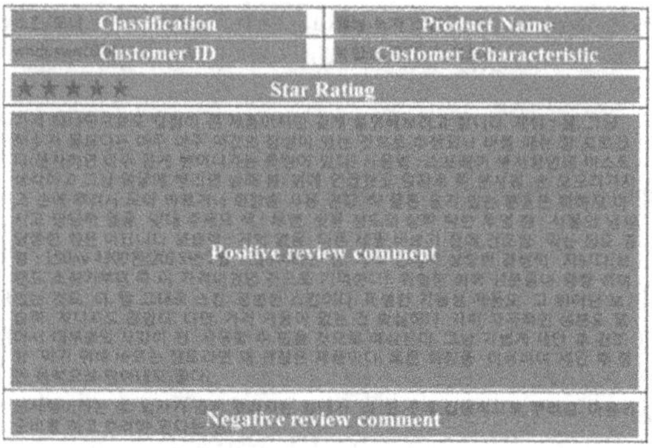

Fig. 2. Composition of online review data

In order to remove the noise of the text data, we used a function to extract the main morpheme in the R program. The types of noise of data removed in this study are as follows. The first type is a word that ignores spacing, or has a typo. The other type is low frequency words. Among the words, when the frequency is low even though it is a word contained in the main morpheme recognized by the morphological analysis function of R, it is removed from the existing data because it does not give a significant meaning to the analysis results.

In the preprocessing process of the data, a user data dictionary is constructed by using a review text corpus. Then, words appearing frequently in the extracted data although not in the basic dictionary of R are inputted in advance, and words with low frequency or no meaning are judged as noise. The case study shows how the review data is converted into review information and visualized, so the product name was recorded without disclosing the actual names.

4.2 Analysis Results

This section summarizes the results of applying the proposed analysis framework to the review data. The co-occurrence matrix of the most frequent 20 keywords can be obtained by extracting the keywords from the review data and multiplying the reduced word-document matrix by its transposition.

Then, we can obtain the following result by finding the relation of each word as shown in Fig. 3. The figure shows 8 clusters based on the K-means clustering method. As the semantic value of each topic increases, the keywords with positive values increase their meanings, and those with negative values decrease their meanings. Therefore, the larger the absolute value, the higher the positive or negative correlation between the subject and the keyword.

The result of visualizing the comprehensive review information derived from the association rule analysis can be found in [12]. The size of the circle increases as the support value increases, and the color of the circle becomes dark as the lift value increases.

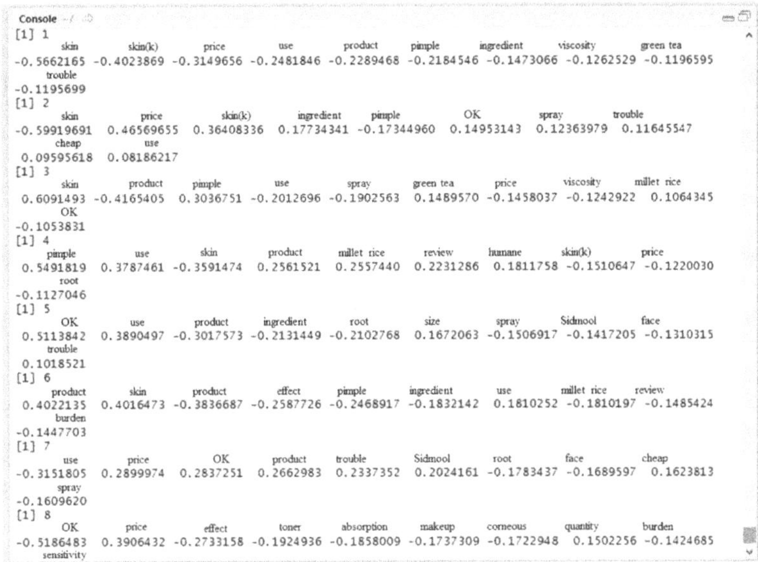

Fig. 3. Clustered review keywords

Differentiated analysis results can be presented according to user characteristics, which enables personalized visualization of review data analysis. The proposed method can provide the result of collecting and analyzing positive opinions about the product for customers with dry skin and the results of analyzing positive opinions of oily skin users respectively.

5 Conclusions

In this research, we presented a new approach to visualization of unstructured online product reviews. The visualization results proposed in this work can be used by customers and also an e-business company. Potential consumers can acquire useful information conveniently through the review data analysis. Also, the company can comprehend customers' opinions on the product and reflect them in developing marketing strategies or renovation plans.

In the future, we will conduct an extension study that applies the proposed methodology to review data in other industries. Since reviews of each industry sector have different forms, further research is needed on a methodology that can be used universally. In addition, we will implement a system that extracts review information extracted based on rule discovery of words, and recommends related products of different kinds by applying association relations found in previous data.

Acknowledgments. This work is supported by Basic Science Research Program through the National Research Foundation of Korea (NRF) funded by the Ministry of Science, ICT & Future Planning (NRF-2017R1D1A1B05029080).

References

1. Collobert, R., Weston, J.: A unified architecture for natural language processing: deep neural networks with multitask learning. In: Proceedings of the 25th International Conference on Machine Learning, Helsinki, Finland, pp. 160–167 (2008)
2. Liu, B., Zhang, L.: A survey of opinion mining and sentiment analysis. In: Mining Text Data, pp. 415–463. Springer, US (2012)
3. Kangale, A., Kumar, S.K., Naeem, M.A., Williams, M., Tiwari, M.K.: Mining consumer reviews to generate ratings of different product attributes while producing feature-based review-summary. Int. J. Syst. Sci. **47**(13), 3272–3286 (2016)
4. Kim, J., Kim, D.: A Study on the method for extracting the purpose-specific customized information from online product reviews based on text mining. J. Soc. e-Bus. Stud. **21**(2), 151–161 (2016)
5. Mooney, R.J., Bunescu, R.: Mining knowledge from text using information extraction. ACM SIGKDD Explor. Newsl. Nat. Lang. Process. Text Min. **7**(1), 3–10 (2005)
6. Hu, M., Liu, B.: Mining and summarizing customer reviews. In: Proceedings of the Tenth ACM SIGKDD International Conference on Knowledge Discovery and Data Mining, pp. 168–177 (2004)
7. Berry, M.W.: Survey of text mining. Comput. Rev. **45**(9), 548 (2004)
8. Rajaraman, K., Tan, A.H.: Topic detection, tracking, and trend analysis using self-organizing neural networks. In: Proceedings of the 5th Pacific-Asia Conference on Knowledge Discovery and Data Mining, pp. 102–107 (2001)
9. Holton, C.: Identifying disgruntled employee systems fraud risk through text mining: a simple solution for a multi-billion dollar problem. Decis. Support Syst. **46**(4), 853–864 (2009)
10. Woolley, A.W., Chabris, C.F., Pentland, A., Hashmi, N., Malone, T.W.: Evidence for a collective intelligence factor in the performance of human groups. Science **330**(6004), 686–688 (2010)
11. Doan, A., Naughton, J.F., Ramakrishnan, R., Baid, A., Chai, X., et al.: Information extraction challenges in managing unstructured data. ACM SIGMOD Rec. **37**(4), 14–20 (2009)
12. Kim, J., Kim, D.: A method for extracting organized information from online product reviews based on text mining. ICIC Express Lett. Part B Appl. **7**(10), 2211–2216 (2016)

A Trail Detection Using Convolutional Neural Network

Jeonghyeok Kim[1], Heezin Lee[2], and Sanggil Kang[1(✉)]

[1] Department of Computer Engineering, INHA University, 100 Inharo Nam-gu, Incheon 22212, South Korea
sgkang@inha.ac.kr

[2] Department of Earth and Planetary Science, University of California-Berkeley, National Center for Airborne Laser Mapping, Berkeley, CA 94720, USA

Abstract. Small-footprint airborne LiDAR scanning systems are effective in modelling forest structures and can also improve trail detection. We propose a trail detection method through a machine learning method from the LiDAR points. To do that, we analyze features for detecting a trail, digitize each feature and combine the results to distinguish between trail and non-trail areas. Our proposed method shows the feasibility of trail detection by using airborne LiDAR points gathered in dense mixed forest.

Keywords: Trail feature · Forest structure · LiDAR · Trail detection · Neural network

1 Introduction

In the case of forest areas, trail detection is important for forest management be-cause it provides corridors for traveling, access for recreation and education, infrastructure for fire protection, military planning, and search-and-rescue operations [1]. Unfortunately mapping un-inventoried forest trails, and maintaining trail position, width, and slope over large areas can be both time-consuming and expensive [2]. A Light Detection and Ranging (LiDAR) scanning systems have been widely used for forest structure analysis, such as geomorphology [3], bare-surface topography, obstacle detection, and corridor composition [4]. LiDAR can be particularly useful for providing accurate measurements of ground surface elevations [5]. Research for trail detection with LiDAR has proceeded as follows. David et al. [6] developed a semi-automated method for detecting trail ways in forested areas using a normalized Digital Surface Model (nDSM) and LiDAR in-tensity images. The method works well in situations in which a canopy does not occlude the trail, and trail features are radio metrically different from the background vegetation. In many forested environments, the canopy occludes the trail, and the intensity of the LiDAR return from the ground is not reliable when com-pared to that of non-trail areas. Kim et al. [7] developed a trail detection method using height above ground level. They set an area in which obstacles do not exist. However, the rationale for setting the range of the area is insufficient [8]. In this letter, we derive three features for detecting trails including flatness, obstacle density, and ground point density. To determine the optimal threshold of trails/non-trails for each

W. Lee et al. (eds.), *Proceedings of the 7th International Conference on Emerging Databases*, Lecture Notes in Electrical Engineering 461,
https://doi.org/10.1007/978-981-10-6520-0_30

feature, we apply a statistical analysis with assumption of normal distribution of digitized feature values.

2 Trail Detection Using Convolutional Neural Network

Figure 1 is the CNN model for trail detection consisting of feature extraction part and fully connected neural network. The feature extraction part extracts three features such as obstacle, flatness, and intensity.

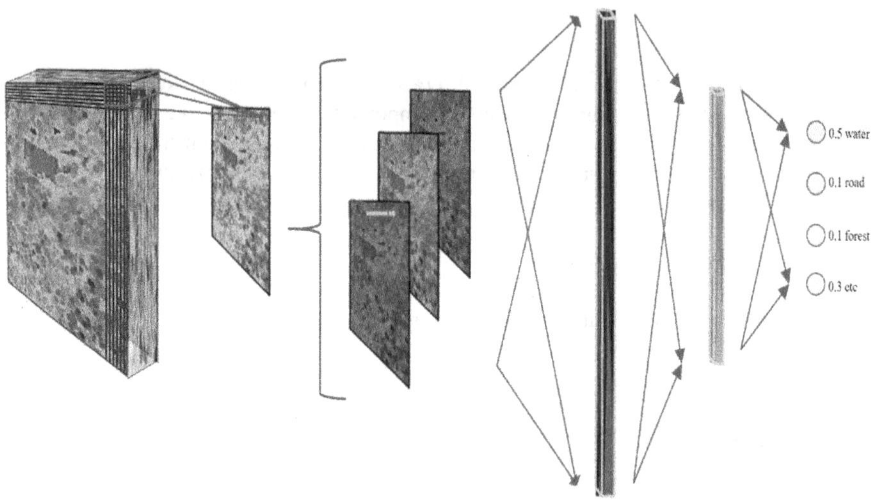

Fig. 1. Convolutional neural network model for trail detection

The obstacle feature determines whether human can go through a space on a path. If there is no LiDAR point in the space which is assumed a predetermined space through which a human can go, the space can be considered as a path. Due to noise in the LiDAR points, even though there exist a few LiDAR points in the space, it cannot be assured as a non-path. However, the less LiDAR points to the center area in the box, the more possible is considered as a path, in general. To take into the consideration, we design the feature detector as dividing it into three sections which are set with equal distance from the center of the sliding window and the weights (w1, w2, w3) are provided to the number of LiDAR points in each section.

The flatness feature represents the distinction of elevation from the ground area to measure how slope on a trail is gentle. If the distribution of elevation of LiDAR points in the space is rough, the space is not enough to walk through so it can be considered as a non-path. The variation of elevations between two adjacent ground areas is specified in order to determine how flat the ground is. The filter used at the convolutional layer for extracting flatness feature. The weights used as a filter during convolutional layer are provided to elevation differences of adjacent areas from center area in the box. To take into the consideration, we design the feature detector as dividing it into two sections which are set with equal distance from the center of the sliding window and the weights (w1, w2) are provided.

The intensity feature represents rigidness of the area, which is measured by LiDAR point intensity reflected from objects. The intensity depends on laser divergence, the diameter of the footprint, the diameter of the receiver, the target diameter, the scan intensity of the laser, and the transmittance of the atmosphere. Based on these characteristic, LiDAR intensity can be utilized for identifying objects. In general, the ground is uniform and rigid so it will have high intensity. Vegetables above the ground are soft so it will have low intensity. The twig or foliage of the tree within aerial area is soft so it will have low intensity. It is difficult to distinguish between path and non-path with only intensity value. Thus, we divide three areas such ground, above ground, and aerial area in the box and then provide weight used as a filter during convolutional layer to the intensity of each area.

Even though some sections are determined as trails from the feature extraction, they are not convinced as a path if they are not connected. Thus, the relative location of the detected features is more important than absolute location. At pooling layer, the spatial size of the detected feature is reduced, the number of parameter weights in CNN and is reduced, and learning overfitting can be controlled. We use max pooling algorithm, which partitions the feature map into a set of non-overlapping 2 by 2 rectangles, and

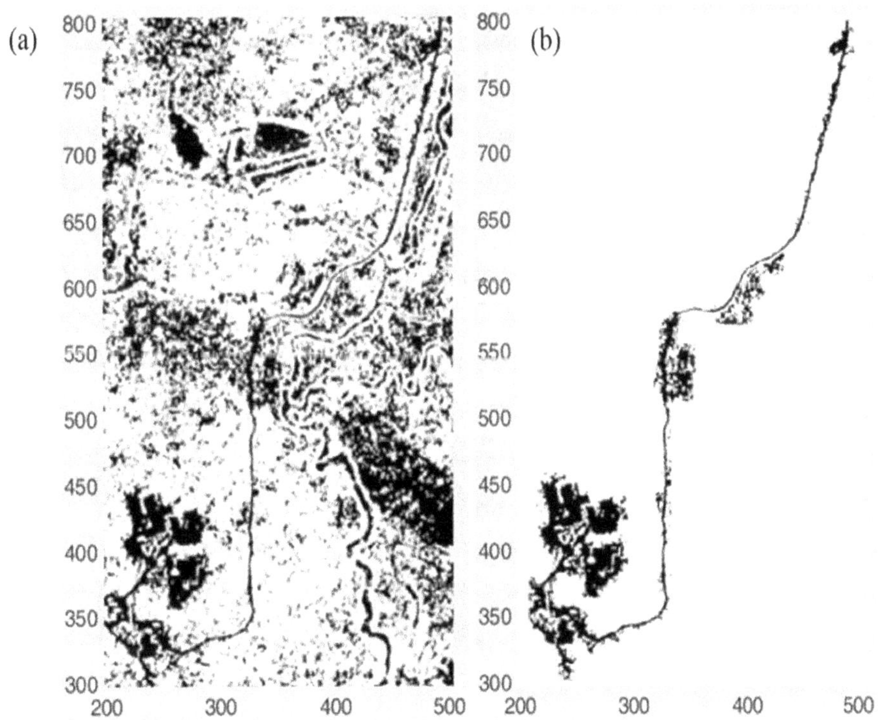

Fig. 2. (a) The 87 percents result in overall area. (b) The connected main trail

then outputs the maximum value out of the detected features for the rectangles. At the final layer, the fully connected neural network trains with features obtained from pooling layer for classifying four categories such as trail, empty space, water, and non-trail. At the output layer of FCNN, softmax function is used for reasoning how the network training penalizes the deviation between the predicted and true labels. As seen the Fig. 2, the main trail can be detected by training the CNN with three features. The black point mean trail and the white mean water, forest and etc.

Convolutional Layers accept three dimensional input data in which three of the dimensions are spatial, and they contains the feature of trail. The layer creates feature maps by convolving the data with learned filters of shape where maps are the spatial dimensions and filter is the number of input feature maps. Convolution can also be applied at a spatial stride. The output is passed through a leaky rectified nonlinearity unit (ReLU) with parameter 0.1. Pooling Layers.

3 Discussion

To evaluate the performance of trail detection, we applied our method to three test sections. Figure 3 shows comparison between the LiDAR points plot and the results in each section. Section A includes a trail across the centre. However, there are several

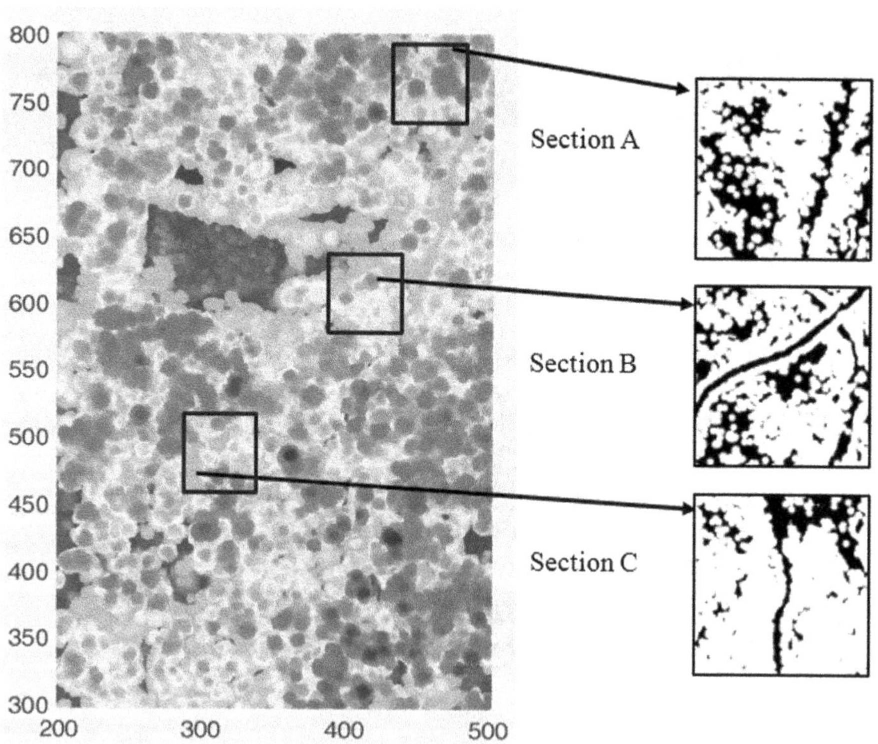

Fig. 3. The left figure is the LiDAR point elevation plot and the right figure shows the results from using our method

vague connections on the trail because the section appears steep gradient with scarps. Overall, the section is 0.05 cm higher in elevation than another test section. This problem is alleviated by the adaptive pre-processing measurements of the structural difference of the section. Section B includes a reasonable trail in the centre. Other trails are disconnected because they are difficult to go through due to the high elevation difference. Section C is a straight trail area in which broadleaf trees are densely distributed. There are numerous LiDAR points on aboveground part by the dense foliage on the canopy. It can be verified that the trail area is clear in the centre. The trail areas are obvious because of densely distributed broadleaf trees.

4 Conclusion

In this letter, we proposed a trail detection method in which we analyzed the three features such as the flatness, obstacle, and intensity using the real LiDAR points gathered. To determine the optimal threshold for each feature, we used the CNN that is composed by three features. Our method was well performed for the environments in which the canopy heavily occludes the trail and the ground elevation change is severe. Also, we focussed on detecting trails in a forest regardless of the type of trails such as bridges, corridors, and empty lots. As a further work, we need to develop the classification of the types of trails. In this case, we need to define the patterns of the LiDAR points according to the types of trails using various statistical data mining techniques.

Acknowledgment. This research was supported by Basic Science Research Program through the National Research Foundation of Ko-rea (NRF) funded by the Ministry of Education (NRF-2016R1D1A1B03932447).

References

1. Erik, A.H., et al.: Estimating forest structure parameters on fort lewis military reservation using airborne laser scanner (LiDAR) data. In: Proceedings of the 2nd International Precision Forestry Symposium, 3, pp. 22–23 (2003)
2. Blundell, B.S., et al.: Terrain gap identification and analysis from lidar data for military mobility. In: Proceedings from the American Society for Photogrammetry and Remote Sensing, 3, pp. 172–184 (2004)
3. Richter, A.M., et al.: Terrestrial laser scanning (LiDAR) as a means of digital documentation in rescue archaeology: two examples from the faynan of jordan. Vir. Syst. Multimedia **18**, 11–13 (2012)
4. Guan, H., et al.: Deep learning-based tree classification using mobile lidar data. Remote Sens. Lett. **6**, 864–873 (2015). doi:10.1080/2150704X.2015.1088668
5. Roberto, S., et al.: Lidar obstacle warning and avoidance system for unmanned aircraft. Int. J. Mech. Aerosp. Indust. Mechatron. Eng. **8**, 702–713 (2014)
6. David, N., et al.: Pathway detection and geometrical description from ALS data in forested mountaneous area. Paris, France, IAPRS 38, Part 3/w8 (2009)
7. Kim, A., Olsen, R.: Detecting trails in lidar point cloud data. In: Laser Radar Technology and Applications, vol. 17. doi:10.1117/12.918631
8. Kampa, K., Slatton, K.C.: An adaptive multiscale filter for segmenting vegetation in ALSM data. In: IEEE Geoscience and Remote Sensing Symposium, vol. 6, pp. 3837–3840 (2004)

Design of Home IoT System Based on Mobile Messaging Applications

Sumin Shin, Jungeun Park, and Chulyun Kim(✉)

Department of IT Engineering, Sookmyung Women's University, Cheongpa-ro 47-gil 100, Yongsan-gu, Seoul 04310, South Korea
ssm6796@naver.com, {binue0820, cykim}@sm.ac.kr

Abstract. With the proliferation of smart devices and sensors, the Internet of Things (IoT) has been rapidly emerged. The term of IoT generally refers to not only network connectivity of smart objects with computing capability but also applications based on the network connectivity. The smart objects generate, exchange and consume their data with only minimal human intervention. Smart Home is a service that connects various home appliances, such as refrigerators, televisions, air conditioners, and so on, in order to improve user experiences and provide combinational operations between them. In this paper, we introduce a design of Home IoT system based on mobile messaging applications with ChatBot, an artificial intelligence based agent to remotely control of home appliances and inquire their status through natural language messages. We propose the standard hierarchy of Home IoT system and demonstrate the proposed system by implementing Arduino based prototype systems.

Keywords: Chatbot · Home IoT · raspberryPi · Artificial intelligence

1 Introduction

Internet of Things (IoT) is rapidly emerging as a core value of ICT market, which has been chosen annually by Gartner, the top 10 strategic technology trends since 2012. In particular, smart sensors have increased due to the proliferation of smart devices, such as mobile phones and the convergence and connectivity between devices has been ensured. The environment rapidly heightened interest in IoT across the field of ICT convergence [1]. Smart Home means a house where various devices in homes such as TVs, refrigerators, and washers can be connected to the network and provide intelligent services [2]. In the early 2000s, 'Home Network' based on a wired Internet grew up with the introduction of high-speed Internet access, while the latest issue, 'Home IoT' can be seen as an expanding area of the existing market with wireless Internet environment and M2M technology [3]. Namely, 'Home IoT' is developed by blending the smart home with the IoT, and the accessibility of the 'Home IoT' is gaining momentum by popularization or smart phone and application as it becomes easier to control the device in both the home and outdoors.

Furthermore, interest in chatbot is rising as major global messaging companies announced that they will adopt chatbot technology based on AI. A chatbot is a computer program which responds like an intelligent entity when conversed with.

W. Lee et al. (eds.), *Proceedings of the 7th International Conference on Emerging Databases*, Lecture Notes in Electrical Engineering 461,
https://doi.org/10.1007/978-981-10-6520-0_31

The chatbot understands one or more human languages by Natural Language Processing [4]. In the past, chatbot used simple patterns matching the previously defined keywords to generate a pre-defined response. But lately, due to advances in artificial intelligence technology, the development of questions and commands was reached in the context of human knowledge.

In this paper, we want to propose a design of home IoT system that allows users to remotely control various devices and sensors in their home via mobile messaging applications. We designed this system with raspberry Pi that is inexpensive, but highly-efficient and easy to operate sensors using GPIO. In the raspberry Pi, sensor nodes such as temperature, humidity, infrared and camera module were installed to monitor smart home environment, and a relay module was also included to implement electronic device controls.

The rest of this paper is as follows. In Sect. 2, we give an existing case similar to the system proposed in this paper and introduce the design of the system and results in Sect. 3. We describe the experiments and results in Sect. 4 and present conclusions and future research directions in Sect. 5 [5].

2 Related Work

2.1 NEST

Nest was a company focused on smart home climate control and fire alarms and was taken over by Google in 2014. Nest's temperature control system learns the behavior of residents' behavioral patterns directly after the installation mode and enters into full-scale indoor climate control within a week [6]. In the case of fire alarm systems, fire alarms are sent in the form of smartphone message, as well as alarms. Different types of IoT products of Nest can exchange information with each other to better understand consumers' intentions or situations. Although It has the advantage that the product or service is intelligent. Various malfunctions have been reported as the shortcomings of the Nest, including two times more expensive than conventional temperature controls, and a variety of malfunctions, including fire alarm malfunction and malfunction of the temperature control (Fig. 1).

Fig. 1. Nest Thermostat & Nest Protect (Smoke + CO Alarm)

2.2 LG HomeChat

LG Homechat commands people in electronic devices to understand what people talk about in natural language, and plays a personal assistant to deliver the response of the devices and the results. It has three key features: communication in natural language, user friendly service orientation, starting as a IoT platform (IoT) platform. Also It provides character and stickers, and personal recommendation functions. However, there are limitations to providing limited services to LG Household appliances (Fig. 2).

Fig. 2. LG HomeChat

2.3 Related Studies

Eliza proposed by Joseph Weizenbaum of MIT Artificial Intelligence Laboratory [7] is an initial form of chatbot that matches user input based on rules and keywords written in scripts. There is Jabberwacky [8] which records conversations with users in a more advanced form and conducts dialogues based on the learning based on them. Jabberwacky uses ALICE, which has a unique markup language called AIML (Artificial Intelligence Markup Language), and heuristic pattern matching, and patterns stored in the knowledge base matched with user input are used to generate output statements.

3 Design of Home IoT System Based on Messaging Applications

3.1 Features and Key Functions

The home IoT system based on the mobile messaging application proposed in this paper is based on 'chatbot' technology, which is based on 'IoT' technology and interaction between users and electronic devices and combines artificial intelligence and messaging application. The system has major functions such as electronic device management, electronic device monitoring, electronic device control, and electronic device notification, all of which are communicated through the user and mobile messaging applications.

3.2 System Configuration and Development Environment

System Configuration. The architecture of this system we propose appears as indicated in Fig. 3. The system consists of A virtual home consisting of a raspberry Pi 3

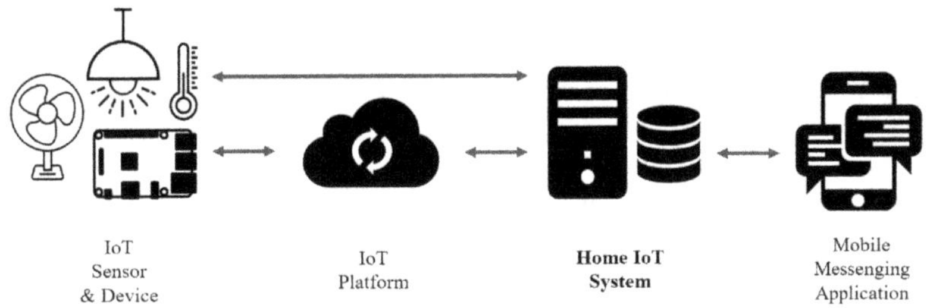

Fig. 3. Configuration of our home IoT system

Kit, sensors (temperature and humidity sensor, infrared sensor and relay module) and a related home appliance, IoT platform capable of collecting data from sensors and providing a simple level of statistical processing, a server that provides parsing of users' requirements and implement data on electronic devices to natural language, Lastly, a messaging Application that users and electronic devices communicate in natural language.

Development environment. The scope and environment of the system is shown in Table 1. The hardware consists of a temperature and humidity sensor, an infrared sensor, a relay module, an electric fan, and a camera installed in the raspberryPi3 and A program was implemented to control the modules and transmit data using PYTHON. As an IoT platform, we use UBIDOTS to record the sensor values and acts as a function of configuring specific events, and the data is delivered in JSON. The server receives a specific data from the IoT platform and generates a notification message to the messaging application, and receives a message requesting the data on the IoT platform. Implementation uses Apache 2 and PHP. It was implemented using Apache 2 and PHP. The chatting application was currently implemented using Naver LINE API, which is widely used, and was implemented using PHP.

Table 1. Scope and environment of design

Hardware	Raspberry Pi 3 (Python), Temperature and humidity sensor, Infrared sensor, Relay module, Camera module, and Electric fan
IoT platform	UBIDOTS (Json)
Home server	Apache2 & PHP (+Javascript)
Mobile messaging application	Naver LINE API (PHP)

3.3 Design of Home IoT System

The Design of the system is largely divided into sensor programming of raspberry Pi, home server building, and chatting programming. We processed user syntax within the server using keyword-based processing and rule-based processing, analyzed the commands that user intended to use, and categorized commands to intensity control, power control, and value modification to answer or response appropriately to users' commands.

The keywords that chatbot recognizes in your message are as shown in Tables 2 and 3. In the case of fan power control, for switching on the fan, it was necessary to perform a message that included 'ON', '켜 (turn on)', '덥다 (it's Hot)', '더워 (it's Hot)', '시작 (Start)', and '작동 (Operation)'. Also for switching off the fan, it was necessary to perform the messages containing 'OFF', '꺼 (turn off)', '춥다 (it's cold)', '추워 (it's cold)' and '그만 (Stop)'. To know the sensor information, we alerted the user to the temperature information and humidity information when user sent messages containing '온도 (temperature)' and '습도 (humidity)'.

Table 2. Keyword to control electric fan power

Turn on	ON	켜	덥다	더워	시작	작동
Turn off	OFF	꺼	춥다	추워	그만	

Table 3. Keyword to obtain sensor information

Temperature	온도	몇 도	**Humidity**	습도

The main features of the system are check the temperature and humidity sensor value, motion sensing through infrared sensors and photo shooting and sending, and fan power controls via the relay module. Depending on the three modes of operation, the following methods of implementation are shown.

Checking the Temperature and Humidity Sensor Value. The temperature/humidity sensor attached to the raspberry pi is measured at a temperature and humidity of 0.5 s and the measured sensor values are stored in the temperature and humidity variable tables in UBIDOTS. When the user asks the system to 'temperature', 'humidity' at line messaging application, the system access to the UBIDOTS to obtain the most up-to-date data and answer it back to the user. Further, the ability to transmit PUSH messages to the user was implemented if the temperature value recorded in the UBIDOTS exceeded a certain value (Fig. 4).

Motion Sensing through Infrared Sensors and Photo Shooting and Sending. When motion sensing is detected by an infrared sensor, the function of the shooting photo is automatically taken up by the camera module. This function consists of an infrared sensor sensing, a shooting photo, and a photo transfer of a raspberry pi. First, the infrared sensors detected motion movements every 0.5 s. Then, if motion sensing was detected, the camera module was activated for five seconds to record the picture.

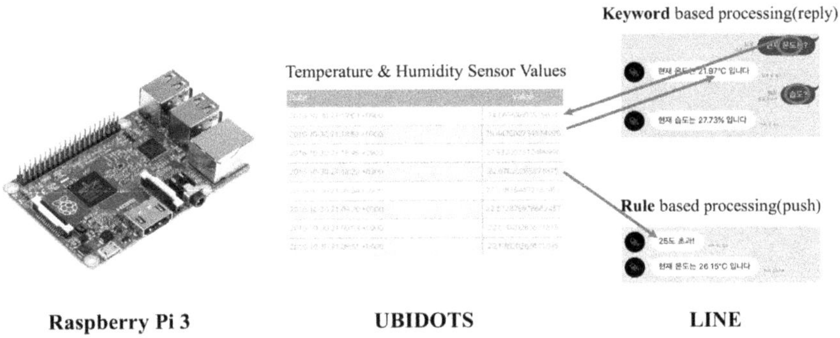

Fig. 4. Feature of the system: checking the temperature and humidity sensor value

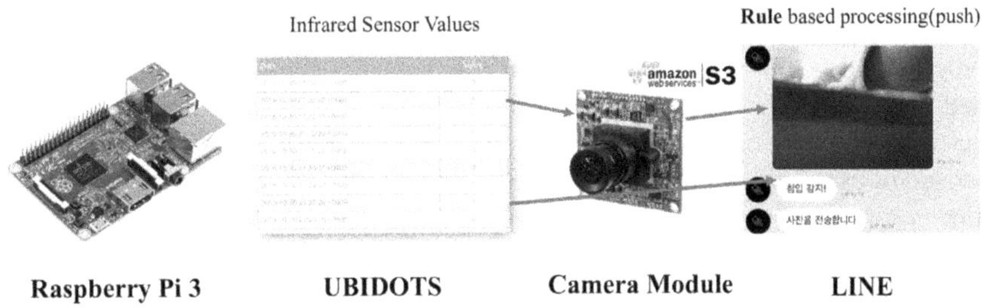

Fig. 5. Feature of system: motion sensing through infrared sensors and photo shooting and sending

The photographed picture is stored inside the raspberry pi as jpg format, and after taking pictures, the image file is uploaded to the Amazon Web Services (AWS) server and the uploaded image is PUSH to the user at line messaging application (Fig. 5).

Fan Power Controls via the Relay Module. Arelay module was used to control the power of all devices that could connect to the outlet as well as the fans via the raspberry pi. The relay module was allowed to be controlled by connecting to the GPIO pins of the raspberry pi. In order to control the switch within the relay module, it was implemented by setting whether or not to send an electrical signal to the GPIO pin of the raspberry pi.

First, when a user sends a message to control power via LINE, the system will update the relay module parameter value at UBIDOTS in accordance with the corresponding message. Then, the system controls the power of the fan according to the variable value in UBIDOTS by accessing UBIDOTS every 2 s. At this point, the relay module and the fan are not connected directly and are connected to the fan via power strip, allowing other devices to be used besides the fan (Fig. 6).

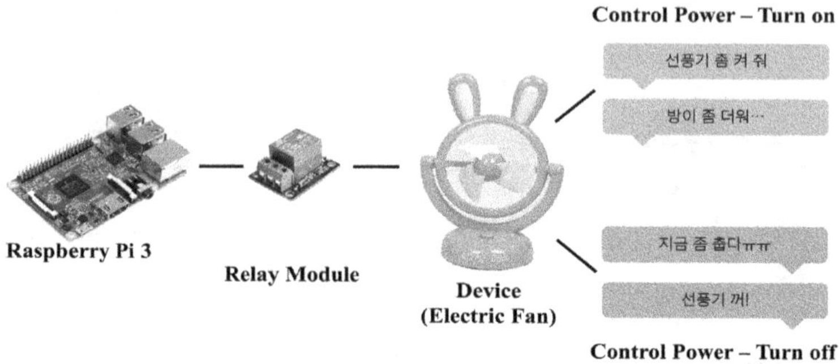

Fig. 6. Feature of the system: fan power controls via the relay module

4 Experimental Results

In order to verify the system proposed in this paper, a simple experiment was conducted for users who had no prior knowledge of the system implemented. The objective of the experiment was to find out how correctly this system worked in the command of users without prior knowledge, and to identify this, the user ordered to list all the messages they would enter when controlling the fan. Because the respondent replied that they will enter '온도' or '몇 도' to obtain temperature information, and '습도' for humidity information, we compared the user response to the fan power control only.

Figure 7 is a user's message on the motion of turning-on the fan, and a total of 15 researchers replied. There were ten messages like '켜 줘', '켜', '켜라' containing '켜', overwhelming compared to other messages. Since the system recognizes 21 of the 25 responses of users as a power control keyword and operates the fan, it has an operating rate of 84% for turning on the electric fan.

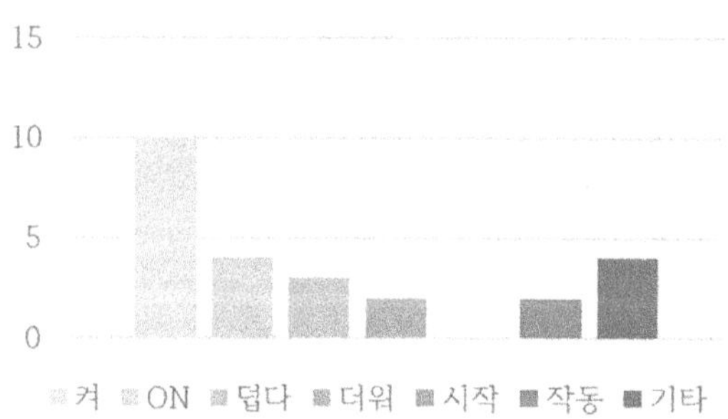

Fig. 7. Type of user language in fan power control – turn on

Figure 8 is a user's message on the motion of turning-off the fan and a total of 15 researchers replied. There were eleven messages like '켜 줘', '켜', '켜라' containing '켜', and no one responds to the 'cold' keywords built into the system. In other words,

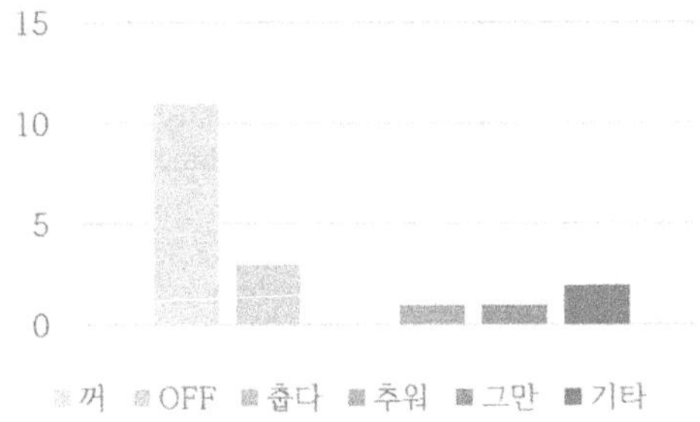

Fig. 8. Type of user language in fan power control – turn off

there were '끄자', '끌래'. Recognizing 16 out of 18 respondents' answers as a power control keyword and stops the operation of the fans, the system has an operating rate of 88% for the operation of the fan.

Based on the above two experiments, the system shows an average of 86% of the operating rate and indicates that most of the users can perform their intended actions for this system.

5 Conclusion

We presented the design of the system using raspberry Pi and LINE application to implement Home IoT based on Mobile Messaging application. The system manages the various sensor information of the raspberry Pi using the independent IoT platform. Not only the LINE we used, but also all mobile messaging applications such as Kakao Talk and Telegram are available. Therefore, it has high reliability and scalability.

The system proposed in this paper is designed to inform the user if there is a user request or a certain situation arose, so that they can control or respond to them. Future studies is including technologies such as GPS, Machine Learning, and Natural Language Processing to the proposed system to establish custom automatic control systems, such as running a boiler, knowing that the user's location is near the house. We will also implement a more developed version of chatbot through the use of a user's conversations and common knowledge establishments. If these technologies are introduced, a more complete system will be implemented.

References

1. Rose, K., Eldridge, S., Chapin, L.: The Internet of Things: An Overview Understanding the Issues and Challenges of a More Connected World. The Internet Society (ISOC), 22 October 2015
2. Bea, S., Kim, J.: Korea Institute of Science and Technology Evaluation & Planning (KISTEP), Changing the paradigm of Internet of Things (IoT) development and security, 13 July 2016
3. Jang, H., Lee, S.: Prospects for Smart Home Development Based on the Internet. Institute for Information & Communications Technology Promotion, 9 September 2015

4. Kim, Y.: 6 Components of Smart Home (Home IoT) Ecosystem. KT Economics & Management Research Institute, 26 November 2014
5. Khanna, A., Pandey, B., Vashishta, K., Kalia, K., Pradeepkumar, B., Das, T.: A study of today's A.I. through Chatbots and rediscovery of machine intelligence. Int. J. ue Serv. Sci. Technol. **8**(7), 277–284 (2015)
6. KB Financial Group Management Research Institute, KB knowledge Vitamin, A Study on the Introduction of Mobile messenger Chatbot and Its Implications, 11 May 2016
7. Weizenbaum, J.: Eliza—A computer program for the study of natural language communication between man and machine. Commun. ACM **9**(1), 36–45 (1966)
8. Güven, G., Franchi, S.: Dialogues with colorful personalities of early AI. Constructions Mind. Artif. Intell. Humanit. Spec. Issue Stanford Humanit. Rev. **4**, 161–170 (1995). http://web.stanford.edu/group/SHR/4-2/text/dialogues.html

A Design of Group Recommendation Mechanism Considering Opportunity Cost and Personal Activity Using Spark Framework

Byungho Yoon[1], Kiejin Park[2(✉)], and Suk-kyoon Kang[1]

[1] Department of Industrial Engineering, Ajou University, Suwon, Korea
{cafesky7, tjrrbsl120}@ajou.ac.kr
[2] Department of Integrative Systems Engineering, Ajou University, Suwon, Korea
kiejin@ajou.ac.kr

Abstract. Group Recommendation is a method of recommending a specific item (e.g. product, service) to a group formed of several members. Least Misery, one of the representative group recommendation method, can recommend items considering group dissatisfaction, but has a drawback of low recommendation accuracy and the other method, Average methods has high accuracy but cannot consider the group dissatisfaction. In this paper, we developed a group recommendation method that improves the recommendation accuracy by measuring the Opportunity Cost (Opportunity Cost is the largest value of the remaining item that is discarded when you select a specific item) and personal activity, taking into account group dissatisfaction. Hadoop-Spark Framework was used in the experiment to distribute large scale of data safely and process efficiently. In Experiment result the proposed group recommendation method improved the recommendation accuracy by 27% while considering the dissatisfaction of the group members compared to the Least Misery method.

Keywords: Group recommendation · Least Misery · Opportunity cost · Hadoop · Spark

1 Introduction

The Recommendation method is a method of recommending specific items (e.g., goods, services, etc.) that are expected to be preferred by users in various fields. Collaborative filtering (CF) is a popular recommendation method, and users with similar preferences can be obtained through evaluation information of items purchased in the past by a specific user and another user. Through this, we can predict the preference of item that has not purchased before from the specific user, and based on this prediction, the system can recommend new item to user [1]. In addition, the recommendation method is classified into two categories. (1) Personal recommendation is to recommend specific items that each user is expected to prefer. (2) Group recommendation is to recommend items that can minimize the dissatisfaction of the members to a group formed by many members.

W. Lee et al. (eds.), *Proceedings of the 7th International Conference on Emerging Databases*, Lecture Notes in Electrical Engineering 461,
https://doi.org/10.1007/978-981-10-6520-0_32

The group recommendation aims to prevent unfortunate member occurrence (e.g., a situation in which a vegetarian group selects a stake) by selecting a group. Representative methods for group recommendation include Least Misery, Most Pleasure, and Average methods. And among them, the Least Misery method can recommend items that consider group dissatisfaction, but has drawbacks of low recommendation accuracy [2].

So, in this paper, we have developed a group recommendation method that improves the recommendation accuracy considering the dissatisfaction level of the group, reflecting the opportunity cost concept and personal activity.

This paper is composed of 5 sections. In Sect. 2, we describe Hadoop and Spark in relation to group recommendation. In Sect. 3, we describe a group recommendation method considering personal activity and opportunity cost. Section 4 presents experimental and performance evaluations, and Sect. 5 presents conclusions and future work.

2 Related Research

2.1 Group Recommendation Process

The group recommendation process can be divided into three stages: Group search, Group modeling, and Group rating prediction [3, 4].

1. The group search is a step of forming an appropriate group considering the purpose of recommendation, and generally groups users that have similar preferences into the same group.
2. The group modeling is a step of applying the group recommendation method to the formed group in the previous step, and is generated as the group preference information based on the preference information of the individual members of the group.
3. Finally, the group rating prediction generates a predicted rating for each item in a specific group based on group formed in the previous step and group recommendation method, and recommends a specific item to the group based on the predicted rating.

Table 1. Examples of group modeling.

	Item 1	Item 2	Item 3	Item 4
User 1	2	4	4	1
User 2	5	3	4	5
User 3	3	3	2	4
User 4	1	1	4	5
Least misery	1	1	2	1
Most pleasure	5	4	4	5
Average	2.75	2.75	3.5	3.75

To show pros and cons of group recommendation process, one example is given by Table 1. In the case of applying each group modeling method to Item 1, Least Misery has the smallest value 1 of the members, Most Pleasure is the highest value of the members 5, and Average is 2.75, which represents the group ratings. Through the results of each method, The Least Misery method is characterized by low recommendation accuracy because the lowest rating of the members is simply the rating of the group in order to consider the dissatisfaction of the members. In addition, Average is the average rating of the group members as the rating of the group. Therefore, the recommendation accuracy is higher than the least misery, but it is difficult to consider the dissatisfaction of the group members.

2.2 Opportunity Cost and Personal Activity

Opportunity cost means the highest value of any item the user give up. Choosing to take advantage of opportunity costs means making reasonable decisions when there are several alternatives to choose from, among which the losses resulting from a particular alternative are improved and then the alternative with the least loss is selected [5]. In addition, in order to measure the influence of each member on group decision making and to reflect it on the rating, various methods have been studied. In [6], in case of determining a place to travel, they applied the weights to the preference calculation by changing the weights according to the roles of the members, but it is difficult to calculate if the role of the members in the group is unknown. In [7], we propose a method to apply weights differently through the social connection between each member and can be usefully applied in cases where the relationship can be clearly grasped (e.g., Facebook), but in other cases, It is difficult to define how much friendship can be seen as a relationship.

2.3 Collaborative Filtering (CF)

Collaborative filtering begins with the idea that users with similar preferences for a particular item will show similar preferences for other items. The similarity measure between users is measured through the preference information of the past purchase item of a specific user, It is a method of predicting the preference of purchase items and recommending item to users.

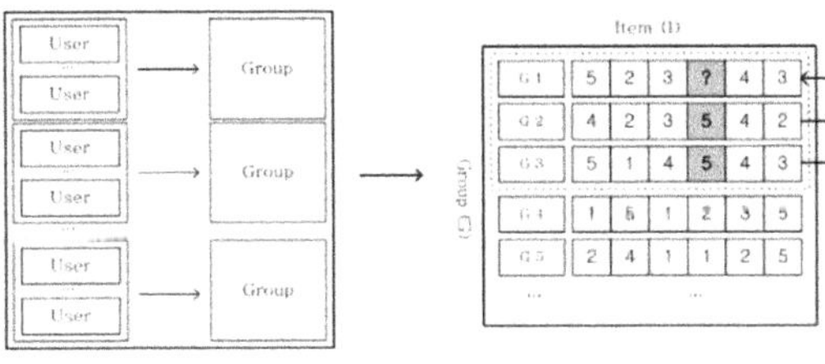

Fig. 1. Examples of group recommendation method using CF.

Figure 1 shows an example of group recommendation through CF, and the preferences existing for each user are generated by group preference through group modeling. As a result, a matrix composed of preferences for groups and items can be obtained, and through this, it is possible to calculate the similarity between the target group (G 1) and the group (G 2, 3) having similar tastes. Therefore, the unavailable item (Item 4) of Group 1 can be calculated through the similarity and similarity of Item 4 of Groups 2 and 3.

2.4 Matrix Factorization (MF) and Alternating Least Squares(ALS)

Among the collaborative filtering methods, Matrix Factorization (MF) was introduced through Netflix Prize, a global recommendation algorithm competition. It is a method of predicting the rating of an item that has not been evaluated based on the user's preference information, and recognized as a leading method because of the high recommendation accuracy [8].

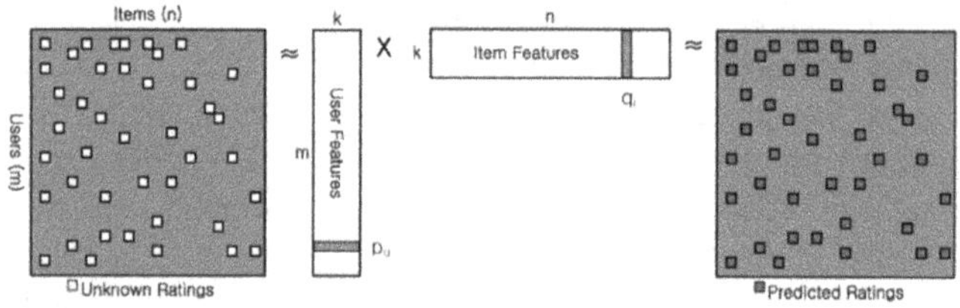

Fig. 2. Predicting item rating using MF.

MF is one of the matrix decomposition methods. It is a sparse type matrix with blank spaces (= means the item that user unrated), and can be expressed as an m × n matrix as shown in Fig. 2. By using MF, we can extract the latent feature matrix that consists of k-feature vectors (see Eq. 1). Then, the optimization process finds p_u, q_i that minimizes the difference between the actual rating and the predicted rating, and generates a predicted rating similar to the actual rating, thereby extracting the predicted rating for all ratings.

$$\min_{q*,p*} \sum_{u,i} \left(r_{ui} - q_i^t p_u\right)^2 \tag{1}$$

Alternating Least Squares (ALS) is one of the methods to optimize the MF, and this method is popular because the calculation is process in a distributed fashion and this can be used efficiently in a distributed computing framework which is the basis of recent big data analysis [9].

2.5 Hadoop and Spark Framework

Spark is an in-memory processing framework that, in the case of iterative processing, loads data into memory and drives high-speed data analysis.

Through this framework, it can solve the disk bottleneck caused by MapReduce, which is a large-scale data processing method in Hadoop, and can efficiently execute data processing requiring iterative processing such as machine learning [10]. In addition, Hadoop is a cluster system composed of a number of slave nodes for storing and processing data and a single master node for managing them, which can store and process large amounts of data. In Fig. 3, YARN (Yet Another Resource Negotiator) manages the entire resources, tasks, scheduling and defects of the Hadoop Cluster and parallelizes the work on each slave node through the Spark Driver Program of the Master Node. In this paper the prototype was constructed for performance evaluation of the proposed method based on the Hadoop-Spark framework.

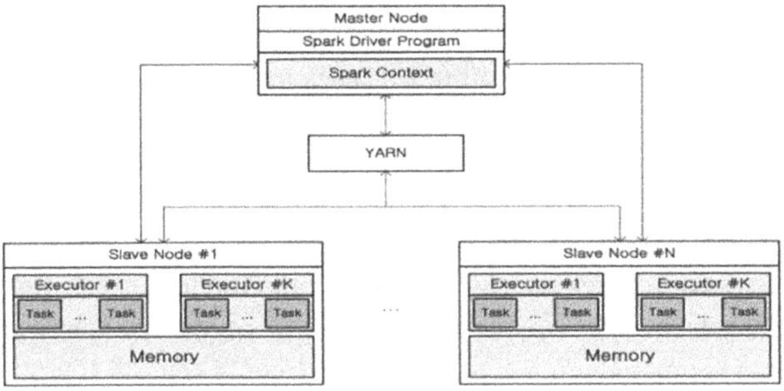

Fig. 3. Hadoop-based Spark framework architecture.

3 Group Recommendation Method Considering Opportunity Cost and Personal Activity

3.1 Group Search: K-Means Clustering

In the group search, K-Means Clustering was used to construct groups with similar preference members. And this method measures the difference between rating data of a given individual user by Euclidean distance and forms k clusters with minimized error between ratings, (shown as Eq. 2).

$$\arg\min \sum_{i=1}^{K} \sum_{x \in s_i} ||x - \mu_i||^2, \text{ i} = 1, 2, \ldots \text{K} \tag{2}$$

In K-Means method, is the number of k that denotes the number of clusters is one of the hyper-parameter to set by hand and the performance of the model can be very different depending on the choice of k. So, in this paper, the Elbow Method [11] is adopted to determine the appropriate k value. And this is a method of gradually increasing the number of K-Means clusters and finding an appropriate k value through the result. In this method, the Elbow Point, which is the interval in which the center of each cluster (μ_i) and the data belonging to the cluster, which is the sum of the errors, is gradually decreased and gradually decreases, is set as the k value.

3.2 Group Modeling: Opportunity Cost and Loss

In this paper, we apply K-Means to the personal rating data to generate k groups, which are expressed as G = $\{g_1, g_2, \ldots, g_k\}$, k = 1, 2, …, And K respectively.

$$r_{g_k i} = \frac{\sum_{l=1}^{L} r_{u_{lk} i}}{L} \quad \mathrm{I} = 1, 2, 3 \cdots \mathrm{I} \tag{3}$$

The rating of each group is taken as the average rating of the group members, and the rating $(r_{g_k i})$ of the *i*-th item in group k is given by Eq. 3, where $r_{u_{lk} i}$ is the rating assigned to item *i* by the member ℓ of the kth group. At this time, to get the opportunity cost for each member of the group, first we search for the item with the highest value in the group rating, and the personal rating of the item after the search is considered the opportunity cost of each member.

The loss can be obtained by subtracting the opportunity cost from the rating of each member of the dissatisfaction degree by selecting a specific item. The larger the value, the greater the dissatisfaction felt by the group selection. Therefore, the larger the loss is, the smaller the weight is reflected in the rating, and the larger the weight is, the more the recommendation considering the dissatisfaction of the members in the group is made possible.

$$l_{g_k i} = \frac{\sum_{l=1}^{L} \left(r_{u_{lk} i} - c_{u_{lk}}\right) \cdot PA_{u_{lk}}}{L} \tag{4}$$

The loss is defined by the average of the member's personal activity multiplied by the value of each member's item *i* minus the opportunity cost of each member. The opportunity cost of the ℓth member of the kth group is expressed as Eq. 4 when $l_{g_k i}$ denote the loss for a particular item *i*.

3.3 Group Modeling: Personal Activity

To measure the influence of each member of the group, we define Repeatability and Helpfulness criteria. Repeatability is measured by the number of items evaluated by each member in the group and refers to the degree of participation in group activities. |GI| is the number of items evaluated by the m-th member of the group G, and $|S_{u_m}|$ is the number of items evaluated by the m-th member of the group G = $\{u_1, u_2, \ldots, u_m\}$, and based on this rates E_{u_m} denotes the repeatability of the evaluation can be derived as Eq. 5.

$$E_{u_m} = \frac{|S_{u_m}|}{|GI|} \tag{5}$$

- The Helpfulness indicates the consent of other members of the text content evaluated by a particular member in a group, and can be measured as the sum of the number of different member agreements on the evaluation of each item. However, because the number of values can be excessive, the recommended result can be distorted if there is a large impact on the overall result.

$$H_{u_m} = \frac{h_{u_m} - H_{\min}}{H_{\max} - H_{\min}} \times (\max - \min) + \min \tag{6}$$

In order to solve this problem, the utility of the evaluation is normalized, and then the sum of the number of consent can be expressed by h_{u_m} for the evaluation of the members belonging to the group G and the Helpfulness of the evaluation is expressed as H_{u_m} (see Eq. 6).

$$PA_{u_m} = sqrt(E_{u_m} \times H_{u_m}) \tag{7}$$

In addition, the personal activity (PA_{u_m}) of the members of Group G is defined as Eq. 7, which is the product of the number of items evaluated and the consent of other members of the evaluation and personal activity means how member u_m is active in the group, And how well u_m represents the personality of group.

3.4 Metric for Performance Evaluation

RMSE (Root Mean Square Error) is used to evaluate the performance of the proposed group recommendation method, which is a typical performance metric for recommendation accuracy measurement [12]. RMSE is a measure of the difference between the actual group's rating and the expected rating through the group recommendation method.

$$\sqrt{\frac{\sum_{j=1}^{n} (\hat{r}_{ui} - r_{ui})^2}{N}} \tag{8}$$

4 Performance Evaluation

Experiment was conducted with Amazon Review Dataset [13] to evaluate the performance of the proposed method. The IDs (reviewerID), item number (Asin), overall rating of items, evaluation text, a Helpful column was used in this experiment as shown in Table 1.

Table 2. Shows the sample of Amazon review dataset

ReviewerID	Asin	Overall	Helpful
A2XU46XXNV19C8	439893577	3.0	[59, 59]
ACI4789O50TU5	1556345542	4.0	[1, 1]
A32M15RZXWZ5GS	979959330	4.0	[0, 0]

The experimental cluster environment consists of 1 master and 6 slaves, and detailed cluster specification is shown in Table 3.

Table 3. Experimental environment.

	Master	Slave
CPU	Intel® Core™ i7 or i5 CPU (Skylake)	
Memory size (Total)	16 GB	64 GB*6(384 GB)
Storage size (Total)	6 TB	8 TB*6 (48 TB)
OS	Ubuntu 16.04 LTS	Ubuntu 14.04 LTS
Spark version	2.01	2.01
Hadoop version	2.72	2.72

The experiment process is as follows. First, we adopted the K-means clustering to the Amazon Review datasets to extract the similar taste group (Table 3).

The WSSSE was calculated by applying the Elbow Method to determine the number of clusters of K-Means. The result is shown in Fig. 4, and the k-value is set to 20.

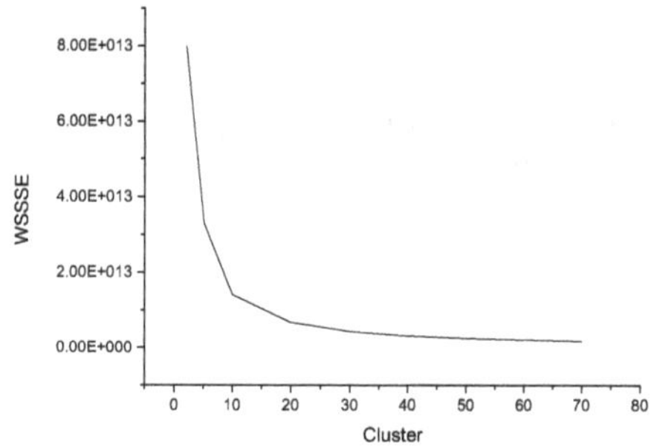

Fig. 4. WSSSE calculation for each cluster.

Group ratings for each item were created with an average rating for each item in a particular group member. After that, the opportunity cost and individual activity were measured according to the method proposed in Sect. 3 of this paper, and the results were reflected in the group ratings. A group recommendation experiment prototype was constructed by applying ALS-based MF, by using Scala language.

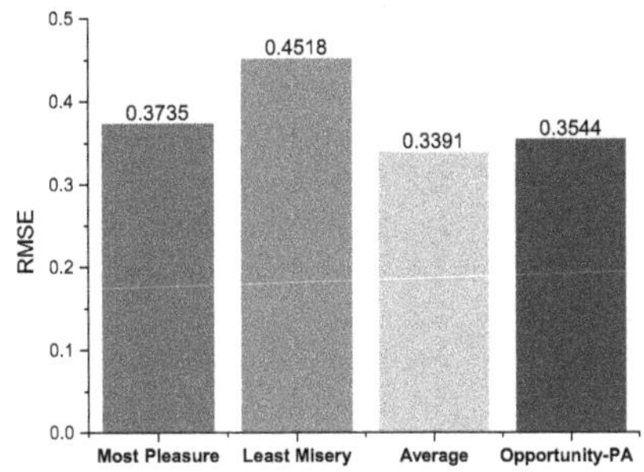

Fig. 5. RMSE comparison of proposed group recommendation and other group recommendation methods.

We compare the proposed group recommendation method with Least Misery, Most Pleasure, Average and RMSE proposed in Fig. 5. As a result, we can find that the proposed method is about 27% better than Least Misery and 5% better than Most Pleasure On the other hand, the recommendation accuracy was about 4% lower than the average method. This means that it is advantageous to apply the proposed model to the group recommendation considering the dissatisfaction level of the group members as a whole, and the average method if not.

5 Conclusion

In this paper, we propose a group recommendation method for solving the disadvantages of Least Misery among group recommendation methods considering opportunity cost concept and personal activity. While Least Misery can take into account satisfaction, the prediction accuracy is low and the average is high, but it is difficult to reflect the dissatisfaction. We attempted to solve this by applying the concept of opportunity cost (loss). Also, considering the different influences of the individuals reflected in the group decision making in the real world, we tried to reflect this through individual activity. As a result, it was confirmed that the recommended recommendation accuracy is better than Least Misery and Most Pleasure among group recommendation methods.

We also used In-Memory based Spark Framework to solve the bottleneck caused by iterative processing of data in MapReduce Framework.

In future research, we will apply the proposed group recommendation method to measure the change of prediction accuracy according to the applied group size of the proposed method, the proposed group recommendation method will be applied to the group (small, medium, and large) classified according to the number of group members.

Acknowledgement. This work is supported by Ajou University Research Fund.

References

1. Linden, G., Smith, B., et al.: Amazon.com recommendations: item-to-item collaborative filtering. IEEE Internet Comput. **7**(1), 76–80 (2003)
2. Boratto, L., Carta, S., et al.: Discovery and representation of the preferences of automatically detected groups: exploiting the link between group modeling and clustering. Future Gener. Comput. Syst. **64**, 165–174 (2016)
3. Boratto, L., Carta, S.: State-of-the-Art in group recommendation and new approaches for automatic identification of groups. Inf. Retrieval Mining Distrib. Environ. **324**, 1–20 (2011)
4. Ntoutsi, I., Stefanidis, K., et al.: gRecs: a group recommendation system based on user clustering. Database Syst. Adv. Appl. **7239**, 299–303 (2012)
5. Hong Yang, S., Long, B., et al.: Collaborative competitive filtering: learning recommender using context of user choice. In: The 34th International ACM SIGIR Conference on Research and Development in Information Retrieval, pp. 295–304 (2011)
6. Ardissono, L., Goy, A., et al.: Intrigue: personalized recommendation of tourist attractions for desktop and hand-held devices. Appl. Artif. Intell. **17**(8–9), 687–714 (2003)
7. Gartrell, M., Xing, X., et al.: Enhancing group recommendation by incorporating social relationship interactions. In: The 16th ACM International Conference on Supporting Group Work, pp. 97–106 (2010)
8. Lu, J., Wu, D., et al.: Recommender system application developments: a survey. Decis. Support Syst. **74**, 12–32 (2015)
9. Koren, Y., Bell, R., et al.: Matrix factorization methods for recommender systems. J. Comput. **42**(8), 30–37 (2009)
10. Park, K., Baek, C., et al.: A development of streaming big data analysis system using in-memory cluster computing framework: spark. In: Advanced Multimedia and Ubiquitous Engineering. Lecture Notes in Electrical Engineering, LNEE, vol. 393, pp. 157–163. Springer (2016)
11. Kodinariya, T., Makwana, P.: Review on determining number of cluster in k-means clustering. Adv. Res. Comput. Sci. Manage. Stud. **1**(6), 90–95 (2013)
12. Herlocker, J., Konstan, J., et al.: Evaluating collaborative filtering recommender systems. ACM Trans. Inf. Syst. **22**(1), 5–53 (2004)
13. Amazon Review Data. http://jmcauley.ucsd.edu/data/amazon/

EEUM: Explorable and Expandable User-Interactive Model for Browsing Bibliographic Information Networks

Suan Lee, YoungSeok You, SungJin Park, and Jinho Kim(✉)

Department of Computer Science, Kangwon National University, Kangwon, Chuncheon, Republic of Korea
{suanlee,yys,paksj,jhkim}@kangwon.ac.kr

Abstract. 8 A suitable user interactive model is required to navigate efficiently in information network for users. In this paper, we have developed EEUM (Explorable and Expandable User-interactive Model) that can be used conveniently and efficiently for users in bibliographic information networks. The system shows the demonstration of efficient search, exploration, and analysis of information network using EEUM. EEUM allows users to find influential authors or papers in any research field. Also, users can see all relationships between several authors and papers at a glance. Users are able to analyze after searching and exploring (or navigating) bibliographic information networks efficiently by using EEUM.

Keywords: Information networks · Graph database · Data visualization · User-interactive model

1 Introduction

An information network is structured to interact with various types of objects. It can be represented as a graph, which consists of $G = (V, E)$, where V is a set of vertices (or nodes) and E is a set of edges (or links). Each edge has two vertices. Node can be linked by one or more links. Each node represents as a object (or entity), and each edge represents as a link, which is a relationship between two nodes. For example, each entity in bibliographic information networks is identified as author, paper, conferences, or other related type of bibliography. Also, each link is identified as 'write' or 'cite'.

Numerous researchers lately are working on information network. Professor Jiawei Han and his team are leading the research, and various research results have been published. They have conducted various studies to perform data mining in bibliographic information networks [1,2], and they have developed a program called, BibNetMiner [3]. In addition, he has conducted research of ranking-based clustering and classification using star network schema in heterogeneous information networks [4,5]. He also has conducted on top-k similarity

W. Lee et al. (eds.), *Proceedings of the 7th International Conference on Emerging Databases*, Lecture Notes in Electrical Engineering 461,
https://doi.org/10.1007/978-981-10-6520-0_33

search in heterogeneous information networks [6], stochastic topic models [7], and co-author relationship prediction [8].

By taking advantages of the previous results of information network researches, we have developed the EEUM system[1] that allows users to search and to explore (or navigate) efficiently bibliographic information networks using WebVOWL [9]. EEUM provides a web-based user interactive interface as easy-viewing of information networks. An information network can be expanded or removed by user's preferences, and multiple filtering functions allow efficient analysis. Influence on information network can be proven by using various notations and centralities of each entity.

2 System Overview

Structure of our proposed EEUM system has shown as Fig. 1. First, EEUM provides users able to explore on information network efficiently through browsing modules. Also, EEUM provides users to organize and to display information network through graph modules. At server modules, graph database stores all data and returns query's result of bibliographic information network.

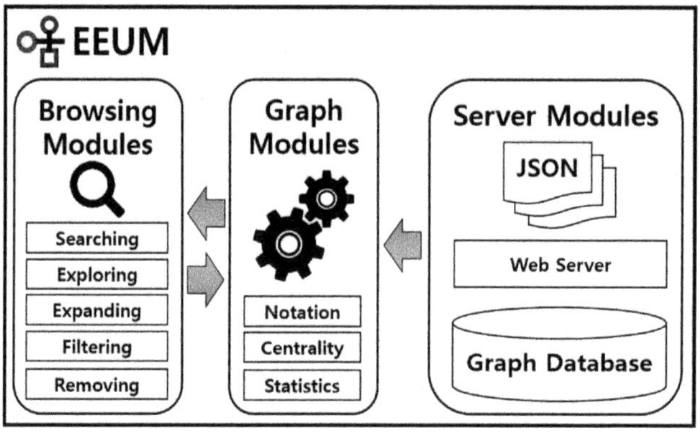

Fig. 1. EEUM system architecture

Figure 2 shows the web-based user interface of the EEUM. Users are able to explore and to expand information networks through the graph structure, and they are able to analyze data efficiently through several functions.

2.1 Browsing Modules

Searching. EEUM provides a module that can search efficiently in bibliographic information networks. This module basically allows searching for author and

[1] EEUM system: http://eeum.suanlab.com.

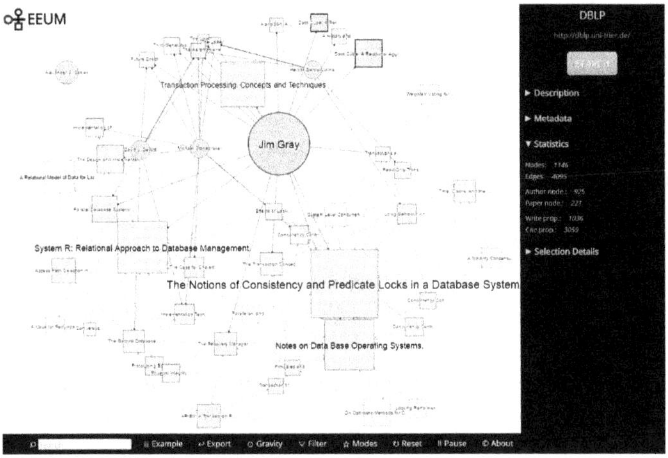

Fig. 2. Web-based user interface

similarity searching for keywords such as title and journal. It features auto-complete for over 1.8 million authors and over 1600 journals. Users can set conditions to search either for the start node only or for the entire node only. In the search module, user can specify year range of the published date to search, and can specify number of hops, the portion of each path between source and destination.

Exploring. EEUM allows users to explore efficiently in bibliographic information networks. Users can explore a structure of graphs and all relationships between various nodes. They can see and read a detailed information by selecting paper-node or author-node. By selecting a paper-node, EEUM provides users a detailed information of the node such as title, type, EE, volume, journal, pages, year, URL, and key. As similar for selecting an author-node, it provides users a detailed information of the node such as author's full name, type, author's DBLP website, and a list of his/her written papers.

Expanding. Users would acquire to expand their associated networks rather than continuously searching with conditions for exploring bibliographic information networks. In expanding module, EEUM allows users to expand their assoicated networks for acquiring a new network.

For example, all nodes of authored and co-authored papers along with co-authors will be appeared to information networks with links when expanding of an author-node. Also, as similar to a paper-node, all nodes of papers (including cited and cited-by papers) and author(s) that are related to the node are will be appeared as well. Therefore, users can expand networks by selecting 'expand' from context menu of each node.

Removing. Users can remove specific nodes that are displayed on bibliographic information networks. In order to remove, clicking right-mouse button on a node will display menu. Then selected node will be removed by selecting 'remove' from menu. Removing an author-node performs one of two methods for user's choice. (1) All nodes that are linked with selected node including relations, which are links in this paper, will be removed, or (2) a selected node, the node's relations, and its authored paper-nodes which contain no other relations will be removed. However, removing a paper-node performs that only a selected node and the node's relations will be removed. Users can efficiently search and explore on information networks after removing user's unwanted some nodes.

Filtering. Bibliographic information networks may included uncountable nodes and relations on web-based user interface when user starts to search, and it can be complexed to user. Therefore, EEUM provides several filtering functions that can be useful in bibliographic information networks. These functions are the followings:

- Cited Paper Node(s): show/hide all node(s) of cited paper and its relation.
- Cite Relation(s): show/hide all 'cite' links.
- Write Relation(s): show/hide all 'write' links.
- Degree of Collapsing: show/hide number of nodes as many as the number of relations based on the all of the above filters. When number of all nodes exceeds a specific number (100 as default), collapsing will performed automatically, which is called auto-collapsing. This performs to display only upper nodes after hiding each node's relations in ascending order.

2.2 Graph Modules

Notation. EEUM presents several objects with nodes and relations in bibliographic information networks. Notation and description of each object is shown in Table 1. EEUM shows all relationships between various objects, and it allows users to analyze by exploring in information network.

Centrality. At EEUM, node's size is adjusted depending on centrality in bibliographic information networks. In case of an author who wrote and published numerous papers, the author-node's centrality increases as well as a size of the node. As similar to paper-node, when number of times that a single paper has cited by people, a centrality increases. Figure 3 shows that all author-nodes and paper-nodes are set depending on its centrality of all authors that are displayed on user interface, and similarlity to all paper-nodes.

Statistics. At graph module, EEUM provides statistical information which will be displayed numbers, which are total number of all nodes, number of all edges, author-nodes, paper-nodes, 'write' links, and 'cite' links, separately. Number of all author-nodes is a sum of all authors including selected author-node, and

Table 1. A description of each object's notation

Object	Type	Definition	Notation
Paper	node	Object with information of a paper; Linked directly to author or selected author (node)	
Cited Paper	node	Object with information of Paper; No link directly to author or selected author	
Selected Paper	node	Object when paper node or cited(by) paper node is selected, or when papers title matches with keyword that is searched by user	
Author	node	Object with information of an author	
Selected Author	node	Object when author node is selected, or when authors name matches with keyword that is searched by user	
Publishing	relation	Relation that is linked to paper published by Author	
Citation	relation	Relation that is linked to paper cited by other paper	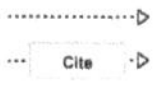

a number of all paper-nodes is a sum of all papers including selected paper-node, and all cited and cited-by paper-nodes. Users are able to easily see the statistical information of bibliographic information networks that they are currently searching.

2.3 Server Modules

Web Server. Web server retrieves user's query results from graph database. First, EEUM generates graph data based on the query results that were initially retrieved from user. Graph data of the query results is stored and managed as a JSON[2] file. Generated graph data can be reused when an identical query is requested. Web server is linked to the graph database which is currently based

[2] https://en.wikipedia.org/wiki/JSON.

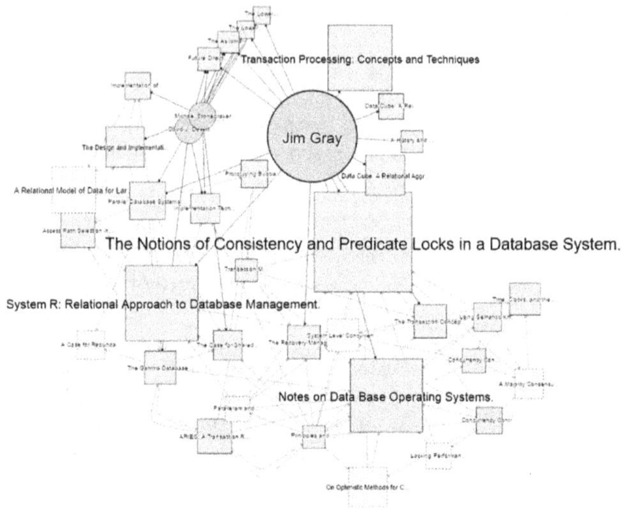

Fig. 3. Example of the size difference of nodes according to centrality (nodes with less than 10 relations are collapsed)

on JSON file, and web server can efficiently shows users the necessary parts of the bibliographic information networks.

As a user explores on information network with the query result, network can be expanded, or the node and relation can be removed. When expanding information network is required, user can append the result of query to graph data. When removing a node or a relation is required, user can remove the node and its relations from graph data without querying to graph database.

Graph Database. At EEUM, we use graph database Neo4j[3] for efficient storage and usage of graph data, and all data of bibliographic information networks are stored in a graph database. Neo4j is an ACID compliant transaction database with graph storage and processing capabilities. While system saves directional data as $[Node - Relation \rightarrow Node]$, data is stored as one of two types of data, $[Author - Write \rightarrow Paper]$ or $[Paper - Cite \rightarrow Paper]$.

Author-node contains an author's name, and paper-node contains various information such as title, type, EE, volume, journal, pages, year, URL, and key. Relation is identified as 'write' or 'cite', and each relation has information of 'from node' and 'to node' to acknowledge directionality.

In this paper, we use DBLP dataset to construct bibliographic information networks. As previously stated, we store DBLP dataset to graph database.

[3] https://neo4j.com/.

3 Experimental Case Studies

EEUM shows the process of exploring and analyzing DBLP dataset. In this paper, we show demonstration in two scenarios.

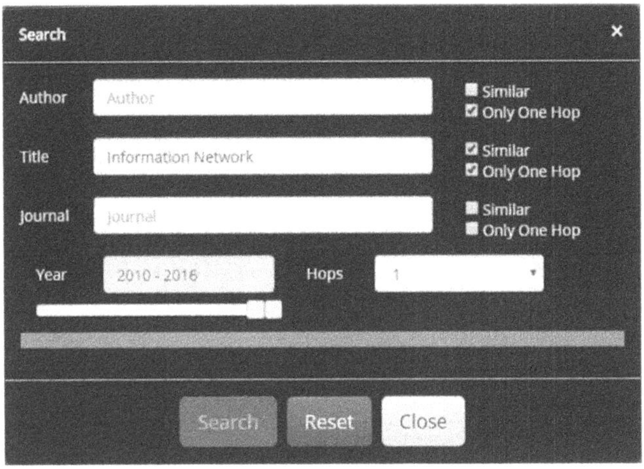

Fig. 4. Search model interface in EEUM system

Scenario A [Information Network]. It is supposed that user explores entire information network, and performs searching for influential author in applicable field. Firstly, Fig. 4 shows as a screenshot of searching the network for title including 'information network' keyword.

Figure 4 shows the results from the search, "Jiawei Han". In the result, "Jiawei Han" is an author who is influential in a field according to the centrality as shown in figure. In the Fig. 5, two large networks are shown. One network contains numerous papers and co-authors that are related to "Jiawei Han", and other networks centers on "Philip S. Yu". If user desires to see only articles that are related to 'Information Network' written by "Jiawei Han", then user is able to search after inputting "Jiawei Han" as author and 'Information Network' as title. A network after retrieving above conditions is shown as Fig. 5. Figure 5 shows single network with numerous co-authors distributed around at centered node, "Jiawei Han". One of co-authors, "Yizhou Sun", is located closed to "Jiawei Han" in network. This can be seen in form of a graph that "Jiawei Han" and "Yizhou Sun" work together on paper on a field of information network.

Scenario B [Data Cube]. It is supposed that user understands the influence on "Jim Gray" and 'Data Cube' in bibliographic information networks. User searches author as "Jim Gray" and title as 'Data Cube'. To analyze the influence of the paper based on the results, user expands network to obtain information of co-author(s) and cited(by) paper(s) for data cube. Figure 6 shows how much

Fig. 5. Results for information network research (degree of collapsing: 4)

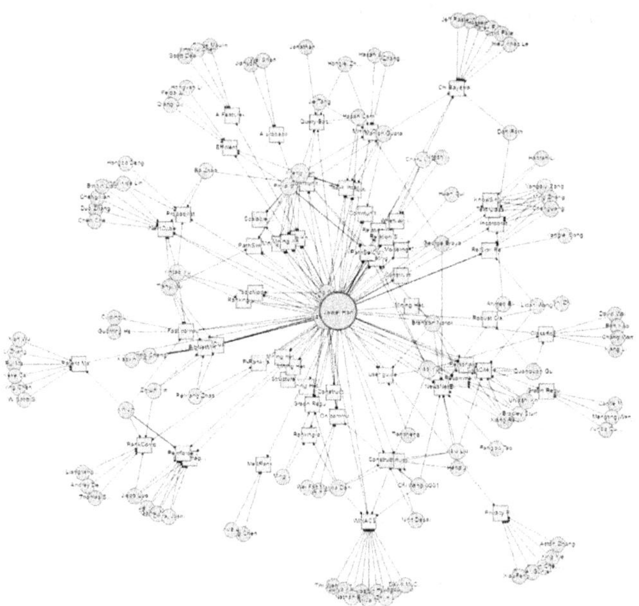

Fig. 6. Results for "Jiawei Han" in 'information network' research

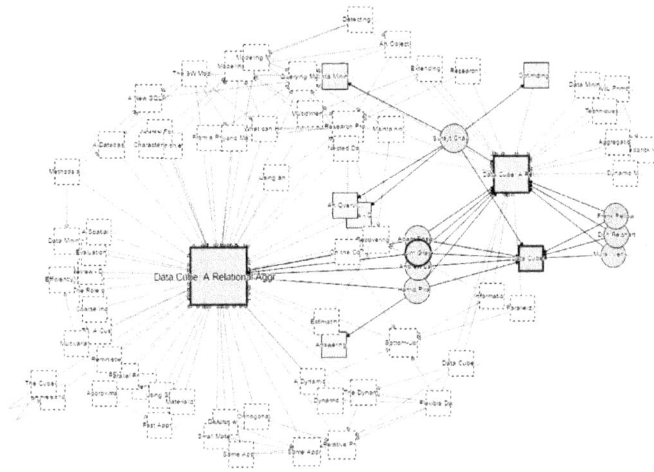

Fig. 7. Information network result of "Jim Gray" author and "data cube" paper

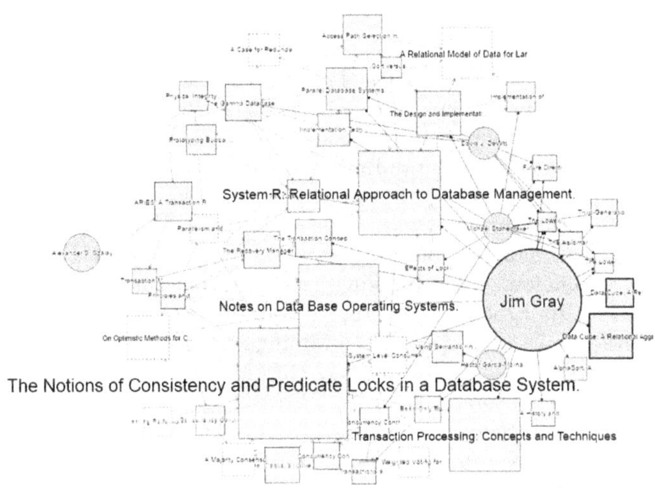

Fig. 8. Information network about "Jim Gray"

influence of data cube papers. Figure 7 shows as expansion of network when user would like to understand how much "Jim Gray", one of authors for the Data Cube Thesis, has influential. In Fig. 7, papers and co-authors of "Jim Gray" are expanded to Fig. 8. In Fig. 8, user can determine that "Jim Gray" is the author who wrote numerous influential papers as information network shows.

4 Conclusion

We have designed and implemented explorable and expandable user-interactive model for browsing bibliographic information networks. Our system allows users to efficiently search, search and analyze information networks based on a graph database. Users can use EEUM to find influential authors or papers in the bibliographic information network. Currently, we are working on combining EEUM systems with various information networks and efficiently summarizing and analyzing the graphs. In the future, we will carry out studies to efficiently analyze large-scale information networks.

Acknowledgments. This work was supported by the Industrial Technology Innovation Program through the Korea Evaluation Institute of Industrial Technology (Keit) funded by the Ministry of Trade, Industry and Energy (Project#: 10052797, Project name: The development of the real-like business simulation platform enable by case-based business games).

References

1. Sun, Y., Han, J.: Mining heterogeneous information networks: a structural analysis approach. ACM SIGKDD Explor. Newsl. **14**, 20–28 (2013)
2. Sun, Y., Han, J.: Mining heterogeneous information networks: principles and methodologies. Synth. Lect. Data Min. Knowl. Discov. **3**, 1–159 (2012)
3. Sun, Y., Wu, T., Yin, Z., Cheng, H., Han, J., Yin, X., Zhao, P.: BibNetMiner: mining bibliographic information networks. In: Proceedings of the 2008 ACM SIGMOD International Conference on Management of Data (2008)
4. Ji, M., Han, J., Danilevsky, M.: Ranking-based classification of heterogeneous information networks. In: Proceedings of the 17th ACM SIGKDD International Conference on Knowledge Discovery and Data Mining (2011)
5. Sun, Y., Yu, Y., Han, J.: Ranking-based clustering of heterogeneous information networks with star network schema. In: Proceedings of the 15th ACM SIGKDD International Conference on Knowledge Discovery and Data Mining (2009)
6. Sun, Y., Han, J., Yan, X., Yu, P.S., Wu, T.: Pathsim: Meta path-based top-k similarity search in heterogeneous information networks. In: Proceedings of VLDB Endowment, vol. 4, pp. 992–1003 (2011)
7. Deng, H., Han, J., Zhao, B., Yu, Y., Lin, C.X.: Probabilistic topic models with biased propagation on heterogeneous information networks. In: Proceedings of the 17th ACM SIGKDD International Conference on Knowledge Discovery and Data Mining (2011)
8. Sun, Y., Barber, R., Gupta, M., Aggarwal, C.C., Han, J.: Co-author relationship prediction in heterogeneous bibliographic networks. In: 2011 International Conference on Advances in Social Networks Analysis and Mining (ASONAM) (2011)
9. Lohmann, S., Negru, S., Haag, F., Ertl, T.: Visualizing ontologies with VOWL. Sem. Web **7**, 399–419 (2016)

Proximity and Direction-Based Subgroup Familiarity-Analysis Model

Jung-In Choi(✉) and Hwan-Seung Yong

Department of Computer Science and Engineering, Ewha Womans University, Seoul 120-750, South Korea
Jungin.choi.db@gmail.com, hsyong@ewha.ac.kr

Abstract. In this paper, we have reported an effective model for familiarity analysis in indoor environments based on proximity and direction. We employ the positioning data of users; thus, we avoid recording the action or any conversation pertaining to the users. We use the beacon signal to find a user's location and choose a subgroup, which is a temporary group obtained using the location of the users. The proposed method analyzes the familiarity using two different methods. The proximity-based method is used to calculate the familiarity based on the time for which the user has stayed in the subgroup. The direction-based method is used to calculate the familiarity based on the direction of each user in the subgroup. This study addressed situations arising in an event or a group activity in indoors to analyze the degree of familiarity by determining the location of a user.

Keywords: Bluetooth low-energy beacon · Familiarity analysis · Indoor positioning · Subgroup analysis

1 Introduction

The Internet of Things (IoT) is expected to be overpopulated by numerous objects with intensive interactions, heterogeneous communications, and millions of services [1]. A large amount of data will be generated because of the considerable increase in the number of users of wearable devices and social network. With IoT devices, the familiarity between users or devices can figure and it helps to analyze user's property and to provide personalized services. So many researches study familiarity-analysis method and previous studies have used social networking services (SNS), sensors installed in buildings, etc. [2]. However a user's privacy could be put at risk, many studies have not used personal information but instead have used the users' location-tracking data [3]. Accordingly, many studies have used sensor data and they use a personal device, which helps transfer the user's location data to the server [4, 5]. Others use attached sensor data employ video, image, and voice data [6]. Unfortunately, these data and personal device contain a user's information. Hence, this is still a matter of privacy.

In this study, we assume that each user carries a beacon, and the receiving device is installed on the building. When the device receives the beacon signal, each device calculates the distance between the device and the beacons to find a user's position and

W. Lee et al. (eds.), *Proceedings of the 7th International Conference on Emerging Databases*, Lecture Notes in Electrical Engineering 461,
https://doi.org/10.1007/978-981-10-6520-0_34

transfer the position to the server. From the transmitted data, the server detects the subgroups. A subgroup is a temporary group in a specific location like classroom. A subgroup is determined by user's location and its density within a group. Using the subgroup, this study figure the familiarity score for each user based on two methods: proximity and direction analyses. Unlike other study, we do not use any information other than the location. Moreover, the users need not send any information from their personal devices.

In this paper, we explain the overall structure of the system in Sect. 2. The subgroup-analysis model using the Bluetooth low-energy beacon is explained in Sect. 2.1. The proximity and direction-based familiarity-analysis models are described in Sects. 2.2 and 2.3, respectively. In addition, we show the scenario and experiment results of the proximity and direction-based methods in Sect. 2.4. Finally, the conclusions of this study are presented in Sect. 3.

2 Real-Time Familiarity Analysis Model

We consider that if the users enter a specific location such as a movie theater or a banquet hall, the users are members of a certain group. If the user is in the movie theater, the user might be an audience watching a film. Although the users are performing the same group activity, it does not imply that the users are familiar with each other. We assume that if the users are familiar with each other, they will have a conversation and be in close proximity. Hence, before we analyze the degree of familiarity between the users, we will detect the subgroups, which are defined by their location within the group [7].

A subgroup and tracking data of the user are the basic elements of our analysis model. We recognized the subgroups and tracked a user's route through the familiarity analysis process. We describe our suggested models as follows. Figure 1 shows the process of the proposed system using the models.

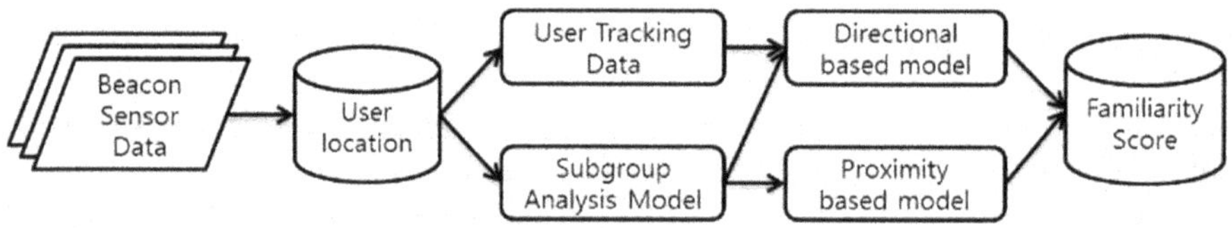

Fig. 1. Process of Familiarity Analysis Model.

While a user carrying a beacon enters an indoor environment, the receiving devices receive the beacon signal. Each receiving device calculates the distance between the user and the receiving device. If a user is within the certain observation area of the receiving device, it is assumed that the user is active around that receiving device. The threshold for the observation area is different from the installation distance. If the interval of each device is 1 m, the threshold is 50 cm. The receiving device will be deemed as being closest to the user if the receiving distance is found to be less than 50 cm.

We recommend that the distance between the receiving devices be in the range of 75 cm to 3 m [7]. This range is the most efficient to obtain a high accuracy with respect to the distance. The receiving device is used to calculate the beacon signals using the distance. The user's location is transferred to the server within the observation area of the receiving device. If the user moves quickly or an error is observed in the calculation, more than one receiving device appears to be the closest to the user. In these cases, the calculation is repeated and the distance is recalculated every 30 s.

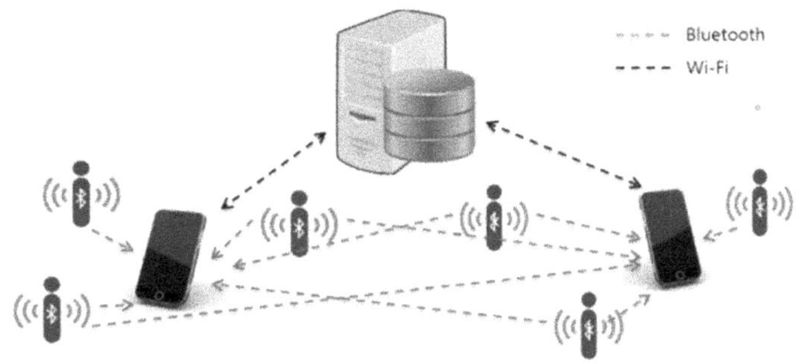

Fig. 2. Sensor Data Flow

Figure 2 shows the flow of the sensor data. The beacon only transmits its signal and ID through the Bluetooth. Moreover, each receiving device transmits the distance data to the server through Wi-Fi. The server is used to determine the subgroups and user routes using the received data. The user routes are based on the user location, which is calculated using a triangulation method [8].

Our familiarity analysis model is based on location. For identify users in an indoor environment, we use the beacon wherein the Bluetooth signals are used to transmit a timestamp, major number, minor number, transmission power, RSSI (dB), and proximity to the receiving device. We can identify a beacon with the major number because it is a unique number such as device ID. The offering proximity data is the distance between the beacon and the receiving device, which is measured using the free-space Friis model. However, the accuracy of the proximity data is 68% [9, 10]. Hence, we use the Average_30s_Distance equation which modified the Friis equation [7]. Using this equation, the difference between the actual and calculated distances is found to be within 10 cm [11]. This is not a problem for the recognition of familiarity. We cannot detect the user's exact location because the calculated distance does not contain direction. Therefore, we use the triangulation equation [8] to determine the user's location.

The proximity-based familiarity analysis is performed using subgroups. The subgroup is calculated using a map, user's location, and density. If the users are in the same subgroup, the familiarity between the users increases every 5 min. The direction-based familiarity analysis is performed using the routes taken by the users. If two users are in the same subgroup and look at each other, they are assigned a familiarity score. In addition, if the users are in same route when they are moving, the familiarity between them increases every 10 min. The results of each familiarity score will be added to final familiarity score in the server.

2.1 Subgroup Analysis Method

The receiving devices receive the beacon signal through Bluetooth. Each receiving device installed under the desk transmits the user's position to the server, and subsequently, the server determines the subgroups. In Table 1, we define four types of subgroups based on their ranges. If the range is over 5 m, the group is public and the range score is 1. A social group, a friendly group, and very-friendly group are assigned ranges of 3 m to 5 m, 1 m to 3 m, and under 1 m, respectively; the groups have range scores of 2, 3, and 4 in regular sequence, respectively. A user may be included in any of the various subgroups. The range score will be used to determine the influence on the subgroup. The number of receiving devices in a certain type of subgroup depends on the interval between the receiving devices. In this study, the interval is 1 m; hence, there is only one receiving device in the very friendly group. The range of the friendly group is 1 m to 3 m; thus, 2 to 9 receiving devices are included. The social group includes 10 to 25 receiving devices. The number of receiving devices in the public group is more than 26.

Table 1. Type of subgroup based on range

Type of subgroup	Range	R_{score}
Public group	5 m and above	1
Social group	3 m to 5 m	2
Friendly group	1 m to 3 m	3
Very-friendly group	Up to 1 m	4

R_{score} is the weight value of each group. It will be used to calculate the entropy score of the subgroup. To calculate the density score D_{score} and density D for each subgroup, we use the following equation.

$$D_{score} = \frac{U}{4S}, \quad D(\%) = \frac{U}{4S} \times 100 \tag{1}$$

In Eq. (1), U is the number of users, and S is the number of receiving devices. For example, if two users are in the observation area of a receiving device, the density score would be 0.5. We consider that four users are sufficient within a distance of 50 cm. Thus, the density score is 1 when four users are within an observation area. However, more than four users can be within the observation area of a receiving device and our system can calculate the density score regardless of the number of users.

To determine whether a subgroup is substantial to be included in the familiarity-analysis model, we calculate its entropy score from the calculated density and the range of the subgroup using Eq. (2). If the density is zero, the user is not within the range and the subgroup entropy (SGE) score is determined to be zero. Otherwise, we multiply the density and range scores and apply natural logarithm as follows.

$$\mathrm{SGE} = \begin{cases} 0 & (D_{score} = 0) \\ \ln(D_{score} \times R_{score}) & (D_{score} > 0) \end{cases} \tag{2}$$

The server is used to calculate the SGE score and this score is used as a basis for determining the influence of the subgroup on the familiarity. The threshold for an influential subgroup is 0.5. Therefore, if the SGE score is above 0.5, we consider it a subgroup for the familiarity-analysis model. For example, if two users are in the observation area of a receiving device, which is a very-friendly group, the equation is $\ln(1 \times 2) \approx 0.69$; thus, it is classified as a subgroup. However, if two users are in the observation areas of two receiving devices, it is not classified as a subgroup. Table 2 lists the entropy scores of the subgroups. The scores qualifying the influential subgroup in the model are in bold.

Table 2. Examples of subgroup entropy score

Number of users	Number of device				
	Very-friendly group	Friendly group			
	1	2	4	6	9
4	**1.39**	0.41	−0.29	−0.67	−1.11
5	**1.61**	**0.64**	−0.07	−0.46	−0.87
10	**2.30**	**1.32**	**0.64**	0.23	−0.17
15	**2.71**	**1.73**	**1.04**	**0.64**	0.23
20	**3.00**	**2.01**	**1.32**	**0.92**	**0.52**

Figure 3 shows an example of the method used to select a subgroup. (a) has two users in one observation area, and (b) has three users in three observation areas. (c) has ten users in four observation areas. Based on the above equations, (a) and (c) are subgroups. In (c), there are possibly two overlapping subgroups, where one has two users and three users in each observation area. In this study, we evaluated these designs and selected the highest range of the subgroup between the overlapped subgroups. However, we did not develop a geometric design but instead limited ourselves to a quadrangle design. The design in (b) which was not selected as a subgroup is recalculated with a higher range. If a user does not belong to the subgroup, the user is considered temporarily isolated.

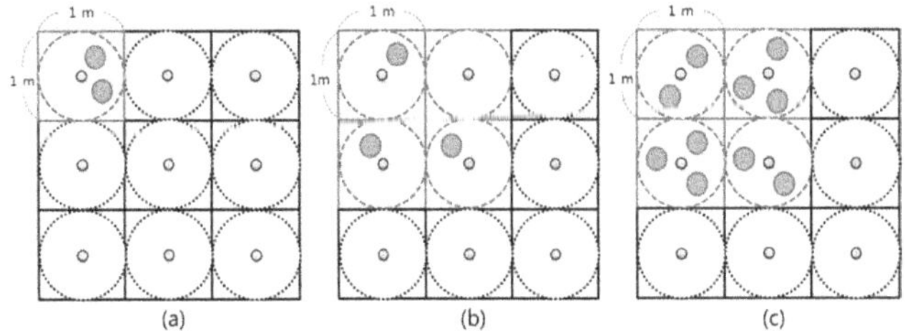

Fig. 3. Example of subgroup analysis

2.2 Proximity-Based Familiarity-Analysis Method

The location and time influences the familiarity. Therefore, we have considered that there is a connection within a subgroup. In this case, the familiarity score between each pair of users increases by 1 point every 5 min.

```
Algorithm 1: Familiarity Calculation Algorithm
Input  : <UserID, SubgroupID, DurationTime>, Familiarity[i][j]
Output: Familiarity score between users
U[ ]=set of UserID;
G[ ]=set of user's SubgroupID;
for i:=0 to U.length do
    for j:=0 to U.length do
        //i and j are number of users
        if i! = j then
            if DurationTime > 3 min and G[i] = G[j] then
                Familiarity[i][j]←Familiarity[i][j] + 1;
            end
        end
    end
end
```

Algorithm 1 describes the method used to calculate the familiarity. Using the beacon's major value, subgroups, and duration time, we calculate the familiarity score using a two-dimensional matrix and recalculate the algorithm every 5 min. UserID is the major value of the beacon, which is unique. When this algorithm is run for the first time, the initial value of the familiarity will be zero. After the first run, the familiarity value uses the saved score that is obtained from the previous calculation. If a user has left the area, the familiarity of the user with others is saved to the server. DurationTime indicates that the user has been included in a certain subgroup for some minutes during the previous 5 min. It enables us to know whether the user is leaving the subgroup. We consider the moment the user leaves as the average time that a user remains in the area within the 5 min duration, not an exact time such as 5, 10, or 15 min. For example, a user is included in A subgroup from 05:04 p.m. to 05:07 p.m. If the system use an exact time, system understand that a user include in the A subgroup during 05:04 p.m. to 05:05 p.m. and during 05:05 p.m. to 05:07 p.m. separately. Hence, we use the duration time then system understand a user include in the A subgroup during 3 min. Thus, if a user stays for more than 3 min, we assign 1 point.

2.3 Direction-Based Familiarity-Analysis Method

The location-based method has a limitation in that the user's direction is not considered. Thus, we add a 10-min tracking data of the users. In Sect. 2.1, we calculated the user's location using the beacon every 30 s. Using the receiving device as reference points, we map the classroom as shown in Fig. 4. Each device is placed in the form of a square shape, and the interval is expressed to set to 1 of the graph value per 10 cm. This is for convenience when calculating the direction of the user. Figure 4 shows the example of the method used to determine whether the users are facing each other through the users' route.

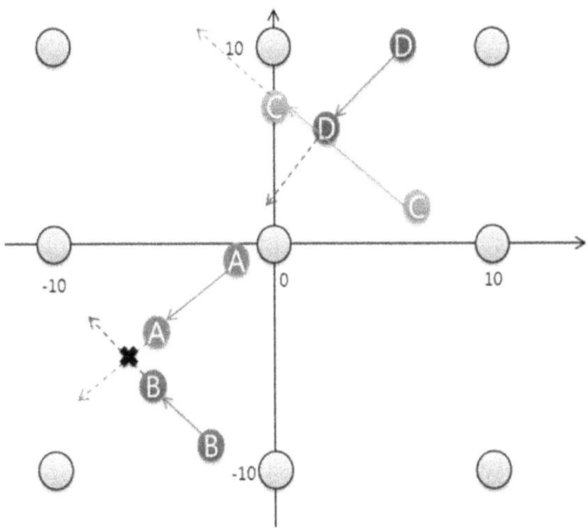

Fig. 4. Example of route tracking of users

When the users are included in the group, the location data for the previous 1 min is recorded to determine the direction of the user's route. We then predict the direction in which the users will move for the next minute. Based on this data, we find the predicted directions of two users in the same subgroup on overlapping. If there is an overlapping data between the users, we assume that users are looking at each other. In Fig. 4, we record the current location and the location before 1 min of users A and B. We then estimate the location of the users after 1 min along the moving routes and follow the arrows as shown in Fig. 4. If the extended dotted lines overlap each other such as users A and B, we consider that they are facing each other. Although the users C and D are close to each other, their extended lines are not overlapping; thus, we assume that they are facing different directions. Naturally, the predicted location will be different from the real location. However, we use this data to determine whether the users are looking at each other.

In addition, if the paths are the same within the same subgroup, we consider that they are highly familiar than others. During the 10 min tracking, we collect 20 location data of the users. If a user is moving along a similar route with another user at the same time, we assume that they are moving together.

Figure 5 shows the example of the 10-min tracking data. The red, blue, and green circles indicate users A, B, and C. Based on our algorithm, three users are familiar with each other. However, user C is moving in a different route than users A and B. It implies that they are together but they are not close to each other. Thus, if a subgroup member has the same route during the 10 min, they obtain 1 more point. Although this type of tracking cannot fully reflect the directionality, unfamiliar users in the subgroup can be determined.

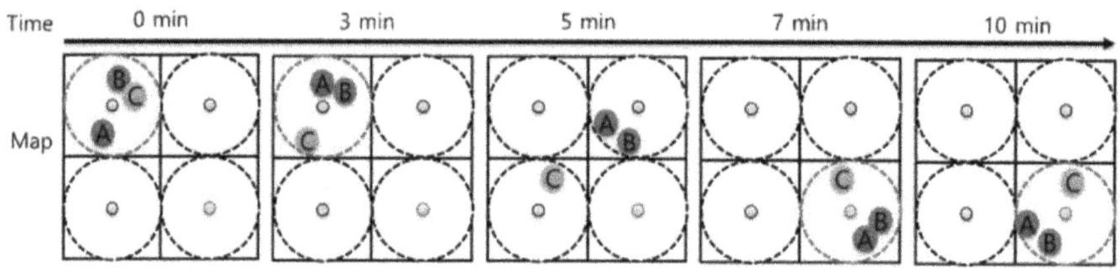

Fig. 5. Example of 10-min tracking

2.4 Experimental Result

To experiment with the proposed model, we need a user movement data. We design a scenario and 10 volunteers move based on our scenario for 15 min. In our scenario, two users remain in one observation area and moves around the receiving device in one direction. Five users move to another observation area every 5 min in various directions. Others move continuously in various directions.

The receiving devices were the Samsung Galaxy S7 smartphones and Nexus 7 tablet. The server was a core i7 PC with 4 GB of RAM. We used a RECO. Inc. beacon sensor. We installed 9 receiving devices under the desk with intervals of 1 m. In the proximity-based familiarity-analysis method, we found two subgroups and each member of the subgroup is assigned a familiarity score of one. Table 3 lists the result of the proximity-based familiarity score.

Table 3. Proximity-based Familiarity Score

	A	B	C	D	E	F	G	H	I	J
A		0	0	0	0	0	0	0	0	0
B	0		0	**2**	**1**	**1**	0	0	0	0
C	0	0		0	0	0	0	0	0	0
D	0	**2**	0		0	0	0	0	0	0
E	0	**1**	0	0		**1**	0	0	0	0
F	0	**1**	0	0	**1**		0	0	0	0
G	0	0	0	0	0	0		0	0	0
H	0	0	0	0	0	0	0		0	0
I	0	0	0	0	0	0	0	0		0
J	0	0	0	0	0	0	0	0	0	

In the direction-based familiarity-analysis method, we track a user's route who is a member of the subgroup. Although the number of users is ten, only three users are part of the subgroup. Two subgroups comprise <B, D> and <B, E, F>. In the <B, D> subgroup, each user moves to different locations. In the <B, E, F> subgroup, each user walks to a particular location. Therefore, B, E, and F are assigned one point each. Table 4 lists the combined result of the familiarity score.

Table 4. Result of Familiarity Score

	A	B	C	D	E	F	G	H	I	J
A		0	0	0	0	0	0	0	0	0
B	0		0	**3**	**3**	**2**	0	0	0	0
C	0	0		0	0	0	0	0	0	0
D	0	**3**	0		0	0	0	0	0	0
E	0	**3**	0	0		**2**	0	0	0	0
F	0	**2**	0	0	**2**		0	0	0	0
G	0	0	0	0	0	0		0	0	0
H	0	0	0	0	0	0	0		0	0
I	0	0	0	0	0	0	0	0		0
J	0	0	0	0	0	0	0	0	0	

The accuracy of the proposed method is over 94%, which is high. This test can only prove that the proposed model is technically possible. In addition, during the test, we found a problem with the proposed model. If two users were both close to the boundary of the observation area but in different observation areas, they cannot obtain a familiarity point. In an actual environment, it is possible that two users are acquainted. Thus, if we use the user's track data as a whole rather than only within a subgroup, we can solve this problem. However, this solution will take more time as more users are involved.

3 Conclusion

In this study, a familiarity analysis model is proposed. The proposed model is based on the proximity and direction, which are the location and route of the users, respectively. Using a Bluetooth signal, we can obtain the location of the users. When a user carries a Bluetooth low-energy beacon, receiving devices installed in an indoor environment receives the signal, and subsequently, the position of the users is calculated. To identify a user's exact location, we use two equations: Average_30s_Distance equation and triangulation equation. Using the locations of the users, we define subgroups. The familiarity score is analyzed using two methods and the results obtained using the methods are combined. The first method is based on the proximity and the second is based on the direction. Each method is used to calculate the familiarity score every 5 min. We experimented with two types of receiving devices, Bluetooth low-energy beacon, server, and 10 volunteers. The result was more than 94% accurate. Although our result shows high accuracy, our experiment was based on an artificial scenario. The experiment does not reflect a real situation and we cannot be sure that it will be effective in practice. To determine realistic errors, we will continue to repeat the test in an actual environment and modify our model. The final objective of our study is to develop a familiarity-analysis model for various situations. This challenge of paper is showing possibility of extended usage of location info at IoT. If we did test for this on grocery store and basket has beacon. This system check each shopper's standing time on each materials and find each shopper's preference. Proposed familiarity analysis method can apply various field.

References

1. Nitti, M., Atzori, L., Cvijikj, I.P.: Friendship selection in the social internet of things: challenges and possible strategies. IEEE Internet Things J. **2**(3), 240–247 (2015)
2. Cranshaw, J., Toch, E., Hong, J., Kittur, A., Sadeh, N.: Bridging the gap between physical location and online social networks. In: Proceedings of the 12th ACM International Conference on Ubiquitous Computing, pp. 119-128. ACM (2010)
3. Li, Q., Zheng, Y., Xie, X., Chen, Y., Liu, W., Ma, W.Y.: Mining user similarity based on location history. In: Proceedings of the 16th ACM SIGSPATIAL International Conference on Advances in Geographic Information Systems, p. 34. ACM (2008)
4. Atzori, L., Iera, A., Morabito, G.: From "smart objects" to "social objects": the next evolutionary step of the internet of things. IEEE Commun. Mag. **52**(1), 97–105 (2014)
5. Wang, W.: Using location-based social media for ranking individual familiarity with places: a case study with foursquare check-in data. In: Progress in Location-Based Services 2014, pp. 171–183. Springer (2015)
6. Laurier, E., Paasi, A., Sage, W.C.: Noticing Talk, Gestures, Movement and Objects in Video Analysis. The SAGE Handbook of Human Geography, London (2013)
7. Choi, J.I., Yong, H.S.: Real-time intragroup familiarity analysis model using beacon based on proximity. In: IEEE Annual Ubiquitous Computing, Electronics & Mobile Communication Conference (UEMCON), pp. 1–5. IEEE (2016)
8. Esteves, J.S., Carvalho, A., Couto, C.: Generalized geometric triangulation algorithm for mobile robot absolute self-localization. In: 2003 IEEE International Symposium on Industrial Electronics, ISIE 2003, vol. 1, pp. 346–351. IEEE (2003)
9. RECO Beacon Homepage. http://reco2.me/. Accessed 10 Jan 2016
10. iBeacon for developers. http://developer.apple.com/ibeacon/. Accessed 10 Jan 2016
11. Choi, J.I., Yong, H.S.: Implementation of close interval indoor positioning using beacon. Navigation **12**, 34–38 (2017)

Music Recommendation with Temporal Dynamics in Multiple Types of User Feedback

Namyun Kim, Won-Young Chae, and Yoon-Joon Lee(✉)

KAIST, Daejeon, South Korea
{namyun.kim, skycwy223, yoonjoon.lee}@kaist.ac.kr

Abstract. In the music streaming service, users can feedback their preference explicitly or implicitly. We propose a link prediction approach for music recommendation with various types of user feedback to alleviate the data sparsity problem and the cold-start problem. Moreover, by reflecting the temporal features of the feedback, we analysis the time-varying user taste. The experiment on real-world dataset demonstrates the effectiveness of our approach in the recommendation quality.

Keywords: Recommender system · Link prediction · Music recommendation

1 Introduction

In many on-line web services, recommendation system has emerged as an indispensable choice [2]. The recommendation system analyzes the user's taste on the items from the collected user feedback, then recommend them.

Traditional recommendation systems can be categorized into two approaches. The first one is the content-based approach. It recommends items based on the characteristics similar to the previous items that the user prefers. The other is collaborative filtering (CF) which recommends items based on the similarity between items or users. CF analyzes the users' feedbacks for items to calculate similarity. It can build a recommendation model for items that are difficult to analyze contents such as music and images. Moreover, it is not free to cold start problem and data sparsity problem, in which recommendation accuracy of early users who lack the amount of user's feedback is poor because it is not enough to collect sufficient amount of user behavior records.

The cold start problem and the data sparsity problem are the traditional challenges in recommendation because they cause negative effects on the recommendation quality. There are some studies conducted to cope with the problems by translating recommendation problem into link prediction task [13]. Various relationships between a user and an item can be represented as a user-item bipartite network. The recommendation problem in the user-item bipartite network can be replaced with a link prediction problem that predicts the link which occurs between the user node and the item node.

However, since the previous link prediction techniques have considered only single kind of user's feedback, they may not fully overcome the data sparsity problem and the cold-start problem. In this paper, we propose a recommendation model that recommends music in user's taste based on link prediction method. Our model considers

W. Lee et al. (eds.), *Proceedings of the 7th International Conference on Emerging Databases*, Lecture Notes in Electrical Engineering 461,
https://doi.org/10.1007/978-981-10-6520-0_35

various types of user feedback (i.e. ratings, repeated listening and skipping current music) and the time-varying temporal feature of the feedback. We adopt a network structure that represents user ratings and listening events. With our approach, we recommend music similar to the user liked or listened before.

In the experiment with real-world dataset, our approach yields up to 78 times better recommendation quality despite of the data sparsity problem. Furthermore, considering temporal feature improves recommendation quality by 1.18 times.

The rest of the paper is structured as follows. In the next section, we describe the traditional recommendation models, then list the previous work to link prediction approach for the recommendation. Section 3 explain our approach in detail. We describe dataset used for experiment and analyze experiment result of our approach in Sects. 4 and 5 respectively, then finally conclude our research in Sect. 6.

2 Related Work

In music recommendation, common techniques are the Content-based Approach and Collaborative Filtering (CF). At first, content-based approach analyzes the profiles of the music and recommends other music which has the same profiles with the user liked music [1, 11]. Therefore, it is possible to construct a recommendation model even if the number of the user feedback is insufficient. However, this approach has the disadvantage of limiting the range of recommendations when it is difficult to analyze the essential content of an item, such as video or music.

Secondly, CF is based on the decisions on the experiences and knowledge that reach each user from a relatively large group of acquaintances. Unlike content-based approach, CF uses only interactions between users and items for recommendation. Due to this characteristic, CF can recommend items that are difficult to analyze contents, such as images and music. However, CF is not free about the following two issues: data sparsity problem and cold start problem.

One of the ways to overcome these issues is link prediction approach. Link prediction problem is a task to predict a new link in a continuously evolving network [12]. Many previous works have been studied to solve the data sparsity problem and improve the accuracy of the recommendation by expressing the sparse data on the network. For example, Dong et al. [3] proposed a transfer-based ranking factor graph model that combines several social patterns with network structure information. Xie et al. [13] proposed a model that consider the relation duality such as similar or dissimilar and like or dislike using the complex number. However, the previous works used only single-type user feedback, which inevitably led to a low accuracy in the recommendation result. Since the music data has a characteristic that user preference changes with time, the temporal feature should be reflected in the music recommendation [5, 6].

In this paper, we propose a novel link prediction method using the various type of user feedback. Moreover, our method considers the temporal feature of user feedback.

3 Methodology

During streaming service, users may show their preference via various feedback. A music recommendation system might consider various kinds of user feedbacks and its temporal feature for recommendation task.

3.1 Network Construction

The user feedback records in music streaming service can be modeled as a user-music bipartite network. From the network, we can get the recommendation of music based on link prediction approach.

We categorize user feedback into two groups: explicit feedback and implicit feedback. The user shows his/her music preference with rating or review. This kind of feedback is called 'explicit feedback'. The traditional recommender system focus on processing explicit feedback for building a recommendation model since it is valuable information for understanding user's taste. Explicit feedback has the benefit of user preference understanding, but collecting explicit feedback is a difficult task. As the result, recommender systems have trouble computing similarity accurately between users or items for recommendation.

Otherwise, recommender system can analyze user's music taste by observing user behavior. The user gives some hints to analyze their taste.

We construct user-music bipartite networks from the explicit and implicit feedback. The user-music bipartite network consists of user and music nodes and links between two types of node. Figure 1 shows an example of building user-music bipartite networks from multiple user feedbacks. In the example, a user feedback, whether it is explicit or implicit, is represented as a weighted link in the network. The weight of link indicates user's preference or confidence to the music. Taking into account the weight of the link, the recommendation system interprets the high weight between the user and the music node as a high interest or preference of the user to the music.

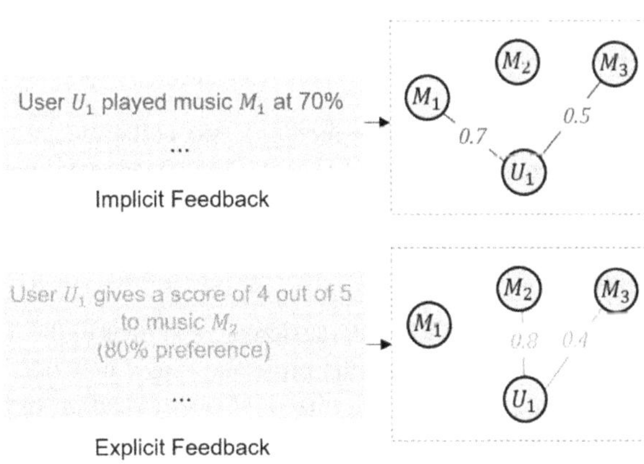

Fig. 1. An example of building user-music bipartite network from user's feedback

3.2 Proposed Method

To address the data sparsity problem, we propose a link prediction based model for recommendation. Given multiple user-music bipartite networks, we recommend music in user's taste by the link prediction method.

Network Projection. Bipartite network projection is a task for compressing a bipartite network into a unipartite network. In Fig. 2, the user-music bipartite network for each user feedback is transferred into a network with music node. If two music nodes have common users in the original bipartite network, they are connected in a projected unipartite network.

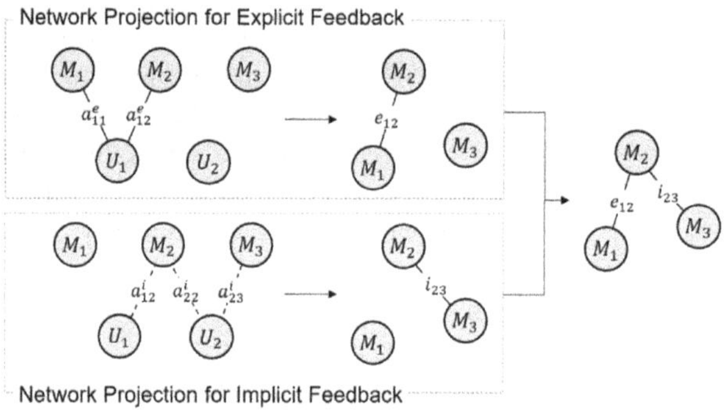

Fig. 2. The user-music bipartite network projection for each type of user feedback

To minimize the data loss during network projection, we compute a link weight in music network from the topological feature in the user-music bipartite network. The weight means the similarity between two music.

Since users' music preference is changed over time [7], temporal feature in user feedback records has to be considered. To do this, first, we construct a music network by summing the projected network from the user-music bipartite network corresponding to time window. Then, we multiply aging factor by the weights of each projected network.

The music networks for each user's feedback are collapsed into a single music network. During the process, the weight for each user's feedback is multiplied by the link weight of corresponding music network. The weight represents the effect of each user's feedback in the viewpoint of user preference. We will discuss how to obtain the optimal weight of each user's feedback.

Link Prediction Algorithm. We find the music that is similar to the user preferred by the random walk with restart (RWR) in the projected music network.

RWR is defined as a sequence of randomized moves by random surfer starting from the certain nodes. The algorithm has been adopted as a similarity measure for recommendation problem [8]. Compared with other similarity measures, RWR can capture the global structure of the network and the composite relationship between nodes [4, 9].

The random surfer repeats one of the following actions, starting from the nodes representing the user liked music.

- **Move to neighbor.** The random surfer moves to one of the neighbors of the current node. The probability that a random surfer moves is directly proportional to the weight of the link between the current node and the neighbor.
- **Back to starting node.** During the network search, the random surfer returns to the starting nodes with predefined restart probability. It limits the range of search space. The restart probability also limits the random surfer from moving too far from the starting nodes.

The relevance score measured by RWR is interpreted as the possibility of link creation. The system recommends the top k ranked music to the user.

Optimization. As mentioned above, we need to find optimal weight for each user's feedback. This optimization process identifies the impacts of each user feedback in user taste analysis.

In this research, we take an evolutionary algorithm to find optimal weight for each user feedback. Compared with other optimization algorithms, the evolutionary algorithm can find global minima in given search space. The overall optimization process is shown in Fig. 3. The algorithm optimizes the weight of user's feedback by an approach inspired by biological evolution such as recombination and mutation.

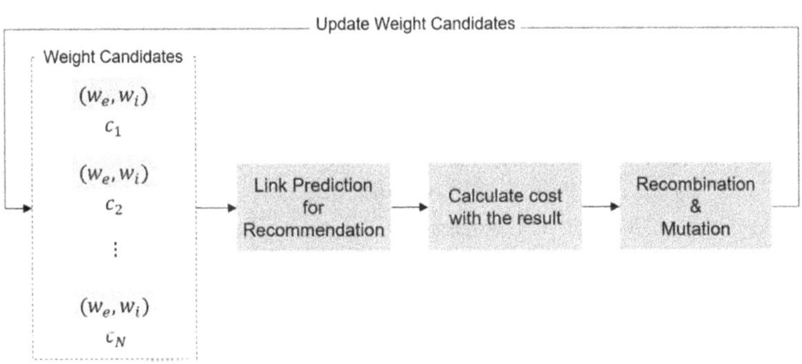

Fig. 3. The overall optimization process

The goal of optimization is to calculate the weight for user's feedback for higher recommendation quality. The cost function for users U in training dataset is defined as below:

$$\text{cost}(c_i, u) = \frac{1}{|U|} \sum\nolimits_{u \in U} \sum\nolimits_{j=1}^{k} r_j^u \tag{1}$$

The Eq. (1) sum all recommendation ranking $\{r_1^u, r_2^u, \ldots, r_k^u\}$ of the music the user u has heard in training data. The link prediction to obtain recommendation results is performed on the projected music network that is calculated with the one of weight candidates c_i. After calculating the cost for all the weight candidates, we update the weight candidates by recombination and mutation among the weights which have small cost. The entire optimization process is repeated until we find the optimal weight in recommendation perspective.

4 Experiment Design

4.1 Datasets

We use one-year user play history and ratings collected from Last.fm in 2014 [10]. Note that there is only 'like' expression in the user rating records. The basic statistics of the dataset is shown in Table 1.

Table 1. Dataset statistics

Data type	The number of data
Users	45,167
Music	4,519,105
Ratings	4,106,341
Listening Events	31,351,954

We sampled the listening events and ratings for 10,000 songs which have at least one interaction with the user. We then partition the sampled dataset into three sets according to timestamp: network construction data, training data, and test data. The dataset in first 10 months is used for network construction, the rest of data sets are used for learning and testing, one month each.

Figure 4 shows the distribution of the correlation between the music and the user in the sampled dataset. The left plot is users' music rating frequency distribution and the

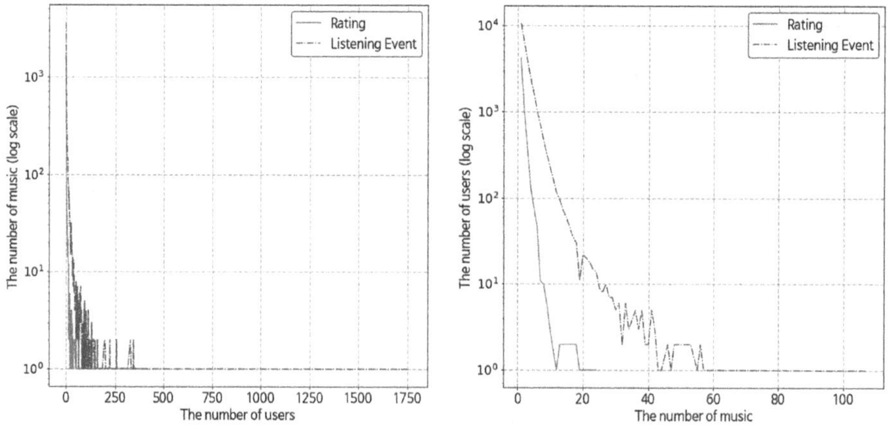

Fig. 4. The distribution of sampled dataset

right one is users' listening frequency distribution. From the both of plots, we can observe that there is the sparsity problem of the dataset; the most of the users only rate or listen small amount of music.

Because of the limitation of the given dataset, we only use two types of user feedback, each represents explicit and implicit feedback; the user ratings and the number of listening events for music.

4.2 Baseline Models

We compare our method against following baseline methods

1. **Naive Item-based CF (I-CF):** This method is naive item-based CF which use user ratings to build recommendation model.
2. **CF with Matrix Factorization (MF-CF):** This method adopts the matrix factorization technique for CF. It also uses user ratings as the input data.
3. **CF Considering Multiple Feedbacks (CCF):** This method uses two model-based collaborative filtering recommendation that uses explicit and implicit behavior as input data.
4. **Simple Link Prediction (LP):** This method adopts link prediction approach on a music network for the recommendation. The music network is projected from the user-music bipartite network built with single kind of user feedback.
5. **Static Link Prediction (S-LP):** This method is modified version of our approach without the consideration on the temporal characteristics of user feedbacks.

4.3 Evaluation Metric

We evaluate various recommendation models for the test dataset with three evaluation metrics; the precision at top 20(P@20), the recall at top 20(R@20) and MAP (Mean Average Precision) [14]. With these metric, we check that how many of top 20 ranked music are actually listened or liked by a user.

5 Experiment

5.1 Data Sparsity Problem

From the experiment dataset, we select 12,858 users as target users. The target users have 491 ratings and 28,718 listening events for 5,045 songs. We show the evaluation of recommendation quality of various methods in Table 2.

In the table, we note that our approach gives a significant improvement in recommendation quality over the other recommendation methods. We further observe that the sparsity of input data affects the performance of recommendation method. The comparison between the two methods, with implicit feedback and with explicit feedback, clearly indicates the sparsity problem. Since the sparsity of explicit feedback is much lower than that of explicit feedback, the experimental results with implicit feedback are much better. Among three algorithms in same types of data, MF-CF

Table 2. The experimental results for the sparsity input data

Feedback type	Methods	Mean P@20	Mean R@20	MAP
Explicit	I-CF	0.004	0.053	0.047
	MF-CF	0.009	0.109	0.047
	LP	0.001	0.009	0.008
Implicit	I-CF	0.008	0.100	0.045
	MF-CF	0.008	0.104	**0.066**
	LP	0.008	0.091	0.040
Multiple	CCF	0.009	0.109	**0.066**
	Ours	**0.014**	**0.166**	0.053

shows better performance than the others. This is because the lack of user feedback drops the accuracy of the similarity measure.

5.2 Cold Start Problem

To evaluate the recommendation results for the cold-start problem, we select 9,486 users as new users who have less than three relationships with music. The experimental results for new users are shown in Table 3.

In the result, we can observe that our approach shows the best performance compared with the others. Note that the most of the evaluation metrics for new users are consistently lower than the results in Table 2. As we can observe in the previous result, the quality of recommendation result depends on the sparsity of input data.

5.3 Temporal Feature Consideration

Next, we examine how the consideration of the temporal feature in music recommendation affects the improvement of recommendation quality. To alleviate the negative effects of the sparsity problem in recommendation quality, we pick up 42 heavy users who have interactions with more than 20 music in the dataset.

In the Table 4, we compare two link prediction approaches: the link prediction method in the static network(S-LP) and our approach. S-LP shows better performance than our approach in P@20 and R@20. However, in MAP, ours better. These results

Table 3. The experimental results for the new users

Feedback type	Methods	Mean P@20	Mean R@20	MAP
Explicit	I-CF	0.004	0.052	0.038
	MF-CF	0.007	0.110	0.041
	LP	0.001	0.009	0.006
Implicit	I-CF	0.007	0.101	0.035
	MF-CF	0.007	0.105	**0.058**
	LP	0.006	0.087	0.033
Multiple	CCF	0.007	0.110	0.056
	Ours	**0.011**	**0.163**	0.043

Table 4. The experimental results for two link prediction approach

Methods	Mean P@20	Mean R@20	MAP
S-LP	**0.0798**	**0.2514**	0.1200
Ours	0.0738	0.2430	**0.1425**

may explain that our approach recommends the music in user's taste with higher rank. Therefore, we can point out that the temporal feature may increases the recommendation quality.

6 Conclusion

In this paper, we propose the link-prediction recommendation model based on music streaming service. With various types of user feedback, our approach tries to alleviate the most well-known challenges in the recommendation: the data sparsity problem and the cold-start problem. To improve recommendation accuracy, we consider additionally the temporal feature of user feedback.

We examine the performance for various recommendation system on the real-world dataset. The experimental result demonstrates that our approach has up to 78 times improvement in recommendation quality despite the data sparsity problem and the cold-start problem. By comparing with the link prediction approach in the static network, we show that considering temporal feature leads 1.18 times better performance.

Acknowledgements. This work was partly supported by KAIST(A0601003029).

References

1. Cano, P., Koppenberger, M., Wack, N.: An industrial-strength content-based music recommendation system. In: Proceedings of the 28th Annual International ACM SIGIR Conference on Research and Development in Information Retrieval, pp. 673–673. ACM (2005)
2. Carrer-Neto, W., et al.: Social knowledge-based recommender system. Application to the movies domain. Expert Syst. Appl. **39**(12), 10990–11000 (2012)
3. Dong, Y., et al.: Link prediction and recommendation across heterogeneous social networks. In: 2012 IEEE 12th International Conference on Data Mining (ICDM), pp. 181–190. IEEE (2012)
4. He, J., et al.: Manifold-ranking based image retrieval. In: Proceedings of the 12th Annual ACM International Conference on Multimedia, pp. 9–16. ACM (2004)
5. Koren, Y.: Collaborative filtering with temporal dynamics. Commun. ACM **53**(4), 89–97 (2010)
6. Munasinghe, L., Ichise, R.: Time score: a new feature for link prediction in social networks. IEICE Trans. Inf. Syst. **95**(3), 821–828 (2012)
7. Park, C.H., Kahng, M.: Temporal dynamics in music listening behavior: a case study of online music service. In: 2010 IEEE/ACIS 9th International Conference on Computer and Information Science (ICIS), pp. 573–578. IEEE (2010)

8. Perozzi, B., Al-Rfou, R., Skiena, S.: Deepwalk: online learning of social representations. In: Proceedings of the 20th ACM SIGKDD International Conference on Knowledge Discovery and Data Mining, pp. 701–710. ACM (2014)
9. Tong, H., Faloutsos, C.: Center-piece subgraphs: problem definition and fast solutions. In: Proceedings of the 12th ACM SIGKDD International Conference on Knowledge Discovery and Data Mining, pp. 404–413. ACM (2006)
10. Turrin, R., et al.: 30music listening and playlists dataset. In: RecSys Posters (2015)
11. van Den Oord, A., Dieleman, S., Schrauwen, B.: Deep content-based music recommendation. In: Advances in Neural Information Processing Systems, pp. 2643–2651 (2013)
12. Wang, P., et al.: Link prediction in social networks: the state-of-the-art. Sci. China Inf. Sci. **58**(1), 1–38 (2015)
13. Xie, F., et al.: A link prediction approach for item recommendation with complex number. Knowl.-Based Syst. **81**, 148–158 (2015)
14. Zhu, M.: Recall, precision and average precision, vol. 2, p. 30. Department of Statistics and Actuarial Science, University of Waterloo, Waterloo (2004)

Effectively and Efficiently Supporting Encrypted OLAP Queries over Big Data: Models, Issues, Challenges

Alfredo Cuzzocrea(✉)

DIA Department, University of Trieste and ICAR-CNR, Trieste, Italy
alfredo.cuzzocrea@dia.units.it

Abstract. Due to emerging technologies like *Clouds*, recently the problem of *encrypting and querying big data* is of great interest trough the community. Here, the main problem consists in devising effective and efficient encryption schemes for big data, and then effective and efficient query algorithms for querying such data in their encrypted form directly. By comparing both lines of research, it emerges that querying encrypted big data plays the major role, as the encryption phase is usually conducted on top of well-recognized state-of-the-art encryption schemes. On the other hand, *OLAP data* are a knowledge-rich class of big data that are extremely important for latest *big data analytics tools*. Inspired by these two authoritative research trends, in this paper we provide the following contributions: (*i*) an overview of most relevant initiatives in the scientific field of *querying encrypted OLAP data*; (*ii*) critical discussion on open issues and research challenges that will dominate the future scene of the investigated research topic.

1 Introduction

With the advent of *Cloud methodologies and paradigms*, *big data security* (e.g., [1–5]) is becoming a hot-topic in database and data warehousing research. Due to emerging technologies like *Clouds*, recently the problem of *encrypting and querying big data* is of great interest trough the community. Here, the main problem consists in devising effective and efficient encryption schemes for big data, and then effective and efficient query algorithms for querying such data in their encrypted form directly. By comparing both lines of research, it emerges that querying encrypted big data plays the major role, as the encryption phase is usually conducted on top of well-recognized state-of-the-art encryption schemes. On the other hand, *OLAP data* [6] are a knowledge-rich class of big data that are extremely important for latest *big data analytics tools*.

Data encryption is a well-focused and mature method for allowing secure data access and publishing over Cloud environments (e.g., [7–9]), in contrast with comparative approaches that adopt even-well-recognized alternatives (e.g., *secure access methods*). According to this consolidated line of research, the main

W. Lee et al. (eds.), *Proceedings of the 7th International Conference on Emerging Databases*, Lecture Notes in Electrical Engineering 461,
https://doi.org/10.1007/978-981-10-6520-0_36

idea consists in encrypting data and then devising ad-hoc query algorithms to process encrypted data *directly* (e.g., [10]). Among several kinds of relevant data targets, OLAP data [6] are, without doubts, a first-class kind of data in emerging *big data analytics scenarios* (e.g., [11–13]).

In this paper, following these evolving trends, we focus on the relevant problem of *querying encrypted OLAP data*. Summarizing, the problem consists in devising ad-hoc algorithms for encrypting OLAP data cubes, and then algorithms for querying so-encrypted cubes, via some relevant properties/principles. Suitable *transformations* from/to the encrypted/decrypted domain must be developed as well.

Figure 1 shows a typical setting of the application scenario we investigate in this paper. Here, the user is interested in querying an (encrypted) OLAP data cube that is stored in a proper Cloud node. Note that the user is not aware that the cube is encrypted and she/he is able to query the cube just by knowing its main *multidimensional-model metadata* [6] (e.g., dimensions, measures, dimensional attributes, and so forth), for big data analytics and decision-making purposes. The reference encrypted OLAP query framework works as follows. First, the input OLAP query is parsed and transformed in a suitable *encrypted query* over encrypted OLAP data. To this end, the *encryption proxy component* (e.g., implemented via suitable Cloud services) takes into consideration: (*i*) the encryption scheme; (*ii*) the reference OLAP query workload; (*iii*) proper OLAP data cube statistics; (*iv*) memory/space constraints. Then, the so-generated encrypted query is issued over the encrypted OLAP data cube directly. The *encrypted query answer* is now returned to the encryption proxy component for its decryption and, finally, to the user who will build meaningfully big data analytics for decision-making purposes. From a research-point-of-view, the most relevant challenge relies in how to effectively and efficiently devise the proper OLAP data cube encryption algorithm and the proper encrypted OLAP

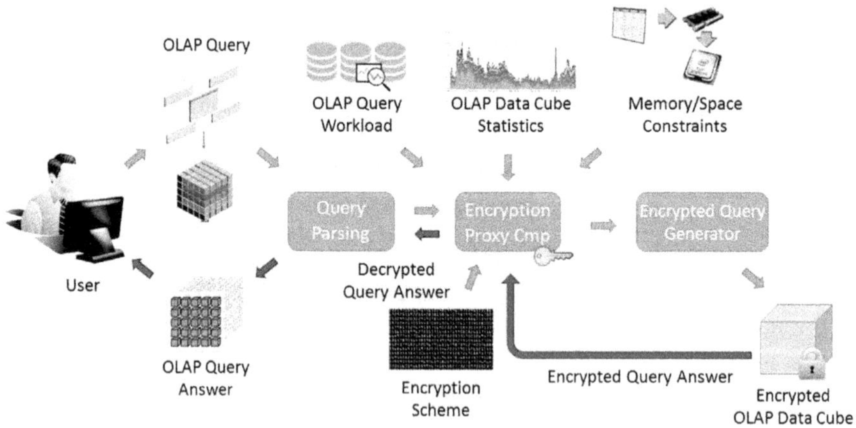

Fig. 1. Querying encrypted OLAP data: application scenario

data cube query algorithm that are the two pillars of the encryption proxy component.

In more details, in this paper we provide the following contributions: (*i*) an overview of most relevant initiatives in the scientific field of *querying encrypted OLAP data*; (*ii*) critical discussion on open issues and research challenges that will dominate the future scene of the investigated research topic. The paper significantly extends the short paper [14], where the embryonic ideas have been initially proposed.

The remaining of this paper is organized as follows. Section 2 provides a brief overview on some state-of-the-art proposals in the context of querying encrypted OLAP data. Section 3 contains a critical discussion on open challenges and future research directions in the investigated topics. Finally, Sect. 4 reports conclusions.

2 State-Of-The-Art Querying Encrypted OLAP Data Proposals: A Brief Overview

Relevant proposals in the context of querying encrypted OLAP data are reported in the following.

[15] presents a novel method for encrypting a Data Warehouse (DW), and the related OLAP system based on the proposed encryption method that is able to query so-encrypted DW data. The proposed algorithm is complex in nature, and performs several encryption tasks depending on statistical properties of target DW data. Authors also conduct several performance tests to validate the proposed OLAP system in terms of query processing performance.

[16] describes a framework for providing encryption-based security over Cloud Data Warehouses (CDW) via an *adaptive* approach. The proposed mechanism applies a separation of concerns approach to obtain such adaptiveness. Proper algorithms that support the introduced mechanism are presented as well.

[17] addresses the specific applicative setting represented by encrypting CDW via *multi-valued encrypted values*, in order to obtain minimal encrypted data redundancy. Authors study grouping (OLAP-like) predicates over encrypted DW and introduce two novel encryption schemes, namely MV-HOM and MVSE-HOM, which specifically support analytical queries over encrypted OLAP data.

[18] argues that *homomorphic encryption* (e.g., [19]), which allows the execution of queries over encrypted data without requiring decryption, has been poorly applied to DW. In order to fulfill this gap, authors propose a framework that defines how a homomorphic-encryption scheme can be used to encrypt numeric OLAP measures, and how SUM-based aggregations of analytic queries are processed over so-obtained encrypted DW. In addition to this, a system architecture for safely processing encrypted DW is presented and described in details.

[20] investigates the specific problem of supporting efficient multidimensional range queries over *attack-resilient databases*. Authors consider outsourced-services' scenarios where the owner make available its proper data to the service provider, which may be curious on them. In order to avoid this problem,

the *Random Space Encryption* (RASP) approach is proposed, with the benefit of providing efficient range search with stronger attack resilience than existing efficiency-focused approaches. Range queries are securely transformed to the encrypted data space and then efficiently processed with a two-stage processing algorithm. A comprehensive experimental campaign completes the analytical contributions of the paper.

[21] applies a novel *conjunctive query scheme* over encrypted multidimensional data in the specific smart grid context. The scheme is called ECQ. Authors focus on emerging smart grids that can collect metering data of users' power consumption where, in order to preserve users' privacy, metering data are mostly encrypted by cryptographic algorithms. They argue that power system data in smart grid has multidimensional attributes. Therefore, querying encrypted multidimensional data along all multiple dimensions is a challenging issue in smart grids. To solve this challenge, the ECQ scheme is introduced. It incorporates the idea of public key encryption and conjunctive keywords search to achieve conjunctive query without data and query privacy leakage. The benefits carried-out by ECQ are truly supported by a detailed security analysis provided by the authors.

Finally, [22] introduces and experimentally assesses MONOMI, a system for supporting the evaluation of (OLAP-like) analytical workloads over sensitive data via encryption methodologies. MONOMI works by encrypting the entire target database and running queries over the encrypted data. It particularly introduces *split client/server query execution*, which can execute arbitrarily complex queries over encrypted data. In addition to this, several techniques that improve performance for such workloads, including *per-row precomputation*, *space-efficient encryption*, *grouped homomorphic addition*, and *pre-filtering*, are introduced.

3 Future Research Directions for Emerging Querying Encrypted OLAP Data Techniques and Algorithms

The problem of querying encrypted OLAP data is relevant in the database and data warehousing research communities. This poses several future research directions to be considered. In the following, we report of some noticeable of them.

Dealing with Complex OLAP Schema. OLAP data cubes very often expose *complex schema* (e.g., [23]). As a consequence, encrypting such cubes become harder and harder. Dealing with complex OLAP schemas is thus a research challenge for future years.

Support for Complex Aggregation Predicates. Conventional proposals do not focus on querying encrypted OLAP data cubes built on top of *complex* aggregation predicates (not just traditional ones like SUM, COUNT, etc.), which instead play a significant role in big data analytics (e.g., [24]).

Support for Filtering Predicates. OLAP queries very often are equipped with ad-hoc *filtering predicates* that limit their scope to specific domains of (encrypted) data. Combining encrypted query execution with filtering predicates is not an easy task (e.g., [25]). This challenge will require lot of attention by the research community.

Extensions to Non-Conventional Query Classes. Usually, OLAP queries are mixed and combined with other interesting non-conventional query classes, such as *iceberg queries* (e.g., [26]) or *skyline queries* (e.g., [27]), as to empower the capabilities of big data analytics tools. When these combinations are issued on top of encrypted OLAP data cubes, critical issues derive, and specialized solutions are necessary as a consequence.

Query Optimization Plans. When encrypted OLAP queries must be executed against a *distributed* collection of (encrypted) OLAP data cubes, then, like in classical distributed query execution paradigms, *query optimization issues* arise (e.g., [28–30]). *Heuristics* seem a promising direction to this end.

Scalable Encryption Methods. While there are several proposals for encryption mechanisms, even quite mature, *scalability* is still a critical requirement for such techniques, especially when they are executed against *big data* (e.g., [31]).

Flexible Transformations From/To Encrypted/Decrypted Multidimensional Domains. As implicitly dictated by the reference encrypted OLAP query framework shown in Fig. 1, a leading component of the target application scenario is represented by the transformations that are necessary to move from/to the encrypted/decrypted *multidimensional* domains, which expose severe challenges due to their inherent complexity. Applying these transformations with *flexibility* (e.g., under OLAP data updates) is a critical requirement for research in this area.

Privacy-Preserving Encryption Mechanisms. When OLAP data cubes are encrypted, their data cells must be universally accessed. This may represent a potential *privacy breach*. Therefore, devising privacy-preserving mechanisms for encrypting OLAP data cubes become mandatory.

Moving to Column-Oriented Databases. A significant achievement for efficiently processing OLAP queries over big data is represented by *column-oriented databases* (e.g., [32]). It has been proven that such databases are capable of efficiently supporting such queries by overcoming limitations of classical row-oriented databases. Relevant questions are now: what happens when encrypted OLAP queries are executed over encrypted column-based OLAP data cubes? Which research innovations will pinpoint this novel querying encrypted OLAP data scheme?

Integration with Hadoop/MapReduce Frameworks. Like for other big-data-oriented methodologies, even OLAP data encryption solutions should be integrated with innovative *Hadoop/MapReduce frameworks* [33,34]. Due to the specific nature of encryption schemes, the latter is not an immediate result to accommodate and it would need engaging efforts during future years.

4 Conclusions

In this paper, we have focused on the emerging querying encrypted OLAP data problem. The main conclusion of our research is represented by the evidence that there is still a lot of work to do in this area, stirred-up by latest *cybersecurity requirements* that are among the most significant ones for the next-generation digital society.

References

1. Cuzzocrea, A.: Privacy and security of big data: current challenges and future research perspectives. In: Proceedings of the First International Workshop on Privacy and Secuirty of Big Data, PSBD@CIKM 2014, pp. 45–47, Shanghai, China, 7 November 2014
2. Cuzzocrea, A., Russo, V.: Privacy preserving OLAP and OLAP security. In: Encyclopedia of Data Warehousing and Mining, 2nd edn. (4 Volumes), pp. 1575–1581 (2009)
3. Bertino, E.: Big data security and privacy. In: 2016 IEEE International Conference on Big Data, BigData 2016, p. 3, Washington DC, USA, 5–8 December 2016
4. Nelson, B., Olovsson, T.: Security and privacy for big data: a systematic literature review. In: 2016 IEEE International Conference on Big Data, BigData 2016, pp. 3693–3702, Washington DC, USA, 5–8 December 2016
5. Moreno, J., Serrano, M.A., Fernández-Medina, E.: Main issues in big data security. Future Internet **8**(3), 44 (2016)
6. Gray, J., Chaudhuri, S., Bosworth, A., Layman, A., Reichart, D., Venkatrao, M., Pellow, F., Pirahesh, H.: Data cube: a relational aggregation operator generalizing group-by, cross-tab, and sub totals. Data Min. Knowl. Discov. **1**(1), 29–53 (1997)
7. Li, J., Ma, R., Guan, H.: TEES: an efficient search scheme over encrypted data on mobile cloud. IEEE Trans. Cloud Comput. **5**(1), 126–139 (2017)
8. Lan, C., Li, H., Yin, S., Teng, L.: A new security cloud storage data encryption scheme based on identity proxy re-encryption. I. J. Netw. Secur. **19**(5), 804–810 (2017)
9. Cui, H., Yuan, X., Wang, C.: Harnessing encrypted data in cloud for secure and efficient mobile image sharing. IEEE Trans. Mob. Comput. **16**(5), 1315–1329 (2017)
10. Arasu, A., Eguro, K., Kaushik, R., Ramamurthy, R.: Querying encrypted data. In: International Conference on Management of Data, SIGMOD 2014, pp. 1259–1261, Snowbird, UT, USA, 22–27 June 2014
11. Cuzzocrea, A., Song, I., Davis, K.C.: Analytics over large-scale multidimensional data: the big data revolution! In: ACM 14th International Workshop on Data Warehousing and OLAP, Proceedings, DOLAP 2011, pp. 101–104, Glasgow, UK, 28 October 2011
12. Cuzzocrea, A.: Analytics over big data: exploring the convergence of datawarehousing, OLAP and data-intensive cloud infrastructures. In: 37th Annual IEEE Computer Software and Applications Conference, COMPSAC 2013, pp. 481–483, Kyoto, Japan, 22–26 July 2013

13. Sakr, S., Elgammal, A.: Towards a comprehensive data analytics framework for smart healthcare services. Big Data Res. **4**, 44–58 (2016)
14. Cuzzocrea, A., Grasso, G.M.: Querying encrypted OLAP data. In: 41st Annual IEEE Computer Software and Applications Conference, COMPSAC 2017, Turin, Italy, 4–8 July 2017
15. Lopes, C.C., Times, V.C., Matwin, S., Ciferri, R.R., de Aguiar Ciferri, C.D.: Processing OLAP queries over an encrypted data warehouse stored in the cloud. In: Data Warehousing and Knowledge Discovery - 16th International Conference, DaWaK 2014, Proceedings, Munich, Germany, pp. 195–207, 2–4 September 2014
16. Guermazi, E., Ayed, M.B., Ben-Abdallah, H.: Adaptive security for cloud data warehouse as a service. In: 14th IEEE/ACIS International Conference on Computer and Information Science, ICIS 2015, pp. 647–650, Las Vegas, NV, USA, 28 June–1 July 2015
17. Lopes, C.C., Times, V.C., Matwin, S.: Towards cloud data warehouses of multi-valued encrypted values. JIDM **5**(3), 335–348 (2014)
18. Lopes, C.C., Times, V.C.: A framework for investigating the performance of sum aggregations over encrypted data warehouses. In: Proceedings of the 30th Annual ACM Symposium on Applied Computing, pp. 1000–1007, Salamanca, Spain, 13–17 April 2015
19. Gentry, C.: Fully homomorphic encryption using ideal lattices. In: Proceedings of the 41st Annual ACM Symposium on Theory of Computing, STOC 2009, pp. 169–178, Bethesda, MD, USA, 31 May–2 June 2009
20. Chen, K., Kavuluru, R., Guo, S.: RASP: efficient multidimensional range query on attack-resilient encrypted databases. In: First ACM Conference on Data and Application Security and Privacy, CODASPY 2011, Proceedings, pp. 249–260, San Antonio, TX, USA, 21–23 February 2011
21. Wen, M., Lu, R., Lei, J., Liang, X., Li, H., Shen, X.: ECQ: an efficient conjunctive query scheme over encrypted multidimensional data in smart grid. In: 2013 IEEE Global Communications Conference, GLOBECOM 2013, pp. 796–801, Atlanta, GA, USA, 9–13 December 2013
22. Tu, S., Kaashoek, M.F., Madden, S., Zeldovich, N.: Processing analytical queries over encrypted data. PVLDB **6**(5), 289–300 (2013)
23. Eavis, T., Taleb, A.: Mapgraph: efficient methods for complex olap hierarchies. In: Proceedings of the Sixteenth ACM Conference on Information and Knowledge Management, CIKM 2007, pp. 465–474, Lisbon, Portugal, 6–10 November 2007
24. Cuzzocrea, A.: Aggregation and multidimensional analysis of big data for large-scale scientific applications: models, issues, analytics, and beyond. In: Proceedings of the 27th International Conference on Scientific and Statistical Database Management, SSDBM 2015, pp. 23:1–23:6, La Jolla, CA, USA, 29 June–1 July 2015
25. Bothe, S., Cuzzocrea, A., Karras, P., Vlachou, A.: Skyline query processing over encrypted data: an attribute-order-preserving-free approach. In: Proceedings of the First International Workshop on Privacy and Secuirty of Big Data, PSBD@CIKM 2014, pp. 37–43, Shanghai, China, 7 November 2014
26. Fang, M., Shivakumar, N., Garcia-Molina, H., Motwani, R., Ullman, J.D.: Computing iceberg queries efficiently. In: Proceedings of 24th International Conference on Very Large Data Bases, VLDB 1998, pp. 299–310, 24–27 August 1998, New York City, New York, USA (1998)
27. Börzsönyi, S., Kossmann, D., Stocker, K.: The skyline operator. In: Proceedings of the 17th International Conference on Data Engineering, pp. 421–430, 2–6 April 2001, Heidelberg, Germany (2001)

28. Cuzzocrea, A., Furfaro, F., Saccà, D.: Enabling OLAP in mobile environments via intelligent data cube compression techniques. J. Intell. Inf. Syst. **33**(2), 95–143 (2009)
29. Cuzzocrea, A.: Accuracy control in compressed multidimensional data cubes for quality of answer-based OLAP tools. In: 18th International Conference on Scientific and Statistical Database Management, SSDBM 2006, pp. 301–310, 3–5 July 2006, Vienna, Austria, Proceedings (2006)
30. Cuzzocrea, A., Matrangolo, U.: Analytical synopses for approximate query answering in OLAP environments. In: 15th International Conference on Database and Expert Systems Applications, DEXA 2004, Proceedings, pp. 359–370, Zaragoza, Spain, 30 August–3 September 2004
31. Cuzzocrea, A., Saccà, D., Ullman, J.D.: Big data: a research agenda. In: 17th International Database Engineering & Applications Symposium, IDEAS 2013, pp. 198–203, Barcelona, Spain, 9–11 October 2013
32. Abadi, D.J., Boncz, P.A., Harizopoulos, S.: Column oriented database systems. PVLDB **2**(2), 1664–1665 (2009)
33. Shvachko, K., Kuang, H., Radia, S., Chansler, R.: The hadoop distributed file system. In: IEEE 26th Symposium on Mass Storage Systems and Technologies, MSST 2012, pp. 1–10, Lake Tahoe, Nevada, USA, 3–7 May 2010
34. Dean, J., Ghemawat, S.: Mapreduce: simplified data processing on large clusters. Commun. ACM **51**(1), 107–113 (2008)

Author Index

W. Lee et al. (eds.), *Proceedings of the 7th International Conference on Emerging Databases*, Lecture Notes in Electrical Engineering 461,
https://doi.org/10.1007/978-981-10-6520-0

CPSIA information can be obtained
at www.ICGtesting.com
Printed in the USA
LVHW011222280620
659213LV00021B/1352